Optical Properties of Semiconductor Nanocrystals

Low-dimensional semiconductor structures, often referred to as nanocrystals or quantum dots, exhibit fascinating behavior and have a multitude of potential applications, especially in the field of communications. This book examines in detail the optical properties of these structures, giving full coverage of theoretical and experimental results and discusses their technological applications.

The author begins by setting out the basic physics of electron states in nanocrystals (adopting a "cluster-to-crystal" approach) and goes on to discuss the growth of nanocrystals, absorption and emission of light by nanocrystals, optical nonlinearities, interface effects, and photonic crystals. He illustrates the physical principles with references to actual devices such as novel light-emitters and optical switches.

The book covers a rapidly developing, interdisciplinary field. It will be of great interest to graduate students of photonics or microelectronics, and to researchers in electrical engineering, physics, chemistry, and materials science.

S. V. Gaponenko is the associate director and head of the nanostructure photonics laboratory at the Institute of Molecular and Atomic Physics, National Academy of Sciences of Belarus. A member of the European Physical Society, he has held the position of visiting scientist at the Universities of Kaiserslautern and Karlsruhe, Germany and at the University of Arizona.

Cambridge Studies in Modern Optics

TITLES IN PRINT IN THIS SERIES

Fabry-Perot Interferometers
G. Hernandez

Holographic and Speckle Interferometry (second edition)
R. Jones and C. Wykes

Laser Chemical Processing for Microelectronics
edited by K. G. Ibbs and R. M. Osgood

The Elements of Nonlinear Optics
P. N. Butcher and D. Cotter

Optical Solitons – Theory and Experiment
edited by J. R. Taylor

Particle Field Holography
C. S. Vikram

Ultrafast Fiber Switching Devices and Systems
M. N. Islam

Optical Effects of Ion Implantation
P. D. Townsend, P. J. Chandler, and L. Zhang

Diode-Laser Arrays
edited by D. Botez and D. R. Scifres

The Ray and Wave Theory of Lenses
A. Walther

Design Issues in Optical Processing
edited by J. N. Lee

Atom-Field Interactions and Dressed Atoms
G. Compagno, R. Passante, and F. Persico

Compact Sources of Ultrashort Pulses
edited by I. Duling

The Physics of Laser-Atom Interactions
D. Suter

Optical Holography – Principles, Techniques and Applications (second edition)
P. Hariharan

Theoretical Problems in Cavity Nonlinear Optics
P. Mandel

Measuring the Quantum State of Light
U. Leonhardt

Optical Properties of Semiconductor Nanocrystals
S. V. Gaponenko

Optical Properties of Semiconductor Nanocrystals

S. V. GAPONENKO

PUBLISHED BY THE PRESS SYNDICATE OF THE UNIVERSITY OF CAMBRIDGE
The Pitt Building, Trumpington Street, Cambridge, CB2 1RP, United Kingdom

CAMBRIDGE UNIVERSITY PRESS
The Edinburgh Building, Cambridge CB2 2RU, UK http: //www.cup.cam.ac.uk
40 West 20th Street, New York, NY 10011-4211, USA http: //www.cup.org
10 Stamford Road, Oakleigh, Melbourne 3166, Australia

First published 1998

Typeset in Times Roman 10/13, in LaTeX 2_ε [TB]

A catalog record for this book is available from the British Library

Library of Congress Cataloging in Publication Data

Gaponenko, S. V. (Sergey V.), 1958–
Optical properties of semiconductor nanocrystals / S.V. Gaponenko.
p. cm. – (Cambridge studies in modern optics)
Includes bibliographical references and index.
ISBN 0-521-58241-5 (hb)

Transferred to digital printing 2003

1. Semiconductors – Optical properties. 2. Crystals – Optical properties. 3. Nanostructures – Optical properties. 4. Quantum electronics. I. Title. II. Series: Cambridge studies in modern optics (Unnumbered)
QC611.6.06G36 1998
537.6′226 – dc21 97-35237
CIP

To my parents, my wife, and my son

"Dum taxat rerum magnarum parva potest res
Exemplare dare et vestigia notitiai."

Lucreti, De Rerum Natura

So far at any rate as so small an example
Can give any hint of infinite events.

(Translation by C. H. Sisson)

Contents

Preface

Electronic states and probabilities of optical transitions in molecules and crystals are determined by the properties of atoms and their spatial arrangement. An electron in an atom possesses a discrete set of states, resulting in a corresponding set of narrow absorption and emission lines. Elementary excitations in an electron subsystem of a crystal, that is, electrons and holes, possess many properties of a gas of free particles. In semiconductors, broad bands of the allowed electron and hole states separated by a forbidden gap give rise to characteristic absorption and emission features completely dissimilar to atomic spectra. It is therefore reasonable to pose a question: What happens on the way from atom to crystal? The answer to this question can be found in the studies of small particles with the number of atoms ranging from a few atoms to several hundreds of thousands atoms. The evolution of the properties of matter from atom to crystal can be described in terms of the two steps: *from atom to cluster* and *from cluster to crystal.*

The main distinctive feature of clusters is the discrete set of the number of atoms organized in a cluster. These so-called magic numbers determine unambiguously the spatial configuration, electronic spectra, and optical properties of clusters. Sometimes a transition from a given magic number to the neighboring one results in a drastic change in energy levels and optical transition probabilities. As the particle size grows, the properties can be described in terms of the particle size and shape instead of dealing with the particular number of atoms and spatial configuration. This type of microstructures can be referred to as *mesoscopic structures* as their size is always larger than the crystal lattice constant but comparable to the de Broglie wavelength of the elementary excitations. They are often called "quantum crystallites," "quantum dots," or "quasi-zero-dimensional structures." As the size of these crystallites ranges from one to tens of nanometers, the word "nanocrystals" is widely used as well. This term refers to the crystallites' size only, whereas the other terms hint at the interpretation of their electron properties in terms of quantum confinement effects.

From the standpoint of a solid state physicist, nanocrystals are just a kind of a low-dimensional structure complementary to quantum wells (two-dimensional structures) and quantum wires (one-dimensional structures). However, the finiteness of quasi-zero-dimensional species results in a number of specific features that are not inherent in the two- and one-dimensional structures. Quantum wells and quantum wires still possess a translational symmetry in one or two dimensions, and a statistically large number of electronic excitations can be created. In nanocrystals the translational symmetry is totally broken, and only a finite number of electrons and holes can be created within the same nanocrystal. Therefore, the concepts of electron-hole gas and quasi-momentum are not applicable to nanocrystals. Additionally, a finite number of atoms in nanocrystals promotes a variety of photoinduced phenomena like persistent and permanent photophysical and photochemical phenomena that are known in atomic and molecular physics but do not occur in solids. Finally, nanocrystals are fabricated by means of techniques borrowed from glass technology, colloidal chemistry, and other fields that have nothing in common with crystal growth.

From the viewpoint of molecular physics, a nanocrystal can be considered as a kind of large molecule. Similar to molecular ensembles, nanocrystals dispersed in a transparent host environment (liquid or solid) exhibit a variety of guest-host phenomena known for molecular structures. Moreover, every nanocrystal ensemble has inhomogeneously broadened absorption and emission spectra due to distribution of sizes, defect concentration, shape fluctuations, environmental inhomogeneities, and other features. Therefore, the most efficient way to examine the properties of a single nanocrystal that are smeared by inhomogeneous broadening is to use numerous selective techniques developed in molecular and atomic spectroscopy.

Additionally, as the size of crystallites and their concentration increase, the heterogeneous medium "matrix-crystallites" becomes a subject of the optics of ultradisperse media, thus introducing additional aspects to the optical properties of nanocrystal ensembles.

Because of these features, studies of the optical properties of nanocrystals form a new field bordering solid state physics, optics, molecular physics, and chemistry.

Despite the fact that matrices colored with semiconductor nanocrystals have been known for centuries as stained glass, systematic study of their physical properties began not long ago. Probably the first investigations of quasi-zero-dimensional structures were the pioneering works by Froelich (1937) and Kubo (1962), in which nontrivial properties of small metal particles were predicted due to a discreteness of electron spectra. The systematic studies of size-dependent optical properties of semiconductor nanocrystals have been stimulated by impressive advances in the quantum confinement approach for fine semiconductor layers (quantum wells) and needle-like structures (quantum wires).

The St. Petersburg school in Russia, which included solid state physics, optical spectroscopy, and glass technology (Ekimov et al. 1980, 1982; Efros and Efros 1982) and independently the Murray Hill group in the United States (Rossetti et al. 1983) were the first to outline the size-dependent properties of nanocrystals due to the quantum confinement effect. Since then, great progress in the field has been achieved due to extensive studies performed by thousands of researchers throughout the world. The advances in the theory of semiconductor quantum dots have been described thoroughly in the nice book by Banyai and Koch (1993). The present book is meant to summarize the progress in experimental studies of semiconductor nanocrystals.

Chapters 1–4 contain a brief description of the theoretical results of electron states in an idealized nanocrystal, a sketch of the growth techniques and structural properties, and a survey of the selective optical techniques and relevant optical effects known for other spectrally inhomogeneous media. These chapters are designed to provide an introductory overview, which seems reasonable considering the interdisciplinary nature of the field. Chapter 5 contains the systematic analysis of the size-dependent absorption and emission processes that can be described in terms of creation or annihilation of a single electron-hole pair within the same nanocrystal. The materials considered are II-VI (CdSe, CdS, ...), III-V (GaAs, InAs), and I-VII (CuCl, CuBr, AgBr) compounds, and nanocrystals of group IV elements (Si and Ge). In Chapter 6, a variety of many-body effects are considered, resulting in the intensity-dependent, or nonlinear optical, phenomena. A variety of crystallite-matrix interface processes that are responsible for the majority of photo-induced persistent and permanent effects such as stable spectral hole-burning or photodarkening, are the subject of Chapter 7. In Chapter 8 we consider the recent advances in the fabrication and description of spatially ordered ensembles of nano- and microcrystals, which is a challenging start towards artificial materials like three-dimensional superlattices of crystallites. The most intriguing kind of these structures is the so-called photonic crystal, which is to photons as an ordinary crystal is to electrons. In some respects, this field combined with nanocrystal optics leads to photonic engineering, providing structures with desirable spectrum, lifetime, and the propagation conditions.

The presentation style of this book was chosen to provide an introduction to, and an overview of, the field in a form understandable for senior and graduate students specialized in physics and chemistry and interested in solid state optics and engineering.

Writing of this book became possible owing to research performed over the world in the period of 1982–1997. During this time the author was at the B. I. Stepanov Institute of Physics of the National Academy of Sciences of Belarus at Minsk. I wish to express my sincere gratitude to my academic teacher Prof. V. P. Gribkovskii, who encouraged my scientific activity in my student years and promoted all my further initiatives. I am grateful to Prof. P. A. Apanasevich and to all my colleagues at the Stepanov Institute, who managed to maintain a creative atmosphere in spite of the unfavorable external conditions during the last decade. I am thankful to my co-workers Dr. L. Zimin, Dr. I. Germanenko, Dr. A. Kapitonov, Dr. E. Petrov, Dr. I. Malinovskii, Dr. A. Stupak, Dr. V. Lebed, and Dr. N. Nikeenko and to many colleagues from other institutions with whom the research on semiconductor nanocrystals has been performed. My special thanks are to Prof. C. Klingshirn and Dr. hab. U. Woggon (Kaiserslautern/Karlsruhe) for continuous collaboration during many years. I am very grateful to Prof. L. Brus (Murray Hill/New York), Prof. S. W. Koch (Tucson/Marburg), Prof. V. Tsekhomskii (St. Petersburg), Prof. L. Banyai (Frankfurt/Main), Prof. N. Peyghambarian (Tucson), Dr. A. Efros (Washington), and Prof. T. Itoh (Sendai) for ongoing stimulating discussions on the physics and chemistry of semiconductor nanocrystals. I am indebted to a number of prominent scientists over the world for kind permission to reproduce their excellent results in this book.

During the final stage of this book project the critical reading of selected chapters by Dr. V. Gurin, Dr. E. Petrov, and Dr. M. Artemyev was of great help, as was the assistance of N. Gritsuk who made the compuscript of the text and a large part of the artwork.

Last but not least, I should like to thank the publishing house of Cambridge University, especially Dr. P. Meyler, for the excellent and fruitful cooperation.

S. V. Gaponenko
Minsk, July 1997

1
Electron states in crystal

A lot of features connected with absorption and emission of light in nanocrystals can be understood in terms of the quantum confinement approach. In this approach, a nanocrystal is considered as a three-dimensional potential box in which photon absorption and emission result either in a creation or in an annihilation of some elementary excitations in an electron subsystem. These excitations are described in terms of quasiparticles known for bulk crystals, that is, electrons, holes, and excitons.

This chapter is meant to remind readers of some principal results from elementary quantum mechanics and to provide an elementary introduction to solid state physics, which is essential for the following chapters. We then depart from elementary "particle-in-a-box" problems and consider the properties of an electron in a periodic potential. In the next step, we introduce the concepts of effective mass and quasiparticles as elementary excitations of a many-body system. Finally, we give an idea of the low-dimensional structures that constitute, undoubtedly, one of the major fields of research in modern condensed-matter physics.

1.1 A few problems from elementary quantum mechanics

1.1.1 Particle in a potential well

To restate some basic properties of quantum particles that are necessary to consider electrons in a crystal, we start with a particle in a one-dimensional potential well (Fig. 1.1). The relevant time-independent Schroedinger equation can be written as

$$-\frac{\hbar^2}{2m}\frac{\partial^2}{\partial x^2}\psi(x) + U(x)\psi(x) = E\psi(x), \tag{1.1}$$

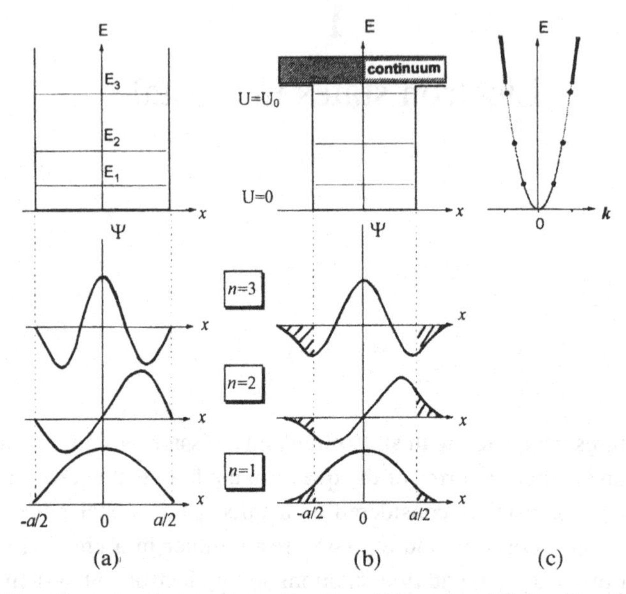

Fig. 1.1. One-dimensional potential well with infinite (a) and finite (b) walls, the first three states, corresponding to $n = 1, 2$, and 3, and the dispersion law in the case of the finite well (c). In the case of infinite walls, the energy states obey a series $E_n \sim n^2$ and the wave functions vanish at the walls. The total number of states is infinite. The probability of finding a particle inside the well is exactly equal to unity. In the case of finite walls, the states with energy higher than u_0 correspond to infinite motion and form a continuum. At least one state always exists within the well. The total number of discrete states is determined by the well width and height. The parameters in Fig. 1.1(b) correspond to the three states inside the well. Unlike case (a), the wavefunctions extend to the classically forbidden regions $|x| > a/2$. The probability of finding a particle inside the well is always less than unity and decreases with increasing E_n. A relation between E and k (dispersion law) in the case of a free particle has the form $E = \hbar^2 k^2/2m$ [dashed curve in Fig. 1.1(c)]. In the case of the finite potential well, a part of the dispersion curve relevant to confined states is replaced by discrete points [solid line and points in Fig. 1.1(c)].

where m is the particle mass, E is the particle energy, and the potential $U(x)$ is considered as a rectangular well with infinitely high walls, that is,

$$U(x) = \begin{cases} 0 & \text{for } |x| \leq a/2 \\ \infty & \text{for } |x| > a/2. \end{cases} \tag{1.2}$$

In Eq. (1.2) a denotes the well width. It is known from elementary quantum mechanics that Eq. (1.1) has the solutions of even and odd types given by expressions

$$\psi^{(-)} = \frac{\sqrt{2}}{\sqrt{a}} \cos\frac{1}{\hbar}\sqrt{2mE} \qquad (n = 1, 3, 5, \ldots) \tag{1.3}$$

and

$$\psi^{(+)} = \frac{\sqrt{2}}{\sqrt{a}} \sin \frac{1}{\hbar} \sqrt{2mE} \qquad (n = 2, 4, 6, \ldots). \tag{1.4}$$

The most important result of the problem is a discrete set of energy values given by

$$E_n = \frac{\pi^2 \hbar^2}{2ma^2} n^2. \tag{1.5}$$

In Fig. 1.1(a) the first three $\psi(x)$ functions for $n = 1, 2, 3$, and the positions of the energy levels are shown. The spacing between neighboring levels

$$\Delta E_n = E_{n+1} - E_n = \frac{\pi^2 \hbar^2 (2n + 1)}{2ma^2} \tag{1.6}$$

grows monotonically with n. The wavefunctions for every state vanish at $x \geq a$. The amplitude of all wavefunctions are the same, and the total probability to find a particle inside the box is exactly unity for all states.

Note that Eq. (1.5) gives values of *kinetic* energy. Using the relation between energy, momentum p, and wavenumber k

$$E = \frac{p^2}{2m}, \qquad p = \hbar k, \tag{1.7}$$

we can write the relevant momentum and wavenumber values

$$p_n = \frac{\pi \hbar}{a} n, \qquad k_n = \frac{\pi}{a} n \tag{1.8}$$

that take the discrete values as well.

If a particle exists, the quantity $\psi \psi^*$ must somewhere be nonzero. Thus, the solution satisfying (1.1) and (1.2) with $n = 0$ cannot be allowed, because this would deny the existence of a particle. The minimum energy a particle can have is given by

$$E_1 = \frac{\hbar^2}{2m} \frac{\pi^2}{a^2}. \tag{1.9}$$

This energy is called the particle's zero-point energy. It can be derived as a result of Heisenberg's uncertainty relation

$$\Delta p \Delta x \geq \frac{\hbar}{2}. \tag{1.10}$$

A particle is restricted to a region of space $\Delta x = a$. Hence, according to (1.10) it must have the uncertainty in its momentum $\Delta p \geq \hbar / 2a$. The latter corresponds to a minimum amount of energy

$$\Delta E = \frac{(\Delta p)^2}{2m} = \frac{\hbar^2}{8ma^2}, \tag{1.9'}$$

which resembles E_1 in (1.9) to an accuracy of $\pi^2/4$.

The parity of the particle wavefunction can be predicted from the symmetry of the problem. The symmetry of a potential well

$$U(x) = U(-x)$$

determines the symmetry of the particle density

$$|\psi(x)|^2 = |\psi(-x)|^2$$

whence

$$\psi(x) = \pm\psi(-x)$$

are the two independent solutions. Generally, the symmetry of wavefunctions can often be useful tools in solving the wave equation for a complicated system.

In the case of a finite height of the walls, wavefunction does not vanish at the edge of the well but exponentially falls inside the classically forbidden region $|x| > a/2$ [Fig. 1.1(b)]. A nonzero probability appears to find a particle outside the well. With growing n this probability increases. The number of the states inside the well is controlled by a condition

$$a\sqrt{2mU_0} > \pi\hbar(n-1), \tag{1.11}$$

where U_0 is the height of the well. The condition (1.11) always holds for $n = 1$. Therefore, there is at least one state inside the one-dimensional potential well with any combination of a and U_0. The possible number of states within the well corresponds to the maximum n value for which (1.11) still holds. In the case presented in Fig. 1.1(b) this number is equal to 3. The absolute position of the energy levels is somewhat lower for finite U_0 as compared with $U_0 \to \infty$ because the effective particle wavelength becomes larger. For deep states Eq. (1.5) can be considered as a good approximation. All states with $E_n > U_0$ correspond to infinite motion and form the continuum of states.

To give an idea of absolute values, consider an electron ($m = m_0$) inside an infinite well with $a = 1$ nm. The energy in this case takes the values $E_1 =$ 0.094 eV, $E_2 = 0.376$ eV, and so on. For comparison, note that kT value for room temperature is 0.025 eV. If we consider a transition from E_1 to E_2

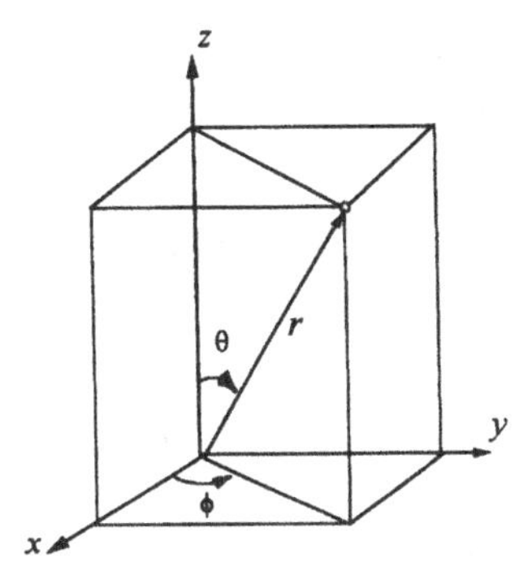

Fig. 1.2. Spherical coordinates.

state stimulated by a photon absorption, that is, $\hbar\omega = E_2 - E_1$, then the relevant photon wavelength in this particular case will be $\lambda = 4394$ nm corresponding to the middle infrared region.[1]

1.1.2 Particle in a spherically symmetric potential

In this case we deal with a Hamiltonian

$$H = -\frac{\hbar^2}{2m}\nabla^2 + U(r), \tag{1.12}$$

where $r = \sqrt{x^2 + y^2 + z^2}$. Taking into account the symmetry of the problem, it is reasonable to consider it in spherical coordinates, r, ϑ, and φ (Fig. 1.2):

$$x = r\sin\vartheta\cos\varphi, \qquad y = r\sin\vartheta\sin\varphi, \qquad z = r\cos\vartheta. \tag{1.13}$$

In spherical coordinates the Hamiltonian (1.12) reads

$$H = -\frac{\hbar^2}{2mr^2}\frac{\partial}{\partial r}\left(r^2\frac{\partial}{\partial r}\right) - \frac{\hbar^2\Lambda}{2mr^2} + U(r) \tag{1.14}$$

where Λ operator is

$$\Lambda = \frac{1}{\sin\vartheta}\left[\frac{\partial}{\partial\vartheta}\left(\sin\vartheta\frac{\partial}{\partial\vartheta}\right) + \frac{1}{\sin\vartheta}\frac{\partial^2}{\partial\varphi^2}\right]. \tag{1.15}$$

[1] 1 eV is equal to $1.602 \cdot 10^{-19}$ Joules. This unit is widely used in atomic and solid state physics. In molecular spectroscopy, the inverse wavelength λ^{-1} measured in cm^{-1} is often used instead of a photon energy $h\nu$. The relations between these units simply come from the formulas

$$E[\mathrm{J}] = h\nu, \qquad E[\mathrm{eV}] = h\nu/e, \qquad \lambda^{-1} = \nu/c,$$

whence

$$E[\mathrm{eV}] = \frac{hc}{\lambda e} = \frac{1.239\ldots \cdot 10^3}{\lambda[\mathrm{nm}]}, \qquad \lambda^{-1}[\mathrm{cm}^{-1}] = \frac{E[\mathrm{eV}]}{1.239\ldots}10^4.$$

We skip mathematical details and highlight only the principal results that arise from the spherical symmetry of the potential. In this case, the wavefunction is separable into functions of r, ϑ, and φ:

$$\psi = R(r)\Theta(\vartheta)\Phi(\varphi), \tag{1.16}$$

and can be written in the form

$$\psi_{n,l,m}(r, \vartheta, \varphi) = \frac{u_{n,l}(r)}{r} Y_{lm}(\vartheta, \varphi), \tag{1.17}$$

where Y_{lm} are the spherical functions, and $u(r)$ satisfies an equation

$$-\frac{\hbar^2}{2m}\frac{d^2u}{dr^2} + \left[U(r) + \frac{\hbar^2}{2mr^2}l(l+1)\right]u = Eu. \tag{1.18}$$

To obtain the energy values, it is possible now to consider the one-dimensional Eq. (1.18) instead of the equation with the Hamiltonian (1.14). The state of the system is characterized by the three quantum numbers, namely, the principal number n, the orbital number l, and the magnetic number m. The orbital quantum number determines the angular momentum value $\mathbf{L}$:

$$\mathbf{L}^2 = \hbar^2 l(l+1), \qquad l = 0, 1, 2, 3, \ldots \tag{1.19}$$

The magnetic quantum number determines the L component parallel to the z axis:

$$L_z = \hbar m, \qquad m = 0, \pm 1, \pm 2, \ldots \pm l. \tag{1.20}$$

Every state with a certain l value is $(2l+1)$ degenerate accordingly to $2l+1$ values of m. The states corresponding to different l values are usually denoted as s-, p-, d-, f-, and g-states and so forth in alphabetical order. For example, states with zero angular momentum ($l = 0$) are referred to as s-states, states with $l = 1$ are denoted as p-states, and so on. The parity of states corresponds to the parity of the l value, because the radial function is not sensitive to inversion (r remains the same after inversion) and spherical function after inversion transforms as follows:

$$Y_{lm}(\vartheta, \varphi) \rightarrow (-1)^l Y_{lm}(\vartheta, \varphi).$$

The specific values of energy are determined by the $U(r)$ function. Consider a simple case corresponding to a spherically symmetric potential well with an infinite barrier, that is,

$$U(r) = \begin{cases} 0 & \text{for } r \leq a \\ \infty & \text{for } r > a \end{cases}. \tag{1.21}$$

Table 1.1. *Roots of the Bessel functions* χ_{nl}

l	$n = 1$	$n = 2$	$n = 3$
0	3.142 (π)	6.283 (2π)	9.425 (3π)
1	4.493	7.725	10.904
2	5.764	9.095	12.323
3	6.988	10.417	
4	8.183	11.705	
5	9.356		
6	10.513		
7	11.657		

Source: Flugge 1971.

In this case energy values are expressed as follows:

$$E_{nl} = \frac{\hbar^2 \chi_{nl}^2}{2ma^2}, \tag{1.22}$$

where χ_{nl} are roots of the spherical Bessel functions with n being the number of the root and l being the order of the function. χ_{nl} values for several n, l values are listed in Table 1.1. Note that for $l = 0$ these values are equal to πn ($n = 1, 2, 3, \ldots$), and Eq. (1.22) converges with the relevant expression in the case of a one-dimensional box [Eq. (1.5)]. This results from the fact that for $l = 0$ Eq. (1.18) for the radial function $u(r)$ is just Eq. (1.1) with the potential (1.2). To summarize, a particle in a spherical well possesses the set of energy levels $1s$, $2s$, $3s$, ..., coinciding with energies of a particle in a rectangular one-dimensional well, and additional levels $1p$, $1d$, $1f$, ..., $2p$, $2d$, $2f$, ..., that arise due to spherical symmetry of the well (Fig. 1.3).

In the case of the spherical well with the finite potential, U_0, Eq. (1.22) can be considered as a good approximation only if U_0 is large enough, namely for $U_0 \gg h^2/8ma^2$. The right side of this inequality is a consequence of the uncertainty relation [see Eq. (1.9′)]. When

$$U_0 = U_{0\,\min} = \frac{\pi^2 \hbar^2}{8ma^2},$$

exactly one state exists within the well, $E_1 = U_0$. For $U_0 < U_{0\,\min}$, no state exists in the well at all. This is an important difference of the three-dimensional case as compared with the one-dimensional problem.

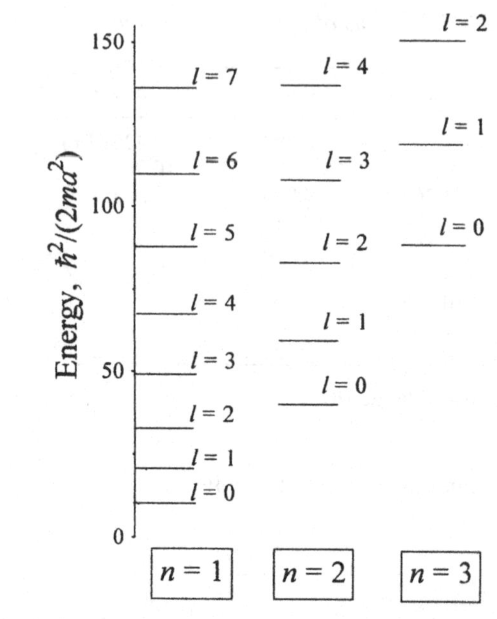

Fig. 1.3. Energy levels of a particle in a spherical well with infinite barrier. Energy is scaled in the dimensionless units of $\chi_{nl}^2 = E_{nl}(\hbar^2/2ma^2)^{-1}$, where χ_{nl} values are the roots of the Bessel functions listed in the Table 1.1. The states are classified by the principal quantum number, n, and by the orbital quantum number, l. Every state is $(2l + 1)$ degenerate. For $l = 0$ (so-called s-states) $\chi_{n0} = \pi n$ holds, and corresponding energies obey a series derived for a particle in a rectangular well [see Fig. 1.1(a)].

1.1.3 Electron in Coulomb potential

For the Coulomb potential

$$U(r) = -\frac{e^2}{r}, \tag{1.23}$$

the equation for the radial part of the wavefunction can be written as

$$\left[\frac{d^2}{d\rho} + \varepsilon + \frac{2}{\rho} - \frac{l(l+1)}{\rho^2}\right] u(\rho) = 0. \tag{1.24}$$

The dimensionless argument and energy

$$\rho = \frac{r}{a^0}, \qquad \varepsilon = \frac{E}{E^0}$$

are expressed in terms of the so-called atomic length unit a^0 and atomic energy unit E^0 given by

$$a^0 = \frac{\hbar^2}{m_0 e^2} \approx 5.292 \cdot 10^{-2}\ \text{nm} \tag{1.25}$$

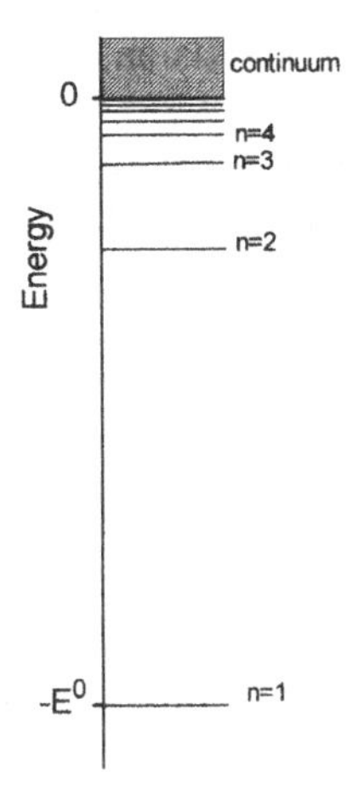

Fig. 1.4. Energy levels of a particle in the Coulomb potential $U(r) = -e^2/r$. For $E > 0$ a particle exhibits infinite motion with a continuous energy spectrum. For $E < 0$ the energy spectrum consists of a discrete set of levels obeying the relation $E_n = -E^0/n^2$, each level being n^2 degenerate.

and

$$E^0 = \frac{e^2}{2a^0} \approx 13.60\ \text{eV} \tag{1.26}$$

with m_0 being the electron mass. The solution of Eq. (1.24) leads to the following result.

Energy levels obey a series

$$\varepsilon = -\frac{1}{(n_r + l + 1)^2} \equiv -\frac{1}{n^2}, \tag{1.27}$$

which is shown in Fig. 1.4. The number $n = n_r + l + 1$ is called the "principal quantum number." It takes positive integer values beginning with 1. The energy is unambiguously determined by a given n value. n_r determines the quantity of nodes of the corresponding wavefunction. It is called the "radial quantum number." For every n value, exactly n states exist differing in l, which runs from 0 to $(n - 1)$. Additionally, for every given l value, $(2l + 1)$ – degeneracy occurs with respect to $m = 0, \pm1, \pm2 \ldots$. Therefore, the total degeneracy is

$$\sum_{l=0}^{n-1}(2l + 1) = n^2.$$

For $n = 1, l = 0$ (1s-state), the wavefunction obeys a spherical symmetry with a^0 corresponding to the most probable distance where an electron can be found. Therefore, the relevant value in real atom-like structures is called "Bohr radius." For $E > 0$ a particle exhibits an infinite motion with a continuous spectrum.

So far, idealized elementary problems have been examined. Now we are in a position to deal with the simplest real quantum mechanical object, that is, with the hydrogen atom consisting of a proton with the mass M_0 and of an electron. The relevant Schroedinger equation is the two-particle equation with the Hamiltonian

$$H = -\frac{\hbar^2}{2M_0}\nabla_p^2 - \frac{\hbar^2}{2m_0}\nabla_e^2 - \frac{e^2}{|\mathbf{r}_p - \mathbf{r}_e|}, \qquad (1.28)$$

where $\mathbf{r}_p$ and $\mathbf{r}_e$ are the radius-vectors of the proton and electron, and p and e indices in the ∇^2- operator denote differentiation with respect to the proton and electron coordinates, respectively. We introduce a relative radius-vector $\mathbf{r}$ and a radius-vector of the center of mass as follows:

$$\mathbf{r} = \mathbf{r}_p - \mathbf{r}_e, \qquad \mathbf{R} = \frac{m_0\mathbf{r}_e + M_0\mathbf{r}_p}{m_0 + M_0} \qquad (1.29)$$

and use the full mass and the reduced mass of the system, M and μ:

$$M = m_0 + M_0, \qquad \mu = \frac{m_0 M_0}{m_0 + M_0}. \qquad (1.30)$$

Hamiltonian (1.28) then reads

$$H = -\frac{\hbar^2}{2M}\nabla_R^2 - \frac{\hbar^2}{2\mu}\nabla_r^2 - \frac{e^2}{r}. \qquad (1.31)$$

One can see that (1.31) diverges into the Hamiltonian of a free particle with the mass M and the Hamiltonian of a particle with the mass μ in the potential $-e^2/r$. The former describes an infinite center-of-mass motion of the two-particle atom, whereas the latter gives rise to internal states. According to (1.27), the energy of these states can be written as

$$E_n = -\frac{Ry}{n^2} \quad \text{for } E < 0 \qquad (1.32)$$

with

$$Ry = \frac{e^2}{2a_B}, \qquad a_B = \frac{\hbar^2}{\mu e^2}. \qquad (1.33)$$

Here Ry is called the "Rydberg constant" and corresponds to the ionization energy of the lowest state , and a_B is the Bohr radius of a hydrogen atom.

The distance between the neighboring levels decreases with n, and for $E > 0$ electron and proton experience an infinite motion.

One can see that the energy spectrum and Bohr radius expressed by (1.33) differ from the relevant values of a single-particle problem by the μ/m_e coefficient. In the case under consideration this coefficient is 0.9995. For this reason expressions (1.25) and (1.26) are widely used instead of the exact values (1.33). This is reasonable in the case of a proton and electron but should be used with care for other hydrogen-like systems. For example, in a positronium atom, consisting of an electron and a positron with equal masses, the explicit values (1.33) should be used.

The problems of a particle in a spherical potential well and of the hydrogen atom are very important for further consideration. The former is used to model an electron and a hole in a nanocrystal, and the latter is essential for excitons in a bulk crystal and in nanocrystals, as well. Furthermore, the example of a two-particle problem is a precursor to the general approach used for many-body systems. It contains a transition from the many-particle problem (proton and electron) to the one-particle problem by means of renormalization of mass (reduced mass μ instead of M_0 and m_0) and a differentiation between the collective behavior (center-of-mass translational motion) and the single-particle motion in some effective field. This approach has far-reaching consequences resulting in the concepts of effective mass and of quasiparticles to be presented in Sections 1.2 and 1.3.

1.1.4 Particle in a periodic potential

Consider a particle in a potential, which satisfies

$$U(x) = U(x + a), \tag{1.34}$$

that is, potential energy is invariant with respect to translation in space by a.

We start with the general properties of wavefunctions satisfying the Schroedinger equation with the potential (1.34). If the argument x is replaced by $(x + a)$,

$$x \rightarrow x + a,$$

one gets an equation

$$-\frac{\hbar^2}{2m}\nabla^2\psi(x + a) + U(x)\psi(x + a) = E\psi(x + a). \tag{1.35}$$

One can see, comparing (1.35) and (1.1), that wavefunctions $\psi(x)$ and $\psi(x+a)$ satisfy the same Schroedinger equation with the same eigenvalue E. If this eigenvalue is nondegenerate (i.e., it has only one eigenfunction), then the wavefunctions $\psi(x)$ and $\psi(x+a)$ may differ in a constant coefficient only,

$$\psi(x+a) = c\psi(x). \tag{1.36}$$

As both eigenfunctions should be normalized, the absolute c value should be

$$|c| = 1.$$

Hence,

$$|\psi(x+a)|^2 = |\psi(x)|^2, \tag{1.37}$$

that is, a particle can be found in the interval Δx near x point with the same probability as near the other point $x+a$, which is equivalent to the x point. Therefore, the average spatial distribution of particles possesses the spatial periodicity of the potential.

Consider the properties of the c_0 value. After two translations, one has

$$\psi\left(x + a_{n_1} + a_{n_2}\right) = c_{n_1} c_{n_2} \psi(x) \tag{1.36'}$$

where

$$a_n = na, \qquad n = 1, 2, 3, \ldots.$$

Taking into account an evident relation

$$a_{n_1} + a_{n_2} = a_{n_1+n_2},$$

one finds that

$$\psi\left(x + a_{n_1} + a_{n_2}\right) \equiv \psi\left(x + a_{n_1+n_2}\right) = c_{n_1+n_2} \psi(x), \tag{1.36''}$$

whence

$$c_{n_1} c_{n_2} = c_{n_1+n_2}. \tag{1.38}$$

This equation has the solution

$$c_n = e^{ika_n}, \tag{1.39}$$

in which k may take any value.

To summarize, wavefunctions that satisfy the Schroedinger equation with a periodic potential can differ from the function that is periodic with the period a only in the phase coefficient of the form $e^{if(x)}$ with $f(x)$ being a linear function of x. Such a wavefunction can be written as

$$\psi(x) = e^{ikx} u_k(x), \qquad u_k(x) = u_k(x + a_n). \tag{1.40}$$

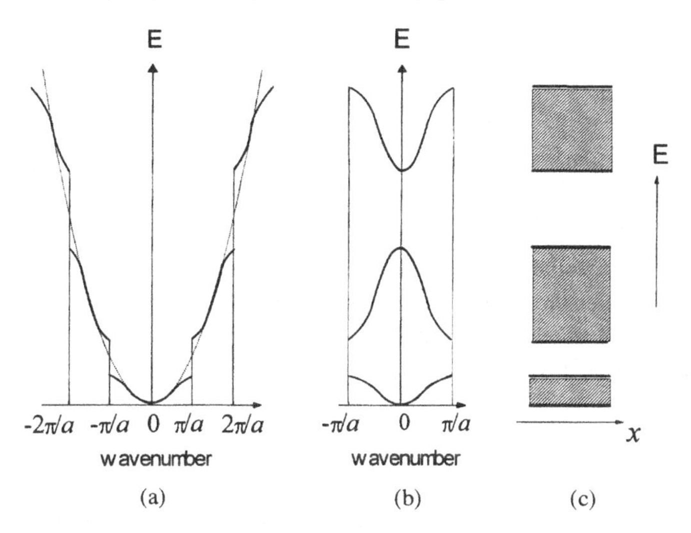

Fig. 1.5. Extended (a) and reduced (b) presentation of the dispersion law of a particle in a one-dimensional periodic potential, and the corresponding energy bands in space (c). The dashed curve in (a) corresponds to the $E = \hbar^2 k^2/2m$ function, which describes the kinetic energy of a free particle. The dispersion curve in the presence of a periodic potential with a period a is shown by the solid lines. It has discontinuities at $k = \pi n/a$, where n is integer. These k values correspond to standing waves that cannot propagate because of multiple reflections from periodic boundaries. Therefore, the energy spectrum breaks into bands separated by forbidden gaps. As the k values differing by $\pi n/a$ appear to be equivalent due to the translational symmetry of space, a reduced dispersion curve (b) can be plotted. It results from (a) by means of a shift of several branches by $2\pi/a$.

Eq. (1.40) means that the eigenfunction of the Hamiltonian with a periodic potential is a plane wave modulated with the same period as the potential. This statement is known as Bloch's theorem.

In what follows, we skip the details and restrict ourselves to the basic results only. The wavenumbers k_1, k_2 differing by a value

$$k_1 - k_2 = \frac{2\pi}{a}n, \qquad n = \pm 1, \pm 2, \pm 3, \ldots, \tag{1.41}$$

appear to be equivalent. This is a direct consequence of a translational symmetry of the space. Therefore, the whole multitude of the k values consists of the equivalent intervals

$$-\frac{\pi}{a} < k < \frac{\pi}{a}; \qquad \frac{\pi}{a} < k < \frac{3\pi}{a}; \qquad \frac{3\pi}{a} < k < \frac{5\pi}{a}; \ldots, \tag{1.42}$$

with the width of $2\pi/a$ each. Each of these intervals contains the full set of the nonequivalent k values and is called the "Brillouin zone." The energy spectrum and the dispersion curve differ from those of a free particle (Fig. 1.5). The dispersion curve has discontinuities at points

$$k_n = \frac{\pi}{a} n; \qquad n = \pm 1, \pm 2, \pm 3, \ldots . \tag{1.43}$$

At this k value the wavefunction is a *standing wave* that arises as a result of multiple reflections from the periodic structure. For every k_n satisfying (1.43), two standing waves exist with different potential energies. This leads to emergence of forbidden energy intervals for which no propagating waves exist. Typically, it is convenient to consider the first Brillouin zone only. Therefore, the extended dispersion curve [Fig. 1.5(a)] can be modified to yield the reduced zone scheme [Fig. 1.5(b)].

The value

$$p = \hbar k \tag{1.44}$$

is called "quasi-momentum." It differs from the momentum by a specific conservation law. It conserves with an accuracy of $2\pi\hbar/a$, which is, again, a direct consequence of a translational symmetry of the space. Although the $E(k)$ relation in the case under consideration differs noticeably from that for a free particle, one can formally express it in the form

$$E(k) = \frac{\hbar^2 k^2}{2m^*(k)}, \tag{1.45}$$

where $m^*(k)$ is a function referred to as "effective mass." In a number of practically important cases, this function can be considered as constant. For every periodic potential, there exist extrema in the band structure. In the vicinity of a given extremum, $E_0(k_0)$, one can write the expansion

$$E(k) = E_0 + (k - k_0)\left.\frac{dE}{dk}\right|_{k=k_0} + \frac{1}{2}(k - k_0)^2 \left.\frac{d^2E}{dk^2}\right|_{k=k_0} + \cdots . \tag{1.46}$$

If the energy is measured from E_0, that is, $E_0 = 0$, and the wavenumber is measured from k_0, that is, $k_0 = 0$, then bearing in mind that $dE(k)/dk = 0$ at the extremum, one has

$$E(k) = \frac{1}{2} k^2 \left.\frac{d^2E}{dk^2}\right|_{k=0} + \cdots . \tag{1.47}$$

Neglecting the contribution from terms higher than k^2, which is justified near an extremum, we just come to Eq. (1.45) with

$$m^{*-1} = \frac{1}{\hbar^2} \left.\frac{d^2E}{dk^2}\right|_{k=0} = \text{const.} \tag{1.48}$$

Note that for a free particle from the relation $E = \hbar^2 k^2/2m$, we have everywhere

$$\frac{1}{\hbar^2}\frac{d^2E}{dk^2} \equiv m^{-1}.$$

Eq. (1.47) with the omitted terms higher than k^2 corresponds to the so-called parabolic band, which is a very helpful approximation in a number of problems dealing with an electron in a periodic crystal lattice.

The effective mass (1.48) determines the reaction of a particle to the external force, F, via a relation

$$m^* a = F, \tag{1.49}$$

where a is the acceleration. Eq. (1.49) coincides formally with Newton's second law. Comparing Fig. 1.5(a) and (b), one can see that, for example, in the vicinity of the minimum point the effective mass is noticeably smaller than the intrinsic inertial mass of a particle. This is evident, because the curvature of the $E(k)$ function, which is just equal to the second derivative, is larger in case (b) near the extremum point than in case (a), shown by a dashed line. Therefore, a particle in a periodic potential sometimes can be "lighter" than in the free space. Sometimes, however, it can be "heavier." Moreover, it can even possess a negative mass. This corresponds to the positive curvature of the $E(k)$ dependence in the vicinity of the maximum. The negative effective mass is not an artifact but an important property peculiar to a particle, which interacts simultaneously with a background periodic potential and with an additional perturbative potential. The negative mass means that momentum of a particle decreases in the presence of an extra potential. This happens because of reflection from the periodic boundaries and can be understood, for example, from the extended dispersion curve in Fig. 1.5(a). The difference of momentum does not vanish but is transferred to the material system responsible for the periodic potential, for example, the ion lattice of the crystal.

To summarize the properties of a particle in a periodic potential, we outline a few principal results. First, a particle is described by a plane wave modulated with a period of the potential. Second, the particle state is characterized by the quasi-momentum. The latter has a set of equivalent intervals, the Brillouin zones, each containing the complete multitude of nonequivalent values. Third, the energy spectrum consists of wide continuous bands separated from each other by forbidden gaps. As a plane wave, a particle in a periodic potential exhibits quasi-free motion without an acceleration. With respect to the external force, the particle's behavior is described in terms of the effective mass. The latter is, basically, a complicated function of energy, but can be considered a constant in the vicinity of a given extremum of the $E(k)$ curve. Generally, the renormalization of mass is simply a result of the interaction of a particle with a given type of the periodic potential.

With this information we shall proceed to electrons in crystal. More detail on the problems considered in this section can be found in textbooks on quantum mechanics (Davydov 1965; Flugge 1971; Landau and Lifshitz 1989; Schiff 1968).

1.2 Schroedinger equation for an electron in a crystal

Consider an ideal crystal with periodic arrangement of atoms. The Hamiltonian of this system should include the kinetic energy of every electron, the kinetic energy of every nucleus, the potential energy of electron-electron interactions, the potential energy of electron-nucleus interactions, and the potential energy of nucleus-nucleus interactions. Therefore, it can be written as

$$H = -\sum_i \frac{\hbar^2}{2m_0}\nabla_i^2 - \sum_a \frac{\hbar^2}{2M}\nabla_a^2 + \frac{1}{2}\sum_{i\neq j} U_1(\mathbf{r}_i - \mathbf{r}_j) + \sum_{i,a} U_2(\mathbf{r}_i - \mathbf{R}_a) + \frac{1}{2}\sum_{a\neq b} U_3(\mathbf{R}_a - \mathbf{R}_b). \tag{1.50}$$

In Eq. (1.50) m_0 and M are the electron and the nucleus masses, $\mathbf{r}$ and $\mathbf{R}$ are the electron and the nucleus radius-vectors. Evidently, it is not possible to solve an equation with Hamiltonian (1.50) for a number of particles in the range of 10^{22}–10^{23}. Therefore, several sequential approximations are developed to deal with this problem.

First, as the nucleus mass, M, is much greater than the electron mass, m_0, nuclei are considered as motionless when electron properties of a crystal are examined. This is known as the *adiabatic approximation* or the Born-Oppenheimer approximation. Using this approximation, the wavefunction can be separated into two functions, depending either on electron coordinates or on nucleus coordinates, to yield two independent Schroedinger equations: one for the nuclear subsystem and another for the electron subsystem. As we are interested in electron properties of a crystal, we write only the latter:

$$-\frac{\hbar^2}{2m_0}\sum_i \nabla_i^2\psi + \frac{1}{2}\sum_{i\neq j} U_1(\mathbf{r}_i - \mathbf{r}_j)\psi + \sum_{i,a} U_2(\mathbf{r}_i - \mathbf{R}_a)\psi = E_R\psi. \tag{1.51}$$

In Eq. (1.51) radius-vectors of nuclei $\mathbf{R}_a$ are parameters but not variables. The wavefunction ψ depends on the whole set of electron coordinates and includes the set of nuclei coordinates as parameters. The parametric dependence of eigenvalues E_R on nuclei coordinates is marked by a proper index.

Second, electrons of inner shells that are tightly bound to nuclei, and electrons of the external shell (valent electrons), are considered in a different way. The former do not determine the electron properties like conductivity, optical transitions, and others and can therefore be considered as lattice components. This means that, instead of nuclei, we deal with ion cores. Therefore, the second term in Eq. (1.51) is only the Coulomb interaction between valent electrons and can be expressed as

$$\frac{1}{2}\sum_{i\neq j} U_1(\mathbf{r}_i - \mathbf{r}_j) = \frac{1}{2}\sum_{i\neq j}\frac{e^2}{|\mathbf{r}_i - \mathbf{r}_j|}. \tag{1.52}$$

Third, under certain conditions the many-particle problem (1.51) can be reduced to a set of one-particle problems by means of the *self-consistent field approximation*. In this procedure, known as the Hartree-Fock method, interactions of each valent electron with all other valent electrons and with all ion cores are accounted for by introducing a periodic potential $u(\mathbf{r})$ that must be adjusted using the symmetry of the lattice and some empirical data to provide the really observable band structure of a given crystal.

According to this program, the Schroedinger equation with Hamiltonian (1.50) reduces to the single-particle equation

$$-\frac{\hbar^2}{2m_0}\nabla^2 + U(r) = E\psi \tag{1.53}$$

with a periodic potential that, in turn, can be reduced to the equation for a free particle by means of the mass renormalization

$$-\frac{\hbar^2}{2m^*}\nabla^2 = E\psi. \tag{1.54}$$

As we have already seen in Section 1.1, the energy spectrum of an electron consists of the bands separated by forbidden gaps. The electronic properties of solids are determined by occupation of the bands and by the absolute values of the forbidden gap between the completely occupied and the partly unoccupied or the free bands. If a crystal has a partly occupied band, it exhibits *metal* properties because electrons in this band provide electrical conductivity. If all the bands at $T = 0$ are either occupied or completely free, material will show *dielectric* properties. Electrons within the occupied band cannot provide any conductivity because of *Pauli's exclusion principle*: only one electron may occupy any given state. Therefore, under an external field an electron in the completely occupied band cannot change its energy because all neighboring

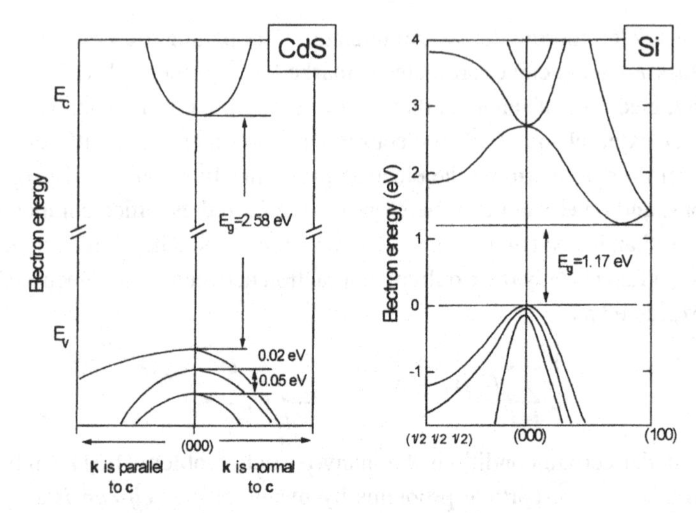

Fig. 1.6. Band structures of the two representative semiconductors, CdS and Si (after Blakemore 1985). In CdS the top of the valence band and the bottom of the conduction band correspond to the same wave number, i.e., CdS is a "direct-gap" semiconductor. In Si the extrema of the conduction and the valence band correspond to the different wavenumbers, i.e., Si is an "indirect-gap" semiconductor.

states are already filled. The highest occupied band is usually referred to as the *valence band*, and the lowest unoccupied band is called the *conduction band*. The interval between the top of the valence band, E_v, and the bottom of the conduction band, E_c, is called the *band gap energy*, E_g:

$$E_g = E_c - E_v. \tag{1.55}$$

Depending on the absolute E_g value, solids that show dielectric properties (i.e., zero conductivity) at $T = 0$ are classified into dielectrics and semiconductors. If E_g is less than 3–4 eV, the conduction band has a non-negligible population at elevated temperatures, and this type of crystal is called a *semiconductor*.

The dispersion curve $E(k)$ for real crystals is rather complicated. The effective mass cannot be considered as a constant, and in a number of cases, can be described as a second rank tensor. However, in a lot of practically important cases, the events within the close vicinity of the E_c and E_v are most important and can be described under the approximation of the constant effective mass, but are sometimes different for different directions. The band structures of the two representative semiconductors, cadmium sulfide and silicon, are given in Fig. 1.6. For CdS crystals the minimal gap between E_c and E_v occurs at the same k value. Crystals of this type are called *direct-gap semiconductors*.

Table 1.2. *Parameters of the most common semiconductors*

	Band gap energy Eg (eV)	Exciton Rydberg Ry^* (meV)	Electron effective mass m_e/m_0	Hole effective mass m_h/m_0[a]	Exciton Bohr radius a_B (nm)
Ge	0.744[b]		$\perp$ 0.19 $\parallel$ 0.92	0.54 (hh) 0.15 (lh)	
Si	1.17[b]	15	$\perp$ 0.081 $\parallel$ 1.6	0.3 (hh) 0.043 (lh)	4.3
GaAs	1.518	5	0.066	0.47 (hh) 0.07 (lh)	12.5
CdTe	1.60		0.1	0.4	
CdSe	1.84	16	0.13	$\perp$ 0.45 $\parallel$ 1.1	4.9
CdS	2.583	29		$\perp$ 0.7 $\parallel$ 2.5	2.8
ZnSe	2.820	19	0.15	0.8 (hh) 0.145 (lh)	3.8
AgBr	2.684[b]	16			4.2
CuBr	3.077	108	0.25	1.4 (hh)	1.2
CuCl	3.395	190	0.4	2.4 (hh)	0.7

[a] hh – heavy hole, lh – light hole
[b] Indirect band gap
Source: After Landholt-Boernstein 1982.

For the Si crystal the minimal energy gap corresponds to the different k values for E_c and E_v. This type of crystal is usually referred to as an *indirect-gap semiconductor.* The band gap energies of the most common semiconductors are given in Table 1.2.

1.3 Concept of quasiparticles: electron, hole, and exciton

Electrons in the conduction band of a crystal, as we have seen in Section 1.2, can be described as particles with charge $-e$, spin 1/2, mass m_e^* (basically variable rather than constant), and quasi-momentum $\hbar k$ with the specific conservation law. One can see that among the above-mentioned parameters, only the charge and the spin remain the same for an electron in a vacuum and in a crystal. Therefore, when speaking about electrons in the conduction band, we mean particles whose properties result from the interactions in a many-body system consisting

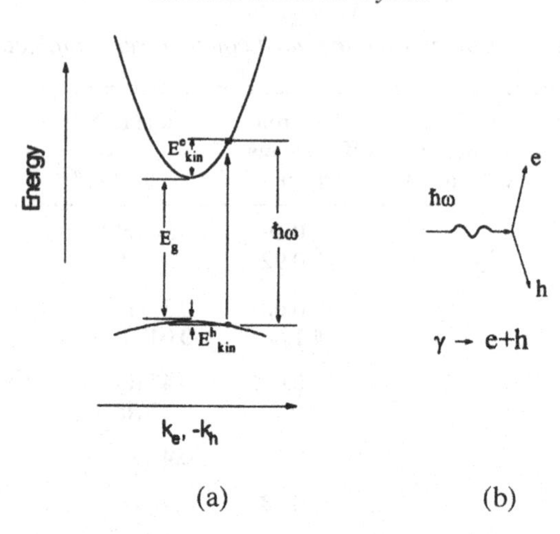

Fig. 1.7. A process of a photon absorption resulting in a creation of one electron-hole pair in different presentations. In a diagram including dispersion curves for conduction and valence band this event can be shown as a vertical transition exhibiting simultaneous energy and momentum conservation (a). This event also may be treated as a conversion of a photon into electron and hole (b).

of the large number of positive nuclei and negative electrons. It is the standard approach in the theory of many-body systems to replace a consideration of the large number of interacting particles by the small number of noninteracting *quasiparticles*. These quasiparticles are described as elementary excitations of the system consisting of a number of real particles. Within the framework of this consideration, an electron in the conduction band is the primary elementary excitation of the electron subsystem of a crystal. The further elementary excitation is a *hole*, which is a quasiparticle relevant to an ensemble of electrons in the valence band from which one electron is removed (e.g., to the conduction band). This excitation is characterized by the positive charge $+e$, spin $1/2$, effective mass m_h^*, and a proper quasi-momentum. In this presentation, the energy of the hole has an opposite sign as compared with the electron energy.

Using the concepts of elementary excitations, we can consider the ground state of a crystal as a *vacuum state* (neither an electron in the conduction band nor a hole in the valence band exists), and the first excited state (one electron in the conduction band and one hole in the valence band) in terms of a creation of one electron-hole pair (e-h pair). A transition from the ground to the first excited state occurs as the result of some external perturbation, for example, photon absorption (Fig. 1.7) with the energy and momentum conservation

$$\hbar\omega = E_g + E_{e\ \mathrm{kin}} + E_{h\ \mathrm{kin}},$$
$$\hbar\mathbf{k} = \hbar\mathbf{k}_e + \hbar\mathbf{k}_h. \qquad (1.56)$$

As the photon momentum is negligibly small, we simply have the vertical transition in the diagram shown in Fig. 1.7(a). This process can be described in another way in the form presented in Fig. 1.7(b). The reverse process, that is, a downward radiative transition equivalent to annihilation of the e-h pair and creation of a photon, is possible as well. These events and concepts have a lot in common with the real vacuum, electrons, and positrons. The only difference is that the positron mass is exactly equal to the electron mass m_0, whereas in a crystal the hole effective mass m_h^* is usually larger than the electron mass m_e^* (see Table 1.2). Being fermions, electrons and holes are described by the *Fermi-Dirac statistic* with the distribution function

$$f(E) = \frac{1}{\exp\dfrac{E - E_F}{kT} + 1} \tag{1.57}$$

ranging from 0 to 1. Here E_F is the chemical potential commonly referred to as the Fermi energy or the Fermi level.

The band gap energy corresponds to the minimal energy that is sufficient for creation of one pair of free charge carriers, that is, electron and hole. This statement can serve as the definition of E_g.

A description based on noninteracting electrons and holes as the only elementary excitations corresponds to the so-called single-particle presentation. In reality, electrons and holes as charged particles do interact via Coulomb potential and form an extra quasiparticle that corresponds to the hydrogen-like bound state of an electron-hole pair and is denoted as an *exciton*. Interacting hole and electron can be described by a Hamiltonian,

$$H = -\frac{\hbar^2}{2m_e^*}\nabla_e^2 - \frac{\hbar^2}{2m_h^*}\nabla_h^2 - \frac{e^2}{\varepsilon|\mathbf{r}_e - \mathbf{r}_h|}, \tag{1.58}$$

which is the same as the Hamiltonian (1.28) of the hydrogen atom with m_e^* and m_h^* instead of m_0 and M, and with the dielectric constant of the crystal $\varepsilon \neq 1$. Therefore, similarly to the hydrogen atom, exciton is characterized by the *exciton Bohr radius*

$$a_B = \frac{\varepsilon\hbar^2}{\mu e^2} = \varepsilon\frac{m_0}{\mu} \times 0.53 \text{ Å}, \tag{1.59}$$

where μ is the electron-hole reduced mass

$$\mu^{-1} = m_e^{*-1} + m_h^{*-1}, \tag{1.60}$$

and by the *exciton Rydberg energy*

$$Ry^* = \frac{e^2}{2\varepsilon a_B} = \frac{\mu e^4}{2\varepsilon^2\hbar^2} = \frac{\mu}{m_0}\frac{1}{\varepsilon^2} \times 13.6\,\text{eV}. \tag{1.61}$$

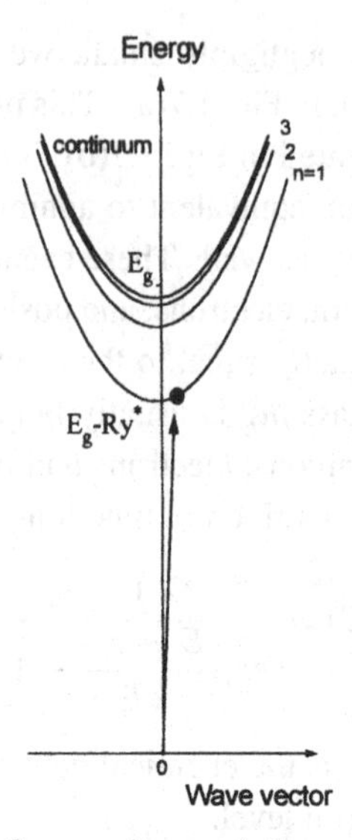

Fig. 1.8. Dispersion curves of an exciton and the optical transition corresponding to a photon absorption and exciton creation. Dispersion curves correspond to the hydrogenlike set of energies $E_n = E_g - Ry^*/n^2$ at $K = 0$ and a parabolic $E(K)$ dependence for every E_n, describing the translational center-of-mass-motion. For $E > E_g$, the exciton spectrum overlaps with the continuum of unbound electron-hole states. An exciton creation can be presented as intercrossing of the exciton and the photon dispersion curves corresponding to the simultaneous energy and momentum conservation. The photon dispersion curve is a straight line in agreement with the formula $E = pc$. This event also can be described as a conversion of a photon into an exciton.

The reduced electron-hole mass is smaller than electron mass m_0, and the *dielectric constant* ε is several times larger than that of a vacuum. This is why the exciton Bohr radius is significantly larger and the exciton Rydberg energy is significantly smaller than the relevant values of the hydrogen atom. Absolute values of a_B for the common semiconductors range in the interval 10–100 Å, and the exciton Rydberg energy takes the values approximately 1–100 meV (Table 1.2).

An exciton exhibits translational center-of-mass motion as a single uncharged particle with the mass $M = m_e^* + m_h^*$. The dispersion relation can be expressed as

$$E_n(\mathbf{K}) = E_g - \frac{Ry^*}{n^2} + \frac{\hbar^2\mathbf{K}^2}{2M}, \tag{1.62}$$

where $\mathbf{K}$ is the exciton wave vector. Eq. (1.62) includes the hydrogen-like set of energy levels, the kinetic energy of the translational motion, and the band gap energy. The exciton energy spectrum consists of subbands (Fig. 1.8) that converge to the dissociation edge corresponding to the free e-h pair. Similarly to the free e-h pairs, excitons can be created by photon absorption. Taking into account that a photon has a negligibly small momentum, exciton creation corresponds to the discrete set of energies

$$E_n = E_g - \frac{Ry^*}{n^2}. \tag{1.63}$$

Exciton gas can be described as a gas of bosons with the energy distribution function obeying the *Bose-Einstein statistic*

$$f(E) = \frac{1}{\exp\frac{E-\varepsilon}{kT} - 1}, \tag{1.64}$$

where ε is the chemical potential. For a given temperature T, the concentration of excitons n_{exc} and of the free electrons and holes $n = n_e = n_h$ are related via the *ionization equilibrium equation* known as the *Saha equation*:

$$n_{\text{exc}} = n^2 \left(\frac{2\pi\hbar^2}{kT} \frac{m_e^* + m_h^*}{m_e^* m_h^*} \right)^{3/2} \exp\frac{Ry^*}{kT}. \tag{1.65}$$

For $kT \gg Ry^*$ most of excitons are ionized and the properties of the electron subsystem of the crystal are determined by the free electrons and holes. At $kT \leq Ry^*$ a significant part of e-h pairs exists in the bound state.

As a result of a creation of excitons and free e-h pairs, the absorption spectrum of direct-gap semiconductor monocrystals contains a pronounced resonance peak at the energy $\hbar\omega = E_g - Ry^*$, a set of smaller peaks at the energies E_n [Eq. (1.63)], and the smooth continuous absorption for $\hbar\omega \geq E_g$ (Fig. 1.9).

Quasiparticles in solids are described in detail in a number of textbooks (Blakemore 1985; Haar 1958; Haug and Koch 1990; Kittel 1986; Klingshirn 1995). Additionally, excitons are the subject of a number of books and reviews (Cho 1979; Davydov 1976; Honerlage, Levy, Grun, et al. 1985; Knox 1963; Rashba and Sturge 1985). The problem of ionization equilibrium is analyzed by Paierls (1979) and Landau and Lifshitz (1988).

1.4 Low-dimensional structures: quantum wells, quantum wires, and quantum dots

In semiconductors the de Broglie wavelength of an electron and a hole, λ_e, λ_h, and the Bohr radius of an exciton, a_B, may be considerably larger than the lattice

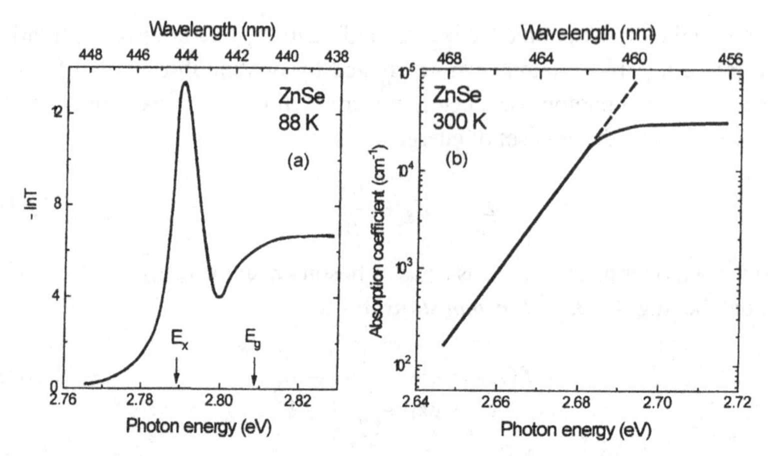

Fig. 1.9. Absorption spectrum of ZnSe single crystal near the fundamental absorption edge at temperatures equal to 88 K (a) and 300 K (b) (Gribkovskii et al. 1990). Zinc selenide possesses the band-gap energy $E_g = 2.809$ eV at $T = 80$ K and 2.67 eV at $T = 300$ K, the exciton Rydberg energy is $Ry^* = 18$ meV. At $kT \ll Ry^*$ the spectrum contains a pronounced peak of the exciton absorption corresponding to $n = 1$ [Eq. (1.63)] with a half-width close to the kT value. The higher subbands are smeared due to thermal broadening. At $kT > Ry^*$ the exciton band is not pronounced but a significant enhancement of absorption at $\hbar\omega < E_g$ due to electron-hole Coulomb interaction occurs. The longwave absorption tail shows an exponential dependence of the absorption coefficient on the photon energy (the Urbach rule) and corresponds to a straight line in a semilogarithmic scale.

constant, a_L. Therefore, it is possible to create a mesoscopic structure, which is in one, two, or three dimensions comparable to or even less than λ_e, λ_h, a_B but still larger than a_L. In these structures elementary excitations will experience quantum confinement resulting in finite motion along the confinement axis and infinite motion in other directions. Modern technological advances provide an opportunity to fabricate low-dimensional structures with size restricted to a few nanometers.

In the case of the size restriction in one dimension, we get a two-dimensional structure, the so-called quantum well. In the case of the two-dimensional confinement the relevant one-dimensional structure is referred to as quantum wire. Finally, if the motion of electrons, holes, and excitons is restricted in all three directions, we come to a quasi-zero-dimensional system, the so-called quantum dot.

In the two- and one-dimensional quantum confined structures, quasiparticles at low concentration can be considered as an ideal gas similar to the three-dimensional crystal. The density of electron and hole states can be expressed

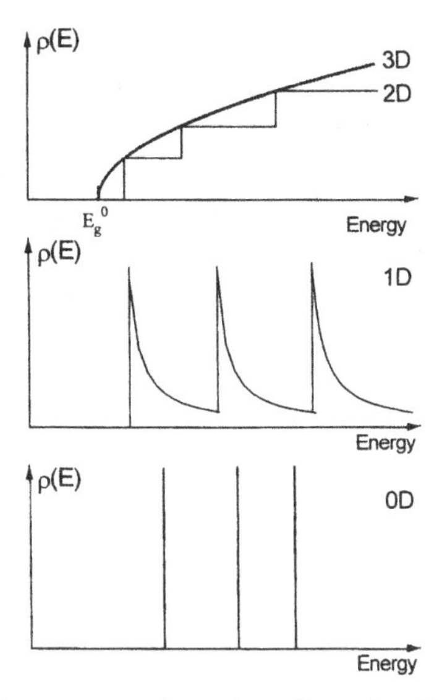

Fig. 1.10. Density of electron states for various dimensionalities. For $d = 1, 2, 3$, the density of states can be expressed by the formula $\rho(E) \propto E^{d/2-1}$. For a quasi-zero-dimensional system, the density of states is described by a set of δ-functions.

in the general form

$$\rho(E) \propto E^{d/2-1} \qquad d = 1, 2, 3 \tag{1.66}$$

where d is the dimensionality, and the energy is measured from the bottom of the conduction band for electrons and from the top of the valence band for holes. In the three-dimensional system, $\rho(E)$ is a smooth square-root function of energy. In the case of $d = 2$ and $d = 1$, a number of discrete subbands appear due to the quantum confinement effect, and the density of states obeys Eq. (1.66) within every subband (Fig. 1.10). For example, in a two-dimensional structure with infinite potential walls, quantization energies are given by a relation similar to Eq. (1.5)

$$E_n = \frac{\pi^2 \hbar^2}{2m_{e,h} l^2} n^2 \qquad n = 1, 2, 3, \ldots \tag{1.67}$$

where l is the size along the confinement direction. The dispersion relation can be written then in the form

$$E(\mathbf{k}) = E_n + \frac{\hbar^2 \left(k_x^2 + k_y^2\right)}{2m_{e,h}}, \tag{1.68}$$

which corresponds to the infinite motion in x, y-directions and the finite motion along the confinement direction coinciding with the z-axis. For $d = 0$ we have to deal with zero-dimensional structure, which differs from quantum wells and quantum wires and is characterized by a discrete δ-function-like density of states, a finite motion of quasiparticles in all directions, and a finite number of atoms and of elementary excitations within the same quantum dot. Moreover, all efforts to create a structure featuring the properties of an ideal quantum dot give structures possessing a number of extra features. These species will be the subject of the rest of the book.

2

Electron states in an ideal nanocrystal

We consider an ideal nanocrystal to be a bit of a crystal with a spherical or cubic shape, the so-called quantum dot. Such species do not exist in nature. Nevertheless, it has been very helpful for the physics of nanocrystals to use these simplified models to trace the basic effects arising from three-dimensional spatial confinement. An extension of the effective mass approximation towards spatially confined structures leads to a particle-in-a-box problem and provides a way to calculate the properties of nanocrystals that are not possible to analyze in other way because of the very large number of atoms involved. This approach fostered the systematic experiments that have determined the major advances in nanocrystal physics. At smaller sizes it converges with the results of the quantum-chemical approach, in which the given number of atoms in the nanocrystal is accounted for explicitly rather than the size.

In this chapter we consider systematically the properties of electron-hole pair states resulting from the effective-mass consideration. We see that an elementary excitation in the electron subsystem of a nanocrystal can be classified as exciton with an extension "exciton in a quantum dot." Afterwards, a survey of quantum-chemical techniques along with the selected examples for semiconductor clusters will be given. Finally, the distinctive size ranges will be outlined to specify the steps of the evolution of properties and of the applicability of the different approaches and concepts to the mesoscopic structures confined in all three dimensions.

2.1 From crystal to cluster: effective mass approximation

On the way from crystal to cluster, it is reasonable to consider the quasiparticles featuring the properties inherent in an infinite crystal, and to include then the finite size of a given crystallite as the relevant potential jump at the boundaries. As the length parameters of quasiparticles (the de Broglie wavelength and exciton Bohr radius) are noticeably larger than the lattice constant for the most common semiconductors, we can consider a crystallite that has a rather large number of atoms and can be treated as a macroscopic crystal with respect to the lattice properties but should be considered as a quantum box for quasiparticles. This statement provides a definition of the term "quantum dot" that is widely used in the theory describing electron properties of nanocrystals in terms of a particle-in-a-box consideration. Therefore, the key point of the effective mass approximation (EMA) in application to nanocrystals is to consider the latter as receptacles of electrons and holes whose effective masses are the same as in the ideal infinite crystal of the same stoichiometry. Hereafter we shall use the term quantum dot to mean the model of a nanocrystal in which the EMA approach is used.

To reveal the principal quantum confinement effects within the framework of the EMA consideration, it is reasonable to deal with the simplest three-dimensional potential well, that is, the spherical potential box with an infinite potential, and to consider electrons and holes with isotropic effective masses. The clear physical results and the elegant analytical expressions can be derived for the two limiting cases, the so-called weak confinement and strong confinement limits, proposed by A. L. Efros and Al. L. Efros (1982).

2.1.1 Weak confinement regime

Weak confinement regime corresponds to the case when the dot radius, a, is small but still a few times larger than the exciton Bohr radius, a_B. In this case the quantization of the exciton center-of-mass motion occurs. Starting from the dispersion law of an exciton in a crystal [Eq. (1.62)], we have to replace the kinetic energy of a free exciton by a solution derived for a particle in a spherical box [Eq. (1.22)]. The energy of an exciton in the weak confinement case is then expressed in the form

$$E_{nml} = E_g - \frac{Ry^*}{n^2} + \frac{\hbar^2 \chi_{ml}^2}{2Ma^2}, \tag{2.1}$$

with the roots of the Bessel function χ_{ml} tabulated in Section 1.1. One can see that the exciton in a quantum dot is characterized by the quantum number n describing internal exciton states arising from the Coulomb electron-hole interaction ($1S$; $2S, 2P$; $3S, 3P, 3D$; ...), and by the two additional numbers, m and l, describing the states connected with the center-of-mass motion in the presence of the external potential barrier ($1s, 1p, 1d \ldots, 2s, 2p, 2d \ldots$, etc.). To distinguish the internal and the external states, we use capital letters for the former and small ones for the latter.

For the lowest state ($n = 1, m = 1, l = 0$) the energy is expressed as

$$E_{1S1s} = E_g - Ry^* + \frac{\pi^2\hbar^2}{2Ma^2}, \tag{2.2}$$

or, put another way,

$$E_{1S1s} = E_g - Ry^*\left[1 - \frac{\mu}{M}\left(\frac{\pi a_B}{a}\right)^2\right] \tag{2.3}$$

where μ is the electron-hole reduced mass (1.60). In Eqs. (2.2) and (2.3) the value $\chi_{10} = \pi$ and the relations (1.59), (1.61) were used. Hence, the first exciton resonance experiences a high-energy shift by the value

$$\Delta E_{1S1s} = \frac{\mu}{M}\left(\frac{\pi a_B}{a}\right)^2 Ry^*, \tag{2.4}$$

which is, however, small compared with Ry^* so far as

$$a \gg a_B \tag{2.5}$$

holds. This is the quantitative justification of the term "weak confinement."

Taking into account that photon absorption can create an exciton with zero angular momentum only, the absorption spectrum will consist of a number of lines corresponding to states with $l = 0$. Therefore, the absorption spectrum can be derived from Eq. (2.1) with $\chi_{m0} = \pi m$ (see Section 1.1):

$$E_{nm} = E_g - \frac{Ry^*}{n_2} + \frac{\hbar^2\pi^2}{2Ma^2}m^2. \tag{2.1$'$}$$

The "free" electron and hole have the energy spectra

$$E^e_{ml} = E_g + \frac{\hbar^2\chi^2_{ml}}{2m_e a^2}, \tag{2.6$'$}$$

$$E^h_{ml} = \frac{\hbar^2\chi^2_{ml}}{2m_h a^2}. \tag{2.6$''$}$$

Therefore, the total excess energy for the lowest electron and hole $1s$ states

$$\Delta E_{1s1s} = E^e_{1s} + E^h_{1s} - E_g = \frac{\hbar^2\pi^2}{2\mu a^2} = \left(\frac{\pi a_B}{a}\right)^2 Ry^* \tag{2.7}$$

is considerably smaller than the Ry^* value as well. Taking into account relations (2.3) and (2.6), consider a difference between the minimal energy necessary for the creation of an unbound e-h pair

$$E_g^{\text{eff}} = E_g + \Delta E_{1s1s} \tag{2.8}$$

and the energy corresponding to the first exciton resonance (2.3) as the effective exciton binding energy Ry^{eff}. It reads

$$Ry^{\text{eff}} = Ry^{*}\left[1 + \left(1 - \frac{\mu}{M}\right)\left(\frac{\pi a_B}{a}\right)^2\right] \tag{2.9}$$

and is larger than Ry^{*}.

2.1.2 Strong confinement limit

Strong confinement limit corresponds to the condition

$$a \ll a_B, \tag{2.10}$$

which means that the confined electron and hole have no bound state corresponding to the hydrogenlike exciton, and the zero-point kinetic energy of electron and hole due to confinement [Eq. (2.6)] is considerably larger than the Ry^{*} value. In this case, the uncorrelated motion of an electron and a hole may be considered as the first approximation, and the Coulomb interaction may be ignored. Then each particle possesses the energy spectrum given by Eq. (2.6). These spectra are sketched in Fig. 2.1. The energy and momentum conservation laws result in selection rules that allow optical transitions that couple electron and hole states with the same principal and orbital quantum numbers. Therefore, the absorption spectrum reduces to a set of discrete bands peaking at the energies

$$E_{nl} = E_g + \frac{\hbar^2}{2\mu a^2}\chi_{nl}^2. \tag{2.11}$$

For this reason, quantum dots in the strong confinement limit are sometimes referred to as artificial atoms or hyperatoms as soon as the quantum dots exhibit a discrete optical spectrum controlled by the size (i.e., by the number of atoms), whereas an atom has a discrete spectrum controlled by the number of nucleons.

However, one should bear in mind that an electron and a hole are confined in space comparable with the extension of the exciton ground state in the ideal infinite crystal. Therefore, an independent treatment of the electron and hole motion is by no means justified, and the problem including the two-particle Hamiltonian with the kinetic energy terms, Coulomb potential, and the confinement potential

$$H = -\frac{\hbar^2}{2m_e}\nabla_e^2 - \frac{\hbar^2}{2m_h}\nabla_h^2 - \frac{e^2}{\varepsilon|\mathbf{r}_e - \mathbf{r}_h|} + U(r) \tag{2.12}$$

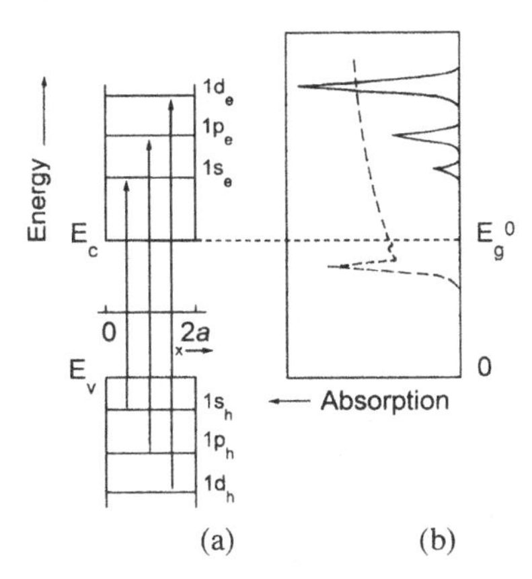

Fig. 2.1. A sketch of optical properties of an ideal spherical quantum dot with isotropic scalar effective masses of noninteracting electron and hole. Electron and hole energy levels (a) obey a series of states inherent for a particle in a spherical box with an infinite barrier. Selection rules allow optical transitions coupling the hole and the electron states with the same quantum numbers. Therefore, the optical absorption spectrum of the parent bulk crystal reduces to a number of discrete bands (b).

should be examined (Brus 1983, 1984). Contrary to the hydrogenlike Hamiltonian (1.28), the appearance of $U(r)$ potential does not allow the center-of-mass motion and the motion of a particle with reduced mass to be considered independently. Several authors have treated this problem with the variational approach (Brus 1986; Kayanuma 1986; Schmidt and Weller 1986) and found that the energy of the ground electron-hole pair state $(1s1s)$ can be expressed in the form

$$E_{1s1s} = E_g + \frac{\pi^2\hbar^2}{2\mu a^2} - 1.786\frac{e^2}{\varepsilon a}, \tag{2.13}$$

in which the term proportional to $e^2/\varepsilon a$ describes the effective Coulomb electron-hole interaction. Comparing this term with the exciton Rydberg energy $Ry^* = e^2/2\varepsilon a$ [Eq. (1.61)] and bearing in mind that our consideration is still with the strong confinement limit ($a \ll a_B$), one can see that Coulomb interaction by no means vanishes in small quantum dots. Moreover, the Coulomb-term contribution to the ground-state energy is even greater than in the bulk monocrystal. This is a principal difference of quantum dots as compared with crystal, quantum well, and quantum wire, where the Coulomb energy of a free electron-hole pair is zero. Therefore, an elementary excitation in a quantum dot can be classified as an exciton with a notation "exciton in quantum dot" (Banyai and Koch 1993; Woggon and Gaponenko 1995). Within the framework of this agreement we will use the term exciton throughout the book even when the relevant state does not obey the hydrogenlike model.

The exciton energy measured as a deviation from the bulk band gap energy E_g in the strong confinement limit can be written in a more general way as an expansion

$$E_{\text{exc}} - E_g = \left(\frac{a_B}{a}\right)^2 Ry^* \left[A_1 + \frac{a}{a_B} A_2 + \left(\frac{a}{a_B}\right)^2 A_3 + \cdots\right] \tag{2.14}$$

with the small parameter $a/a_B \ll 1$. The first coefficient A_1 for various states is described by the roots of the Bessel function [see Eq. (2.6) and Table 1.1]. The second coefficient A_2 corresponds to the Coulomb term in Eq. (2.13) and takes the following values: $A_2 = -1.786$ for the $1s1s$ state, $A_2 = -1.884$ for the $1p1p$ state, and the values between -1.6 and -1.8 for other configurations (Schmidt and Weller 1986). The A_3 coefficient for the $1s1s$ state was found to be $A_3 = -0.248$ (Kayanuma 1986). Summarizing the findings relevant to the ground state, we can write the energy of the first absorption peak as follows:

$$E_{1s1s} = E_g + \pi^2 \left(\frac{a_B}{a}\right)^2 Ry^* - 1.786 \frac{a_B}{a} Ry^* - 0.248 Ry^*. \tag{2.15}$$

The size dependence of the E_{1s1s} energy is plotted in Fig. 2.2 using dimensionless energy units E/Ry^* and length units a/a_B.[1] Note that Eq. (2.15) gives the function describing the shift of the absorption edge from quantum confinement in a form that has no dependence on the material parameters if the energy is scaled in Ry^* units and the length is measured in the units of a_B.

There is one more case that allows an analytical solution of the Schroedinger equation with the confinement potential. It corresponds to a very large ratio of the hole and electron masses, that is, $m_h \gg m_e$ which is equivalent to

$$\mu \approx m_e, \quad a_h \ll a_e; \quad a_e + a_h = a_B \gg a_h \tag{2.16}$$

where

$$a_e = \frac{\varepsilon \hbar^2}{m_e e^2}, \quad a_h = \frac{\varepsilon \hbar^2}{m_h e^2} \tag{2.17}$$

are the electron and the hole Bohr radii. In the case

$$a_h \ll a \ll a_e, a_B \tag{2.18}$$

the hole can be considered as immovable and localized in the center of the dot. This assumption is the same as the Born-Oppenheimer approximation,

[1] Because of the alternating signs of the different terms in Eq. (2.15), at $a/a_B \approx 3$ the relation $E_{1s1s} = E_g$ holds. This fact is by no means confusing because Eq. (2.15) is valid only if $a/a_B \ll 1$, which is not the case.

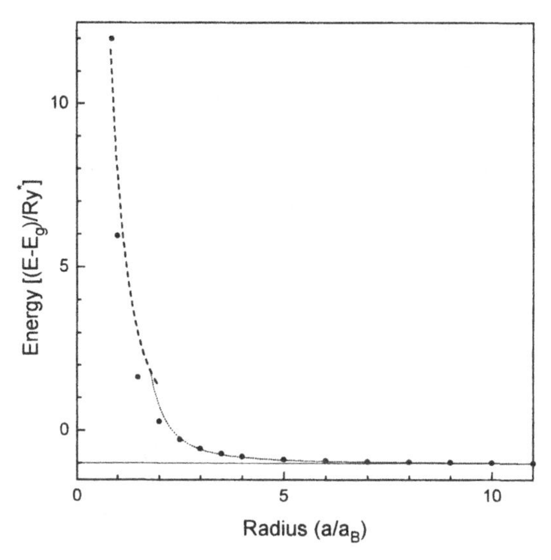

Fig. 2.2. Size dependence of the energy of the first allowed dipole optical transition in an ideal quantum dot according to calculations by Y. Kayanuma (1988) (dots) and the asymptotic functions described by Eq. (2.15) (long-dash line) and Eq. (2.20) (short-dash line). Numerical calculations and the curve plotted according to Eq. (2.20) correspond to $m_e/m_h = 0.3$. The asymptotic dependence described by Eq. (2.15) is independent of the material parameters if the energy and size are measured in units of the exciton Rydberg and the Bohr radius.

and the relevant electron-hole state in a semiconductor quantum dot is called a "donor-like exciton" (Efros and Efros 1982; Kayanuma 1988; Ekimov et al. 1989). In this model, energy states and the absorption spectrum are determined mainly by quantization of the electron motion [Eq. (2.6′)]. However, as a result of the Coulomb electron-hole interaction, each electron level splits into several sublevels (Ekimov et al. 1989). The position of the first absorption maximum can be described by a relation (Kayanuma 1988)

$$E_1 = E_g + 8\left(\frac{a}{a_B}\right)^2 Ry^* \exp\left(-\frac{2a}{a_B}\right). \qquad (2.19)$$

In the limit of a very large hole mass, a problem relevant to an exciton confined in a box has a number of common features with the problem connected with a hydrogen atom inside a cavity. Various theoretical approaches to this problem are extensively discussed by W. Jaskolski (1996). It should be noted that the above consideration in terms of either exciton center-of-mass motion quantization or electron-hole motion quantization does not imply any fundamental physical effect or discontinuity when the size of a dot scales around

$a = a_B$. A presentation of the quantum-size effects in terms of the weak confinement and strong confinement limits is very helpful because it provides true intuitive conclusions and trends based on elementary quantum mechanics, and the concepts developed for macrocrystals appear to be suitable when speaking about properties of nanocrystals. In the real situation, a smooth evolution of quantum dot properties from crystal-like to clusterlike features occurs, and this can be successfully proven within the framework of the effective mass approximation by means of the explicit numerical solution of the Schroedinger problem with Hamiltonian (2.12). This program has been realized by a number of authors (Kayanuma 1988; Mohan and Anderson 1989; Hu, Lindberg, and Koch 1990; Pollock and Koch 1991).

An explicit numerical analysis has revealed basically the same results that can be intuitively foreseen within the framework of the strong confinement and the weak confinement models. The size-dependent blue shift of the absorption onset really shows no sensitivity to the absolute m_e and m_h values and obeys a universal law for all semiconductors in the dimensionless energy and length units, which are similar to the widely used atomic units [Fig. 2.3(a)]. At the same time, for $a \geq a_B$ the $E(a)$ function is rather sensitive to the m_e/m_h ratio [Fig. 2.3(b)]. The simplified relations (2.3) and (2.15) were found to provide a good fit to the exact results in the range $a/a_B > 4$ and $a/a_B < 1$, respectively. The evolution of the absorption spectrum on the way from crystal to cluster is clearly seen in Fig. 2.4, where several first absorption terms are presented by discrete lines with the height proportional to the corresponding oscillator strength. A redistribution of the bulklike hydrogen exciton sequence $E_n = -Ry^*/n^2$ with the oscillator strength $f \propto n^{-3}$ into states corresponding to a relative electron-hole motion in the box is evident. This redistribution occurs along with hybridization and quasi-crossing of the relevant states. For example, the $1S1s$ state [Eq. (2.3)] transforms into the $1s1s$ state of Eq. (2.15).

Although Eq. (2.3) provides satisfactory fitting of numerical results, a better description for $2a_B < a < 4a_B$ has been obtained using another function (Nair, Sinha, and Rustagi 1987; Kayanuma 1988).

$$E_{1S1s} = E_g - Ry^* + \frac{\pi^2\hbar^2}{2M(a - \eta a_B)^2}, \tag{2.20}$$

where the parameter η was found to be close to unity. A possible physical reason for the lower a value in Eq. (2.20) as compared with Eq. (2.3) can be related to a dead layer near the surface of a size close to a_B that becomes inaccessible to the exciton.

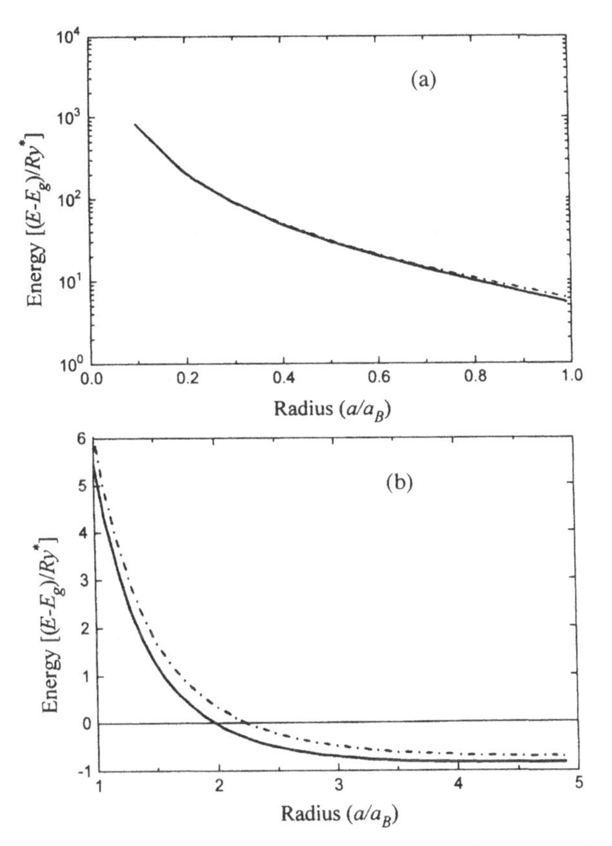

Fig. 2.3. Size dependence of the energy of the first allowed dipole optical transition calculated with proper consideration for the electron-hole Coulomb interaction by means of the original matrix diagonalization technique (from Hu, Lindberg, and Koch 1990). The m_e/m_h value is 0.1 (solid lines) and 0.01 (dash-dot lines). Note that the result for $a < a_B$ is not sensitive to the m_e/m_h value, in agreement with a simplified consideration for the strong confinement limit [Eq. (2.15)].

2.1.3 Surface polarization and finite barrier effects

In real microstructures containing semiconductor nanocrystals, the latter are always embedded in a dielectric medium with a dielectric constant ε_2, normally being less than that of the semiconductor nanocrystal ε_1. Moreover, an interface between the nanocrystal and the matrix in a number of cases cannot be considered as an infinite potential barrier. The difference in dielectric constants (dielectric confinement) gives rise to surface polarization effects (Brus 1984) arising from an interaction of electron and hole inside a crystallite with induced image charges outside. The finite barrier height results in a general

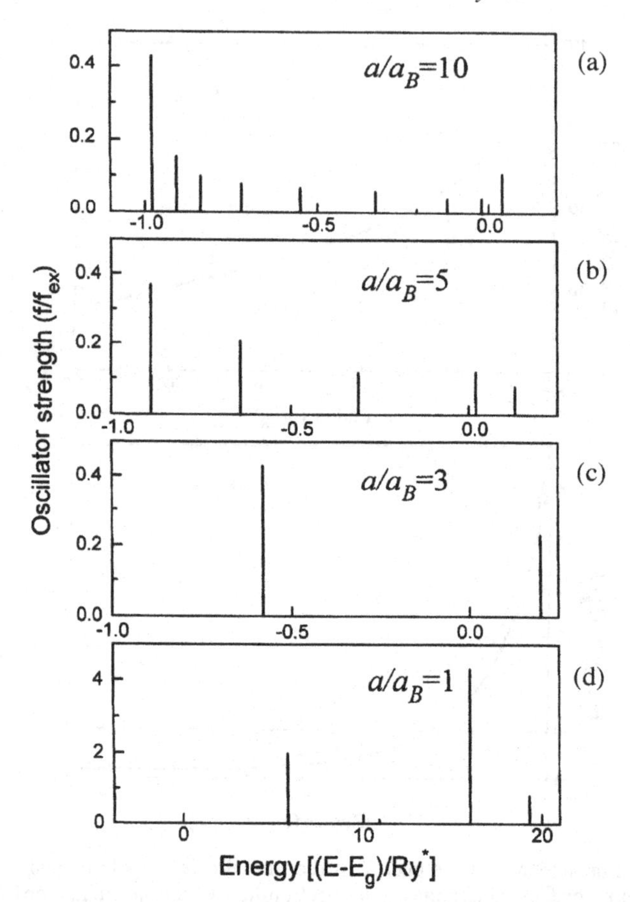

Fig. 2.4. Evolution of the absorption spectrum of a semiconductor from crystallike (a) to clusterlike behavior (d) [according to Y. Kayanuma (1988)]. With decreasing size the hydrogenlike sequence inherent in a bulk crystal (a) reduces to a number of lines corresponding to a relative electron-hole motion in a quantum box. Note a change in the scale in panel (d) as compared with (a)–(c).

energy lowering that can be foreseen on the basis of the elementary quantum mechanics. However, both effects together result sometimes in instabilities that become possible for a certain combination of semiconductor material parameters, matrix properties, and crystallite size.

To take into account the surface polarization effects, an electron-hole Hamiltonian (2.12) should be supplemented with the potential energy of electron and hole interaction with the induced polarization field

$$U_{eh} = U_{ee'} + U_{hh'} + U_{eh'} + U_{he'}, \tag{2.21}$$

where prime denotes the relevant image charge, and the terms can be written as follows (Efremov and Pokutnii 1990):

$$U_{ee'} = \frac{e^2}{2\varepsilon_1 a}\left(\frac{a^2}{a^2 - r_e^2} + \frac{\varepsilon_1}{\varepsilon_2}\right),$$

$$U_{hh'} = \frac{e^2}{2\varepsilon_1 a}\left(\frac{a^2}{a^2 - r_h^2} + \frac{\varepsilon_1}{\varepsilon_2}\right), \tag{2.22}$$

$$U_{eh'} = U_{he'} = -\frac{e^2}{2\varepsilon_1 a}\frac{a}{\sqrt{(r_e r_h)^2/a^2 - 2r_e r_h \cos\Theta + a^2}},$$

with $\Theta = \mathbf{r}_e{}^\wedge\mathbf{r}_h$. The polarization effects are more pronounced in the case

$$\varepsilon_1 \gg \varepsilon_2, \quad m_h \gg m_e, \quad a_h < a < a_e,$$

which has been examined by the variational technique (Efremov and Pokutnii 1990). Interaction of electron and hole with image charges results in a deviation of the energy versus size dependence from that calculated according to Eq. (2.15). A discrepancy becomes nonnegligible for $a \leq 10a_h$. T. Takagahara proposed an analytical expression in the form (2.14) with the coefficients A_2 and A_3 modified to account for the dielectric confinement effect (Takagahara 1993a). Particularly, A_3 was found to vary monotonically from −0.248 to −0.57 when $\varepsilon_1/\varepsilon_2$ changes from 1 to 10.

An influence of the finite potential barrier on the energy states increases with decreasing dot radius and with increasing quantum number for a given dot radius. Numerical solution for Hamiltonian (2.12) with electron-hole Coulomb interaction and the finite potential taken into account for a spherical dot (Tran Thoai, Hu, and Koch 1990) provides absolute values of the effect. For example, for $a = 0.5a_B$ the potential height $U = 40Ry^*$ makes the energy of the ground electron-hole pair state decrease from $E_{1s1s} - E_g \approx 35Ry^*$ down to $E_{1s1s} - E_g \approx 15Ry^*$, that is, more than two times smaller with respect to the bulk band gap energy E_g.

Surface polarization effects in the case of a finite potential barrier may result in self-trapping of carriers at the surface of the dot (Banyai et al. 1992). For decreasing confinement potential at a fixed dot radius, and for decreasing dot radius at a fixed confinement potential, the electron-hole pair state evolves from a volume state in which both particles are mostly inside the dot, to a surface-trapped state in which the radial charge distribution is concentrated near the surface. The trapping effect is more pronounced for the heavier particle, that is, for a hole. Therefore, in the case of a considerable difference in m_e and

m_h, this effect may lead to a charge separation resulting in an enhanced dipole momentum of an exciton in the ground state.

2.1.4 Hole energy levels and optical transitions in real semiconductors

In most semiconductors the valence band near $k = 0$ contains two branches corresponding to the so-called heavy and light holes, and the third branch splits off due to the spin-orbit interaction. Moreover, in a number of materials absolute values of the hole effective mass are different for different directions. For these reasons, a consideration of quantum confinement effects in the valence band of real semiconductors should be made with the complex band structure taken into account. The hole kinetic energy operator should be used in the form of the Luttinger Hamiltonian (see, e.g., Tsidilkovskii 1978), which in the case of a spherically symmetric problem can be expressed in the form proposed by A. Baldereshi and N. Lipari (1973)

$$H_h = \frac{\gamma_1}{2m_0}\left[\hat{\mathbf{P}}^2 - \frac{\mu}{9}\left(\mathbf{P}^{(2)}\mathbf{J}^{(2)}\right)\right], \tag{2.23}$$

where $\mathbf{J}$ is the angular momentum operator, $\mathbf{P}^{(2)}$ and $\mathbf{J}^{(2)}$ are the second rank tensor operators, $\mu = (6\gamma_3 + 4\gamma_2)/5\gamma_1$, and γ_i are called the Luttinger parameters. An analysis of the Schroedinger equation with the electron kinetic energy in the form relevant to a free particle and with the hole kinetic energy in the form (2.23) leads to a significant modification of hole energy levels and of the diagram of optical transitions (Xia 1989; Sweeny and Xu 1989; Sercel and Vahala 1990). The hole wave functions are expressed as linear combinations of the different valence band states and have mixed symmetry. The wave functions are the eigenfunctions of the total angular momentum operator, which has to be conserved in optical transitions. The orbital angular momentum contains not only the quantum number l but also the number $(l + 2)$ because Hamiltonian (2.23) couples states with $\Delta l = 0, +2, -2$. The single-particle quantum numbers should be replaced by new notations to account for the valence band mixing:

$$nl \to n^*(l, l+2)_F,$$

where $F = l + j$ is the total angular momentum number, containing F_z from $-F$ to $+F$, and n^* labels the ground and the excited states. Sometimes the number in the brackets corresponding to the state providing the minor contribution is omitted. Thus, the ground hole state can be denoted as $1S_{3/2}$, where the capital letter S is used instead of the small one to distinguish between the

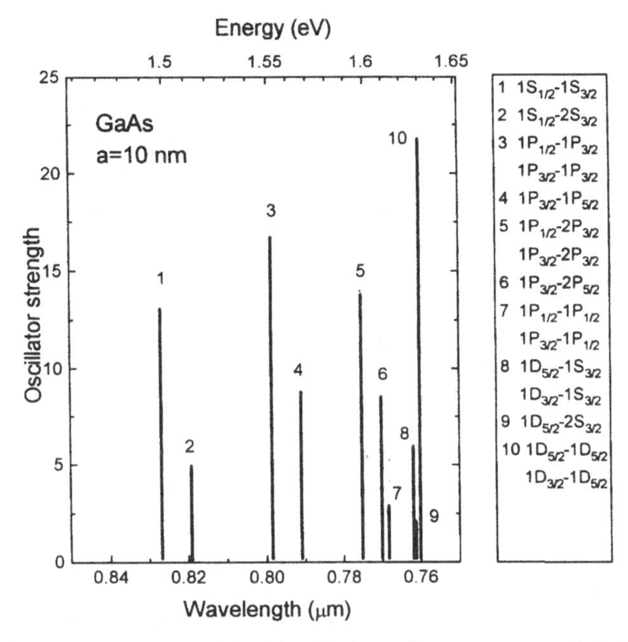

Fig. 2.5. Oscillator strengths of first the 10 absorption resonances of GaAs quantum dot ($a = 10\,\mathrm{nm}$) calculated by J. Pan (1992) within the framework of the effective mass approximation using the Luttinger hole Hamiltonian. The right-hand column shows the hole and electron states involved in each transition. Note a variety of the hole states and of the optical transitions that are absent in the case of the scalar hole effective mass (see Fig. 2.1 for a comparison).

single particle and the mixed state. As a result of valence band mixing, the selection rules allow optical transitions with $\Delta n \neq 0$, and a number of new absorption bands appear. The electron-hole spectra and the optical transition probabilities have been calculated for the case of GaAs (Pan 1992a), CdS (Koch et al. 1992), Si (Takagahara and Takeda 1992), CdSe (Ekimov et al. 1993), and CdTe (Lefebvre et al. 1996). The optical spectrum and the relevant transitions for GaAs quantum dots are presented in Fig. 2.5.

2.1.5 *Size-dependent oscillator strength*

Oscillator strength of exciton absorption per unit volume is proportional to the transition dipole moment and to the probability of finding an electron and a hole at the same site. In the case of a weak confinement we deal with a hydrogenlike state and the electron-hole overlap is size-independent. If the dot radius is still smaller than the photon wavelength corresponding to the exciton resonance

λ_{exc}, that is,

$$a_B \ll a < \lambda_{exc}, \tag{2.24}$$

a superposition of the exciton states having different momentum of the center-of-mass motion $\hbar K$ in the range $0 < \hbar K^2/2M < \hbar\Gamma$ (Γ is the dephasing rate) appears in one exciton band (Feldman et al. 1987). Therefore, the concept of the coherence length a^* and a proper coherence volume $(4/3)\pi(a^*)^3$ has been proposed that satisfies the relation

$$\frac{4}{3}\pi(a^*)^3 \int_0^{\hbar\Gamma} D(E)\, dE = 1 \tag{2.25}$$

where $D(E) = M^{3/2}E^{1/2}/(2^{1/2}\pi^2\hbar^3)$ is the density of states for the translational exciton motion. In the range $a_B \ll a < a^*$, the oscillator strength of the exciton transition *of a crystallite* appears to be proportional to the dot volume V, that is, to the number of atoms N that form a crystallite (Takagahara 1987; Kayanuma 1988; Hanamura 1988). If we consider ensembles of crystallites with varying sizes but with the same volume fraction as the semiconductor phase, the absorption coefficient of the medium per unit length appears to be independent of the dot size inherent in each ensemble. However, the size-dependent oscillator strength *per crystallite* will manifest itself in the size-dependent radiative decay rate and in the enhanced nonlinear susceptibility (Takagahara and Hanamura 1986; Takagahara 1987; Hanamura 1988).

In the strong confinement limit the electron-hole overlap grows monotonically with decreasing size. In this case, for a smaller size the oscillator strength of a given crystallite is nearly size-independent (Brus 1984; Kayanuma 1988). For the first transition the oscillator strength *per crystallite* tends to the value (Kayanuma 1988)

$$f_1 \xrightarrow{a\to 0} f_{ex}\frac{\pi a_B^3}{V} \tag{2.26}$$

where f_{ex} is the exciton oscillator strength of the bulk crystal. Note that the above consideration is true for $a \gg a_L$. Otherwise, the effective mass approximation is not valid.

To summarize, the effective mass approximation provides a description of electronic properties of nanocrystals on the way from crystal-like to the clusterlike behavior in terms of the particle-in-a-box problem. It predicts a number of size-dependent features due to the three-dimensional spatial confinement of quasi-particles. The main manifestation of quantum-size effects is the continuous blue shift of the absorption onset of nanocrystals with decreasing size. To give an idea regarding absolute values of the effect, size-dependent band gaps for

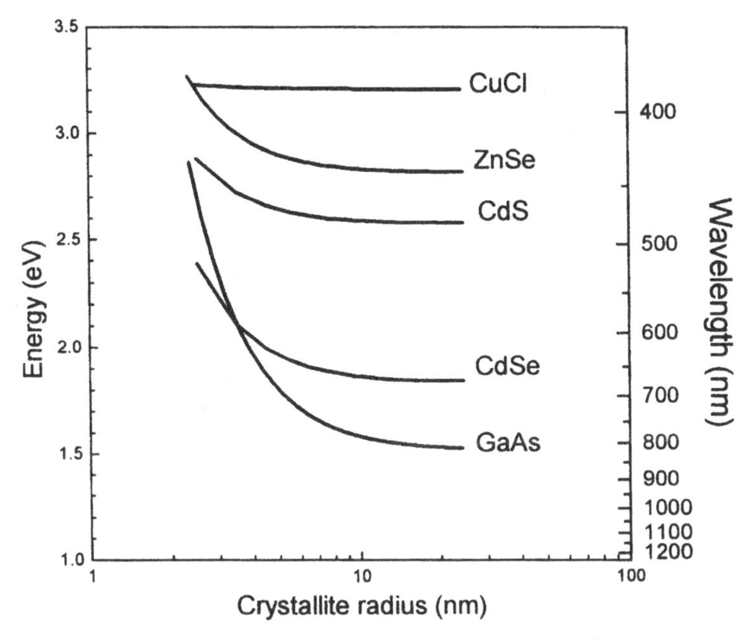

Fig. 2.6. Energy of the absorption onset versus crystallite size calculated according to Eqs. (2.15) and (2.20) for a number of semiconductor materials. The quantum confinement effect results in a monotonic blue shift of the absorption onset with decreasing size. For the same size range, the effect is more pronounced in materials with smaller band-gap energy E_g and larger exciton Bohr radius (GaAs). In the case of CuCl with large $E_g = 3.4$ eV and small $a_B = 0.7$ nm, the weak confinement limit occurs within the size range where effective mass approximation is valid. For crystallite radius exceeding 20 nm the absorption onset corresponds to that of the bulk crystal. For radii less than 2.0 nm the validity of the effective mass approximation becomes questionable.

a number of semiconductor materials are presented in Fig. 2.6. For a quantum dot radius less than 2 nm a crystallite contains less than 10^3 atoms, and the results predicted within the framework of the particle-in-a-box consideration with the constant effective mass should be treated as an estimate only. One possible correction to this approach is to assume the energy dependence of the effective mass, that is, nonparabolicity of the bands, when kinetic energy of confined quasi-particles moves from the band extremum (Nomura and Kobayashi 1991). Another possibility is to use the empirical pseudopotential method in which the exact crystal-field potential experienced by the valence electron is replaced by an effective potential (pseudopotential). This method is used for band structure calculations of bulk crystals and has been successfully applied to nanocrystals (Rama Krishna and Friesner 1991). Results obtained for $a > 2$ nm are close to those predicted by the effective mass approach. For $0.5 < a < 2$ nm the pseudopotential method gives in certain cases a nonmonotonic energy versus size dependence, which is a typical feature of small clusters.

Generally, the potential of the effective mass approach as applied to nanocrystals is not yet exhausted. One of the main advantages of this approach is the ability to reduce a problem connected with nanocrystals to a clear quantum-mechanical form, which can be treated then by means of approaches and techniques developed for other quantum systems and borrowed from other fields besides solid state physics. Recently, a successful application of non-Euclidian quantum mechanics to quantum dots has been demonstrated (Kurochkin 1994). In this approach the potential box is replaced by a free curved space providing an analytical expression for the confined electron-hole pair with the Coulomb interaction without using the strong and weak confinement approximations.

Vice versa, some specific problems considered for quantum dots may be helpful, due to their general formulation, in the examination of different quantum systems in other fields.

In this section we have focused on the fundamental quantum confinement effects predictable by means of the effective mass approach. The theory of quantum dots, including an analysis and description of various calculation techniques, is presented in a book by L. Banyai and S. W. Koch (Banyai and Koch 1993), which is recommended for further reading.

2.2 From cluster to crystal: quantum-chemical approaches

2.2.1 Semiconductor nanocrystals as large molecules

Real nanometer-size semiconductor crystallites, if treated correctly, should be described in the same way as very large molecules. This means that the particular number of atoms and specific spatial configurations should be involved rather than the size. The importance of such an approach becomes crucial for smaller crystallites consisting of less than 100 atoms. In this case, the properties of semiconductor particles[2] should be deduced from the properties of atoms rather than crystals. Therefore, we have to deal with specific types of clusters that can be examined by means of the molecular quantum mechanics often referred to as *quantum chemistry*. Quantum-chemical consideration provides an opportunity to reveal the development of crystal-like properties starting from the atomic and molecular level.

To show the basic steps from atom to cluster and from cluster to crystal, we consider, as an example, a silicon atom, silicon cluster, and silicon crystal (Bawendi, Steigerwald, and Brus 1990). The four-electron sp^3-hybridized Si

[2]The very term semiconductor particles in relation to smaller crystallites is certainly not correct because they do not possess semiconducting features. It is used here only to refer to the proper single crystal whose properties will be reached in the case of an infinitely increasing particle size.

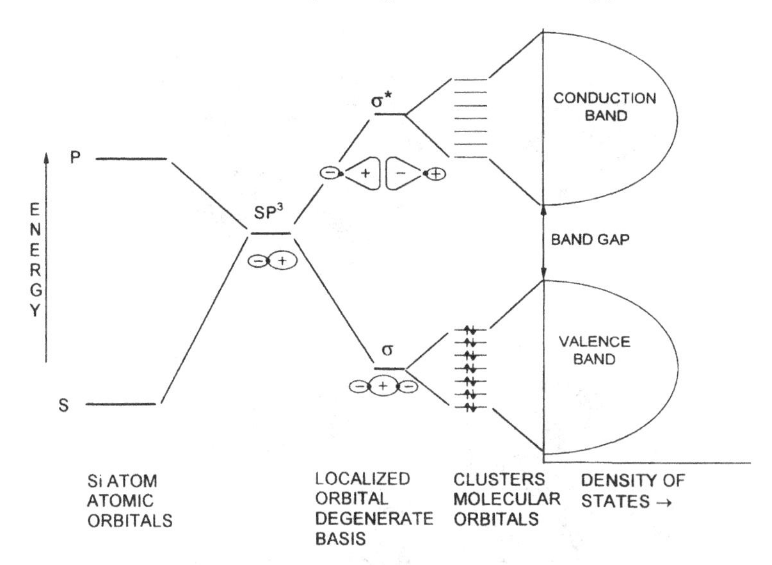

Fig. 2.7. Evolution of silicon atomic orbitals into crystal energy bands [after Bawendi, Steigerwald, and Brus (1990)].

atoms assembled in a cluster can be described using the basis orbitals that are bond, rather than atomic, orbitals between nearest neighbor atoms. They are bonding orbitals σ and antibonding orbitals σ^* (Fig. 2.7). As the number of atoms in the crystallite grows, each of the localized bond orbital sets forms molecular orbitals extended over the crystallite that finally develop into conduction and valence bands. The highest occupied molecular orbital (HOMO) becomes the top of the valence band, and the lowest unoccupied molecular orbital (LUMO) becomes the bottom of the conduction band. The HOMO-LUMO spacing tends to the band gap energy of the bulk monocrystals. The difference of the HOMO energy and vacuum energy determines the *ionization potential*, whereas the difference of the LUMO energy and vacuum energy can be associated with the *electron affinity (redox potential)*.

Particular stable configurations of silicon clusters were calculated by means of Langevin molecular dynamics ($N \leq 7$) (Binggeli and Chelikowsky 1994), tight-binding molecular dynamics simulations ($N = 11$–16) (Bahel and Ramakrishna 1995), and using the stuffed fullerene model ($N = 30$–50) (Pan and Ramakrishna 1994). Several selected clusters are presented in Fig. 2.8. Up to $N = 16$, all configurations correspond to surface clusters. For $N > 30$, unique stable configurations were found at $N = 33$, 39, and 45. These clusters consist of a bulklike core of five atoms surrounded by a fullerenelike surface. The core atoms bind to the twelve surface atoms. The surface relaxes from its ideal fullerene geometry to give rise to crown atoms and dimers. These crown

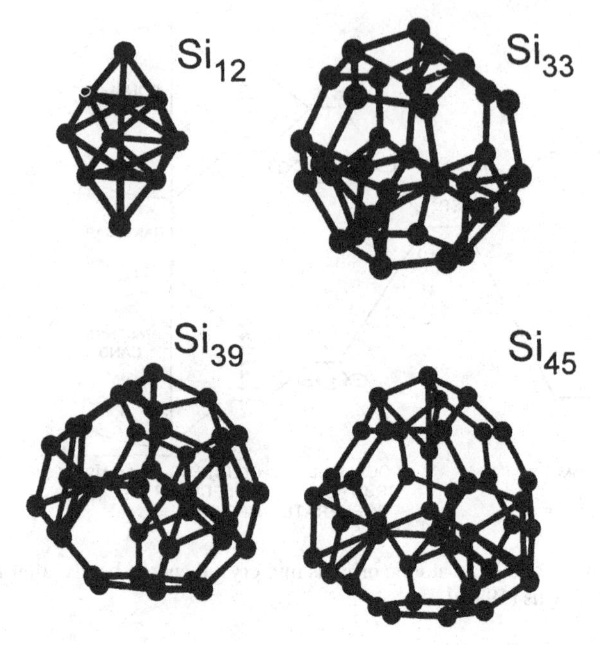

Fig. 2.8. Silicon clusters [after Bahel and Ramakrishna (1995) and Pan and Ramakrishna (1994)].

atoms are threefold coordinated and possess one dangling bond each. The dimers are also threefold coordinated, but the dangling bonds are eliminated by a local π-bonding. Evidently, the energy structure of these and other similar clusters should be analyzed by means of quantum chemistry rather than solid state physics.

2.2.2 General characteristics of quantum-chemical methods

To calculate properties of a molecule or a cluster, one has to deal with the Schroedinger equation with the Hamiltonian similar to that discussed in Section 1.2 in connection with a crystal [see Eq. (1.50)]. It includes kinetic energies of all electrons, kinetic energies of all nuclei, potential energy of electron-electron repulsive interaction, potential energy of the electron-nucleus attractive interaction, and the potential energy of the nucleus-nucleus repulsive interaction. In total, the problem includes N nuclei and n electrons. It can be solved analytically only for $N = n = 1$, which is just a hydrogen atom (see Section 1.1). Therefore, all quantum-chemical approaches are essentially numerical and based on several assumptions and approximations.

The first assumption is the adiabatic or Born-Oppenheimer approximation, which provides an independent consideration of heavy nuclei and light electron motion. To obtain an energy spectrum of the electron subsystem of a cluster or a molecule, the Schroedinger equation should be examined

$$H_e\Psi = E\Psi \tag{2.27}$$

with the total electron Hamiltonian consisting of the single-electron and two-electron components

$$H_e = H_e(1) + H_e(2), \tag{2.28}$$

$$H_e(1) = \sum_{\mu=1}^{n}\left(-\frac{\hbar^2}{2m}\nabla_\mu^2 - \sum_{\nu}^{N}\frac{Z_\nu e^2}{|\mathbf{r}_\mu - \mathbf{R}_\nu|}\right), \tag{2.29}$$

$$H_e(2) = \frac{1}{2}\sum_{\mu,\nu}\frac{e^2}{|\mathbf{r}_\mu - \mathbf{r}_\nu|} \tag{2.30}$$

In Eqs. (2.29) and (2.30) n is the total number of electrons, N is the total number of nuclei, Z_ν is the charge of the ν-nucleus, and $\mathbf{r}_\mu$ and $\mathbf{R}_\nu$ are radius-vectors of the μ-electron and ν-nucleus, respectively. The single-electron Hamiltonian $H_e(1)$ contains the kinetic energy of each electron and the potential energy of the attractive interaction of each electron with all nuclei. The two-electron Hamiltonian $H_e(2)$ includes the potential energy of the repulsive interaction of each electron-electron pair.

Equation (2.27) with Hamiltonian (2.28)–(2.30) forms the basic equation of quantum chemistry. All existing quantum-chemical approaches differ in methods of approximate numerical solution of this equation. First of all, the single-electron approximation is used. The total wave function of system Ψ is either written in the form of a product of single-electron wave functions φ_i, which are called *molecular orbitals* (the *Hartree method*), or in the form of the *Slater determinant* consisting of φ_i, providing an account for the *Pauli principle* (the Hartree-Fock method). Single-electron wave functions are solutions of the equations

$$\left[-\frac{\hbar^2}{2m}\nabla_i^2 - \sum_{\nu=1}^{N}\frac{Z_\nu e^2}{|\mathbf{r_i} - \mathbf{R}_\nu|} + \sum_{\nu=1}^{n-1}\frac{e^2}{|\mathbf{r}_i - \mathbf{r}_\nu|}\right]\varphi_i = E_i\varphi_i. \tag{2.31}$$

The solution of Eq. (2.27) can be found using the *variational principle*, which states that the true wave function corresponds to the minimum average energy of the system

$$\langle E\rangle_{av} = \frac{\int \Psi^* H_e \Psi\, d\tau}{\int \Psi^*\Psi\, d\tau}. \tag{2.32}$$

Molecular orbitals can be considered as linear combinations of single-electron wave functions of atoms (*atomic orbitals*)

$$\Phi_i = \sum_{\nu} c_{\nu i} \chi_{\nu}. \tag{2.33}$$

This approximation is known as the MO-LCAO method (molecular orbitals as linear combinations of atomic orbitals), and this is really an approximation because in a cluster or in a molecule each electron belongs to all atoms.

Thus, to find the wave function and the energy of the electron system it is necessary to solve the Hartree-Fock system of integrodifferential equations. A solution can be found only for electrons in an atom since the latter possesses the central symmetry. In multicenter systems, such as clusters or molecules, a set of known basis functions is used instead of atomic orbitals and then the *coefficients* $c_{\nu i}$ rather than functions χ_ν are optimized. Thus, the Hartree-Fock system of integrodifferential equations reduces to the Hartree-Fock-Roothaan system of *algebraic* equations, which can be written in the form (Flurry 1983)

$$\sum_{\nu} c_{\nu i}(F_{\mu\nu} - E_i S_{\mu\nu}) = 0, \qquad i = 1, 2, \ldots M, \tag{2.34}$$

where M is the total number of the basis functions. Equation (2.34) can be written in the compact matrix form

$$\mathbf{F}\mathbf{c}_i = E_i \mathbf{S}\mathbf{c}_i, \tag{2.34'}$$

where $\mathbf{F}$ is the Fockian operator. In Eq. (2.34) $S_{\mu\nu}$ are overlap integrals

$$S_{\mu\nu} = \int \chi_\mu \chi_\nu \, d\tau, \tag{2.35}$$

that satisfy the condition

$$\sum_{\mu} \sum_{\nu} S_{\mu\nu} c_{\mu i} c_{\nu i} = 1. \tag{2.36}$$

The Fockian elements have the form

$$F_{\mu\nu} = H_{\mu\nu} + 2I_{\mu\nu} + K_{\mu\nu}, \tag{2.37}$$

where $H_{\mu\nu}$ is an operator including the kinetic energy of electrons and the energy of their Coulomb attraction to nuclei, and $I_{\mu\nu}$ and $K_{\mu\nu}$ describe the Coulomb electron-electron repulsion and the electron-electron exchange interaction. Equations (2.34) and (2.34′) form a basis of all nonempirical quantum-chemical techniques, that is, ab initio techniques. In spite of a number of approximations, the ab initio techniques are still very complicated and can be performed only for simplest molecules and clusters (Slater 1974).

2.2.3 Semiempirical techniques

To avoid the difficulties of ab initio calculations in application to complex molecules and clusters, a number of semiempirical techniques within the framework of the Hartree-Fock approach have been developed. These techniques involve some further approximations to reduce the number of integrals in the calculations and introduce some experimental values instead of the calculated ones, providing better agreement with the observable energies (Segal 1977).

In the simple Hückel technique, the following approximations are used.

(i) Coulomb integrals $I_{\mu\nu}$ are considered as being identical and equal to α for all atoms in a cluster;

(ii) Exchange integrals $K_{\mu\nu}$ equal to a constant value β if atoms are coupled by the σ-bond, and zero otherwise. Therefore, σ-electrons are introduced into atomic cores;

(iii) Overlap integrals are taken in the form $S_{ij} = \delta_{ij}$ (δ_{ij} is the Kronecker symbol). The α and β values are treated as empirical parameters.

In the extended Hückel technique, diagonal elements $H_{\mu\mu}$ are attributed to the ionization potential, whereas off-diagonal elements of the Hamiltonian are approximated by a function of $S_{\mu\nu}$, e.g., $H_{\mu\nu} = K S_{\mu\nu}(H_{\mu\mu} + H_{\nu\nu})$ in which the K value is fitted to thermodynamic data for the relevant diatomic complex (Hedvig 1975). $S_{\mu\nu}$ are calculated from a set of atomic basis functions.

In the technique based on complete neglect of differential overlap (CNDO), the $S_{\mu\nu}$ matrix is taken in the unitary form. $H_{\mu\nu}$ elements are calculated with the neglected overlap of any pair of atomic wave functions in all integrals corresponding to the electron-electron repulsion. A number of fitting parameters are used that are adjusted by fitting the computed dimer binding energy to experimental values. The obtained parameters are involved in calculations of clusters. A number of modifications of the CNDO approach have been developed and are described, for instance, in a book by Segal (1977).

2.2.4 Quantum-chemical calculations for semiconductor clusters

The electron structure of semiconductor clusters has been computed by means of the extended Hückel technique (Gurin 1994; 1996) for Cd_nS_m, Cd_nSe_m, and Cd_nTe_m clusters, by means of the tight-binding approach, which is a version of the Hückel technique for ZnS and CdS clusters (Lippens and Lannoo 1989; Einevoll 1992; Ramaniah and Nair 1993; Hill and Whaley 1993), for Si_nH_m clusters (Ren and Dow 1992; Hill and Whaley 1995; Huaxiang, Ling, and Xide 1993) and for PbS clusters (Wang and Herron 1991). Si_nH_m clusters have been treated also by means of the semiempirical LCAO technique (Delerue, Alan, and

Lannoo 1993; Lannoo, Delerue, and Alan 1995). The electronic structure and optical transitions of Zn_nS_m, Mg_nO_m, and Cd_nS_m clusters have been computed using a version of CNDO approach known as CNDO/S (Filatov and Kuzmitskii 1996; Kuzmitskii, Gael, and Filatov 1996). Unlike the effective mass approach, quantum-chemical calculations do not provide functional relationships between the number of atoms in a cluster and its electronic properties. Each specific semiempirical technique uses a large number of parameters that may change considerably the final result in the case of an inadequate choice. Most of the techniques are valid for rather small clusters only within the framework of the reasonable computing time. Typically, a quantum-chemical calculation provides a *single-electron* energy structure that is useful in an analysis of the correlation between the cluster configuration and electron energies and the density of states. However, to calculate the energy of the optical transitions, the energy structure of the cluster in the excited state should be calculated as well. These calculations have been performed using the CNDO/S approach (Kuzmitskii, Gael, and Filatov 1996). In the case of the tight-binding approach, a difference between single-electron energies and optical transition energies was treated (Ramaniah and Nair 1993) in terms of the electron-hole Coulomb interaction using a correction calculated by the perturbation technique [see Eqs. (2.14) and (2.15)]. Such an approach of evaluation of optical transition energies has certain deficiencies. First of all, a correction due to the Coulomb electron-hole interaction appears to be of the same order of magnitude as an absorption blue shift of the first transition with respect to the bulk band gap energy predicted by the tight-binding single-particle calculations. Second, the coefficient A in the Coulomb term $Ae^2/\varepsilon a$ is different for different transitions and should be calculated explicitly. Third, the concepts of the dielectric constant ε and size a are questionable in relation to small clusters. Generally, the quantum-chemical calculations of semiconductor nanocrystals at present can be characterized as a preliminary stage that is expected to be followed by further advances in the nearest future.

We restrict the overview of quantum-chemical results for specific clusters to the consideration of data obtained for CdS-based structures.

Cd_nS_m clusters examined by the extended Hückel technique (Gurin 1994; 1996) were found to exhibit a clear tendency to energy band formation with increasing number of atoms. The chemical bond in these clusters is not reduced to the classical case of sp^3 hybridization of the diamondlike semiconductors. There is a small contribution of d orbitals. Clusters of the same configuration with S replaced by Se and Te obey the same tendency of the energy of the HOMO-LUMO transition ΔE_{HL} as the band gap energy E_g of the parent bulk crystal, that is, for given n and m values $\Delta E_{HL}(Cd_nTe_m) < \Delta E_{HL}(Cd_nSe_m) <$

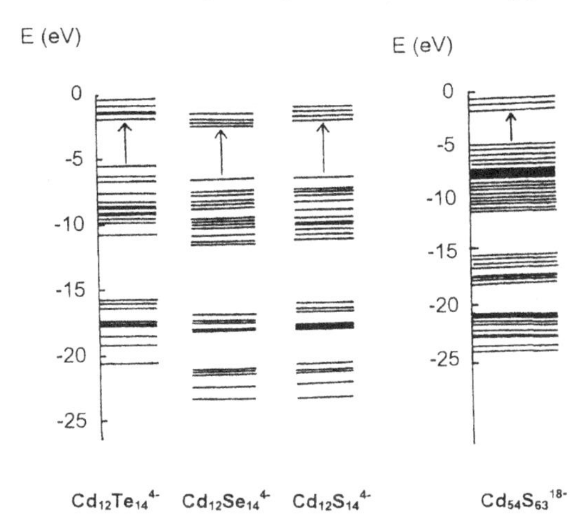

Fig. 2.9. Energy structure of semiconductor clusters computed by the extended Hückel technique (Gurin 1994). Arrows indicate HOMO-LUMO transitions.

$\Delta E_{HL}(Cd_nS_m)$ similar to $E_g(CdTe) < E_g(CdSe) < E_g(CdS)$ (see Fig. 2.9). For the same configuration the increasing number of shells results in a monotonic decrease in the ΔE_{HL} energy (compare data for $Cd_{12}S_{14}$ and $Cd_{54}S_{63}$ in Fig. 2.9). This tendency is the same as that predicted by the effective mass approximation. However, an increase in the total number of atoms N does not result in an unambiguous tendency of the ΔE_{HL} behavior. For example, an increase in the N value leads to a decrease in ΔE_{HL} within the series

(i) $Cd_4S_{10}^{12-} \rightarrow Cd_{10}S_{20}^{20-} \rightarrow Cd_{20}S_{35}^{30-}$;

(ii) $Cd_4S_6^{4-} \rightarrow Cd_{10}S_{16}^{12-} \rightarrow Cd_{20}S_{31}^{22-}$.

However, clusters with close N values belonging to different series have drastically different energies. For example, $\Delta E_{HL}(Cd_4S_{10}) \approx 20$ eV, $\Delta E_{HL}(Cd_4S_6) \approx 10$ eV, and $\Delta E_{HL}(Cd_6S_4) \approx 6$ eV. The nonmonotonic $\Delta E_{HL}(N)$ dependence is a typical feature inherent in clusters that cannot be foreseen within the framework of solid-state theories. The drastic difference in the energy structure for similar total numbers of atoms means that the chemical coordination is more significant than the number of atoms (i.e., the size). In the case under consideration, clusters in series (ii) differ from those in series (i) by removed sulfur atoms with dangling bonds.

A nonmonotonic ΔE_{HL} change with growing N has been revealed also for CdS clusters by means of CNDO/S calculations (Filatov and Kuzmitskii 1996), although the structure of clusters chosen was somewhat different. To

Table 2.1. *Energies and oscillator strengths of the electronic transition in CdS clusters calculated by the CNDO/S technique*

Structure	E_{HOMO}-E_{LUMO}	Transition energy (eV)	Oscillator strength
$(Cd(SH)_4)^{2-}$	8.07	4.28	0
		4.43	4.46
$(Cd_4(SH)_{10})^{2-}$	7.84	4.10	0
		4.12	0.30
$(Cd_{10}S_4(SH)_4)^{4-}$	7.30	4.40	0
		5.12	$3{\cdot}10^{-4}$
$(Cd_{20}S_{13}(SH)_{22})^{8-}$	5.68	3.79	0
		4.16	$5{\cdot}10^{-3}$

Source: After Kuzmitskii, Gael, and Filatov 1996.

model the clusters capped by phenol groups that can be chemically developed (see Chapter 3), the clusters $Cd_nS_mH_l$ have been examined obeying the series $(Cd(SH)_4)^{2-} \rightarrow (Cd_4(SH)_{10})^{2-} \rightarrow (Cd_{10}S_4(SH)_4)^{4-} \rightarrow (Cd_{20}S_{13}(SH)_{22})^{8-}$.

The CNDO/S technique applied provided calculations of excited states of the cluster as well as the ground state. The energy differences of the excited and the ground states rather than ΔE_{HL} value should be attributed to observable energies of optical transitions. It is noteworthy that the first optical transition in small $Cd_nS_mH_l$ clusters was found to be forbidden (Table 2.1). Relative weights and spectral positions of the several optical transitions for the three of clusters are plotted in Fig. 2.10.

The extended Hückel technique and the CNDO/S technique do not provide results for larger clusters because of the tremendous number of calculations. To trace the further tendency in optical spectra of CdS clusters, we plot in Fig. 2.11 the spectra obtained by means of the tight-binding approach (Ramaniah and Nair 1993). One can see that the first optical resonance shifts monotonically towards lower quantum energies when the radius of the cluster grows from 8.7 Å (approximately 100 atoms) to 23.2 Å (approximately 2,000 atoms). At the same time, the spectrum becomes denser, that is, spacing between neighboring resonances becomes smaller.

Finally, in Fig. 2.12, a summary is presented of data for CdS-based clusters and dots evaluated by means of both the solid-state and molecular approaches. One can see a clear agreement of the quantum-chemical and pseudopotential calculations for smaller sizes and a convergence of the results of the effective

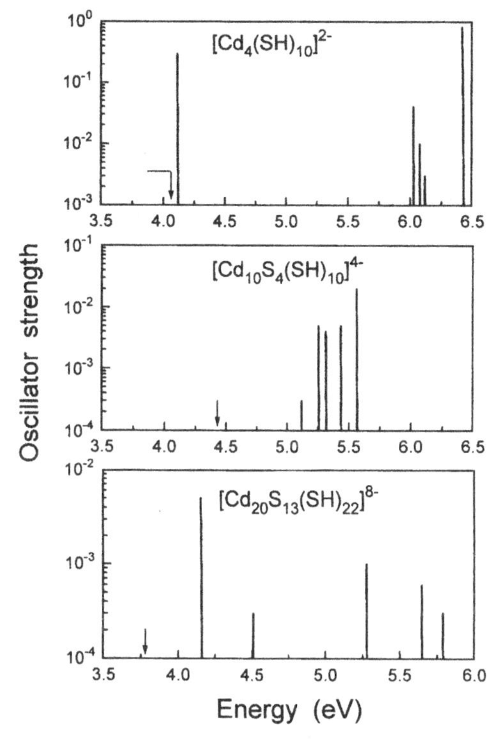

Fig. 2.10. Oscillator strengths of several first optical transitions for several $Cd_nS_mH_l$ clusters computed by CNDO/S technique and plotted according to the calculations of Kuzmitskii, Gael, and Filatov (1996). Transition energies were calculated using data for the excited and ground states. The first transition was found to be forbidden and is indicated for each cluster by an arrow.

mass approximation and the numerical calculations for larger sizes. A discrepancy between the EMA results and the data provided by means of other techniques for smaller particles can be reduced if a finite potential barrier is introduced in a particle-in-a-box problem (Nosaka 1991).

2.3 Size regimes in quasi-zero-dimensional structures

Let us discuss the size ranges for semiconductor nanocrystals related to different models and approaches, which provide an adequate description of their optical properties. The characteristic length parameters to be involved in such a classification are the crystal lattice constant, a_L, the exciton Bohr radius, a_B, and the photon wavelength λ corresponding to the lowest optical transition.

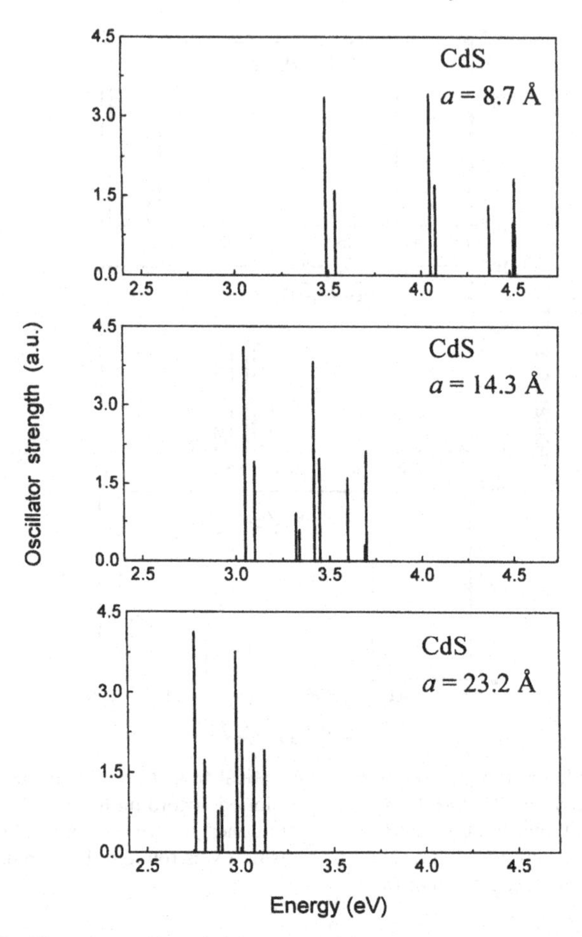

Fig. 2.11. Oscillator strengths of the several first optical transitions of various CdS clusters. The energy of each transition is obtained as $E_i = E_g + \Delta E_i - \Delta E_{\text{Coul}}$, where $E_g = 2.5\,\text{eV}$ is the band gap energy of the cubic CdS crystal, ΔE_i values are the result of the single-particle tight-binding calculations after Ramaniah and Nair (1993), and $\Delta E_{\text{Coul}} = -1.8e^2/\varepsilon a$ is the electron-hole Coulomb interaction energy calculated by the perturbative approach (Brus 1986; Kayanuma 1986; Schmidt and Weller 1986).

First of all, if the size of a nanocrystal is comparable to a_L, an adequate description can be provided only in terms of the quantum-chemical (molecular) approach with the specific number of atoms and configuration taken into account. This size interval can be classified as a *cluster range*. The main distinctive feature of clusters is the absence of a monotonic dependence of the optical transition energies and probabilities versus number of atoms. The size as a characteristic of clusters is by no means justified. A subrange can be

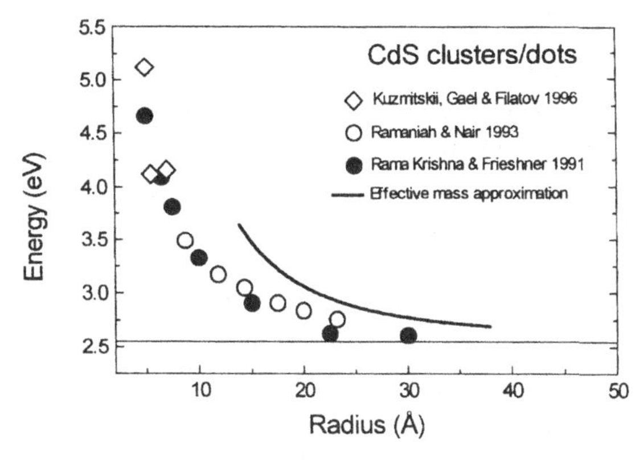

Fig. 2.12. Summary of the data obtained for the energy of the first allowed optical transition versus crystallite radius for CdS clusters (dots). (◇) – results of the CNDO/S technique (Kuzmitskii, Gael, and Filatov 1996); (○) – results obtained by means of the tight-binding approach for single-particle energies with the electron-hole Coulomb interaction accounted for by perturbational calculations (Ramaniah and Nair 1993); (●) – results obtained using the pseudopotential method for single-particle energies (Ramakrishna and Frieshner 1991) and the perturbational approach to account for the Coulomb electron-hole interaction; solid line–analytical expression within the framework of the effective mass approximation for the case of an infinite potential barrier:

$$E_{1s1s} = E_g + \pi^2 \left(\frac{a_B}{a}\right)^2 Ry^* - 1.786 \frac{a_B}{a} Ry^* - 0.248 Ry^*,$$

where $E_g = 2.58$ eV is the band gap energy, $Ry^* = 29$ meV is the exciton Rydberg energy, and $a_B = 28$ Å is the exciton Bohr radius of the bulk CdS crystal. The straight horizontal line shows a position of the first exciton resonance in the bulk CdS crystal. Radii for clusters examined by the CNDO/S technique have been evaluated using the size versus number of atoms relation proposed by Lippens and Lannoo (1989).

outlined corresponding to surface clusters in which every atom belongs to the surface. This type of cluster possesses an enhanced chemical activity.

When $a_L \ll a \ll \lambda$, a nanocrystal can be treated as a quantum box for electron excitations. The solid-state approach in terms of the effective mass approximation predicts a monotonic evolution of optical properties of nanocrystals with size towards properties of the relevant bulk single crystals. A number of analytical relations describing size-dependent properties of nanocrystals have been derived for the cases $a \ll a_B$ and $a \gg a_B$. The exciton Bohr radius value, a_B, divides this range into two subranges providing an interpretation of size-dependent properties in terms of either electron and hole or a hydrogenlike exciton confined motion. Generally throughout this range, the concept of exciton as an interacting electron-hole pair is relevant, and numerical calculations provide an evidence of monotonic optical transition energies and probabilities behavior without any discontinuity around $a = a_B$. It is remarkable that for smaller sizes the results provided by the solid-state and the molecular approaches do converge to one another. This range can be classified as the *quantum dot range*.

As the size of a nanocrystal reaches a value on the order of 100 nm, it becomes comparable to the photon wavelength relevant to resonant optical transitions. This circumstance brings about a number of important aspects. First of all, light scattering should be explicitly considered (Bohren and Huffman 1984). Second, a concept of *photon confinement* should be introduced. A nanocrystal in this case should be treated as a *microcavity* possessing a definite number of electromagnetic modes. This effect is relevant to the modified density of photon states and is currently being actively investigated in relation to various microstructures. Contrary to planar microcavities where photons are restricted only in one dimension, in large nanocrystals photon confinement occurs in all three directions and the term *photonic quantum dot* can be introduced. Furthermore, light absorption and emission in this case will be strongly influenced by *exciton-photon coupling*. In a bulk crystal exciton-photon coupling is described in terms of the new quasiparticle, *polariton*. The concept of a polariton leads to a description of light propagation in terms of the polariton creation and annihilation at the front and the rear boundary of the sample, respectively, whereas photon absorption is interpreted in terms of polariton scattering by lattice imperfections (phonons and impurities). This size range of nanocrystals is less investigated. It can be referred to as *polaritonic range*. A number of interesting phenomena can be foreseen for this range due to exciton-photon coupling under the conditions of the photon confinement.

3
Growth of nanocrystals

Semiconductor nanocrystals can be fabricated using a number of technologies, differing in the environment in which nanocrystals appear, growth conditions, size range, and size distribution, as well as physical and chemical stability and reliability. Nanocrystals can be developed in inorganic glasses and crystals, in liquid solutions and polymers, or on a crystalline surface. In this chapter we provide a brief overview of these techniques and give a synopsis of nanocrystals developed by various techniques.

3.1 Nanocrystals in inorganic matrices

3.1.1 Glass matrices: diffusion-controlled growth

Fabrication of nanocrystals embedded in a glass matrix by means of diffusion-controlled growth is based on commercial technologies developed for fabrication of color cut-off filters and photochromic glasses. Color cut-off filters produced by Corning (United States), Schott (Germany), Rubin (Russia), and Hoya (Japan) are just glasses containing nanometer-size crystallites of mixed II-VI compounds (CdS_xSe_{1-x}). Empirical methods of diffusion-controlled growth of semiconductor nanocrystals in a glass matrix have been known for decades or even centuries (in the case of color stained glasses). Commercial photochromic glasses developed in recent years contain nanocrystals of I-VII compounds (e.g., CuCl, CuBr, AgBr). Typically, silicate or borosilicate matrices are used with the absorption onset near 4 eV (about 300 nm), thus providing optical transmission of the semiconductor inclusions to be studied over the whole visible range.

Growth of crystallites results from the phase transition in a supersaturated viscous solution. The process is controlled by diffusion of ions dissoluted in the matrix and can be performed in the temperature range $T_{glass} < T < T_{melt}$, where T_{glass} is the temperature of the glass transition and T_{melt} is the melting temperature of the matrix. Typically, growth temperatures range between 550°C and 700°C depending on the desirable size of the crystallites and matrix composition.

Diffusion-controlled growth from a supersaturated solution can be described in terms of three distinct precipitation stages, namely *nucleation, normal growth,* and *competitive growth* (Lifshitz and Slyozov 1961; Abraham 1974; Koch 1984; Slyozov and Sagalovich 1987; Lui and Risbud 1990). At the first stage, small nuclei are formed. At the second stage, crystallites exhibit a monotonic growth due to jumps of the atoms across the nucleus-matrix interface. At this stage the supersaturation degree decreases with time and the total volume of semiconductor phase monotonically grows up. Finally, when crystallites are large enough and the degree of supersaturation is negligible (i.e., almost all ions are already incorporated in crystallites) the surface tension plays the main role and the growth dynamics are characterized by a diffusive mass transfer from smaller particles to larger ones. This stage is commonly referred to as Ostwald ripening, competitive growth, diffusion-limited aggregation, or coalescence. Whereas in a real growth process different stages may coexist, in theory the growth dynamics are commonly analyzed separately for each stage, which makes it possible to obtain analytical expressions as well as to give an adequate description for a rather large variety of experimental conditions.

At the nucleation stage the concentration of ions in the matrix can be treated as constant. According to steady-state nucleation theory, the nucleation rate R is given by the expression (Lui and Risbud 1990)

$$R = A \exp\left(-\frac{\Delta G_a}{kT}\right) \exp\left(-\frac{4\pi\sigma a^{*2}}{3kT}\right), \tag{3.1}$$

where ΔG_a is the activation energy associated with jumps of atoms across the nucleus-matrix interface, σ is the interface free energy per unit area, $a^* = 2\sigma/\Delta G_v$ is the critical radius, and ΔG_v is the bulk free energy per unit area. The radius distribution of nuclei *P(a)* obeys the Gaussian function

$$P(a) = P_0 \exp\left(-\frac{4\pi\sigma(a-a^*)^2}{3kT}\right). \tag{3.2}$$

The number of nuclei and the critical radius are controlled by the supersaturation value. The larger the supersaturation, the smaller is the critical radius and the smaller clusters can grow. At the normal growth stage one can consider the number of nuclei to be constant whereas concentration of ions decreases monotonically. Time dependence of the cluster size can be written then as

$$a^2 = a_0^2 + \text{const} \cdot t, \tag{3.3}$$

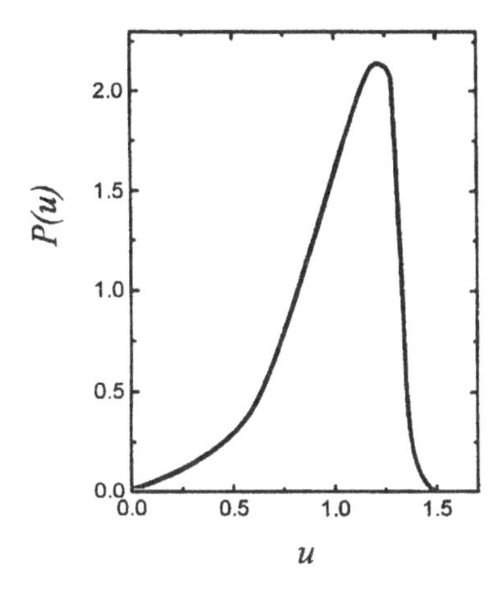

Fig. 3.1. Asymptotic size distribution function corresponding to the competitive growth of crystallites in a glass matrix from a supersaturated solid solution (Eq. 3.4). The function corresponds to the infinitely long time when supersaturation becomes negligible. The function has been derived by Lifshitz and Slyozov (1961).

with the initial radius $a_0 \approx a^*$ (Lui and Risbud 1990). Over long periods of time, a_0 can be neglected as compared with a, and the growth law takes a parabolic form $a^2 \propto t$. Common for both nucleation and the normal growth stage is a steady growth of the total mass of the semiconductor phase.

At the third, competitive growth stage no excessive ions exist in the matrix (i.e., supersaturation is negligible) and the total mass of the semiconductor phase is considered to be constant. Under these conditions surface tension becomes important. An increase in the mean size of crystallites occurs because of the mass transfer from smaller to larger crystallites, because in this case a decrease in the crystallite-matrix surface gives rise to a decrease in the free energy of the system. A critical radius a_{crit} separates crystallites that experience either further growth or dissolution. If $a > a_{\text{crit}}$, a crystallite becomes larger, whereas for $a < a_{\text{crit}}$ crystallites dissociate. Thc size-distribution function in this case takes the form (Lifshitz and Slyozov 1961)

$$P(u) = 3^4 2^{-5/3} e u^2 (u+3)^{-7/3} \left(\frac{3}{2} - u \right)^{-11/3} \exp\left[\left(\frac{2}{3}u - 1 \right)^{-1} \right], \quad (3.4)$$

where $u = a/a_{\text{crit}}$. Unlike Gaussian, the Lifshitz-Slyozov distribution function (3.4) is asymmetric with respect to the mean radius $\bar{a}$ (Fig. 3.1). The peak occurs at $u \approx 1.2$ and the halfwidth of the distribution is about 0.3. The time

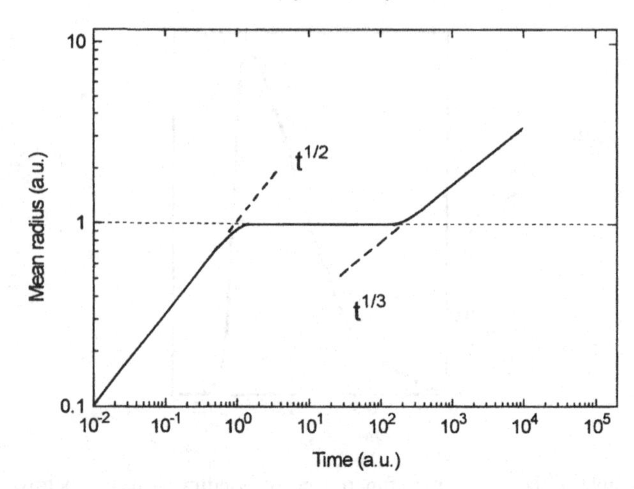

Fig. 3.2. Time dependence of the mean size of particles under conditions of simultaneous nucleation and growth processes in a supersaturated solid solution (after Shepilov 1992). At the earlier stage the function is close to a parabolic law inherent in the ideal normal growth regime. The size distribution at this stage is described by a Gaussian. At the final stage the growth law features cubic dependence, which is typical of competitive growth. The size distribution at the final stage corresponds to the Lifshitz-Slyozov function (3.4) presented in Fig. 3.1. It is important that between these two limiting cases a time interval exists where the mean particle radius remains constant. During this intermediate period, the size distribution evolves from a Gaussian to the Lifshitz-Slyozov function.

dependence of $\bar{a}$ in this case obeys a cubic law:

$$\bar{a} = a_{\text{crit}} = \left(\frac{4\alpha Dt}{9}\right)^{1/3} \tag{3.5}$$

where coefficient $\alpha \sim T^{-1}$, and the diffusion coefficient D can be written as

$$D = D_0 \exp(-\Delta E_{\text{coal}}/kT) \tag{3.6}$$

where ΔE_{coal} is the activation energy. As $\bar{a}^3 \propto t$ holds and the total volume of crystallites remains constant at this stage, the number of crystallites decreases as t^{-1}.

In a real system various stages may coexist. In particular, normal growth can occur along with an increase in the number of crystallites because of the ongoing nucleation process. An attempt to examine the growth behavior in this case by a numerical analysis has been made by Shepilov (1992). In this case time dependence of the mean radius at the initial stage is close to the parabolic law, whereas the final stage follows the cubic dependence. An interval exists in between in which the size distribution changes whereas the mean radius remains constant (Fig. 3.2).

Experimentally, the competitive growth stage is investigated much better than other precipitation regimes, probably because just this stage is used for the commercial production of color and photochromic glasses (Buzhinskii and Bobrova 1962; Tsekhomskii 1978; Bobkova and Sinevich 1984; Borrelli et al. 1987; Vasiliev et al. 1995). The cubic growth law $a^3 \propto t$ was reported for both II-VI and I-VII compounds (Golubkov et al. 1981; Ekimov, Efros, and Onushchenko 1985). The dependence $a^2 \propto t$ inherent in normal growth was not revealed, although a deviation from the cubic law in the earlier period was observed (Potter and Simmons 1988).

Nucleation, normal growth, and competitive growth are characterized by different activation energies ($E_{\text{nucl}} < E_{\text{norm}} < E_{\text{comp}}$), which makes it possible to obtain crystallites of the same size that have experienced somewhat different growth conditions by a proper choice of the temperature/time combination. Also possible is a two-step procedure. For the first step, a low temperature/long time combination provides a development of nuclei with an extremely small size. For the second step, a high temperature/short time combination gives rise to a rapid increase in the size, while the number of crystallites remains constant. The two-step regime provides a better control of the size, size distribution, and concentration of nanocrystals (Borrelli et al. 1987; Woggon et al. 1990).

The normal growth stage was found to yield crystallites of a higher quality and possessing a narrower absorption band as compared with the competitive growth regime (Gaponenko et al. 1993). However, growth duration on the order of 100 hours is necessary to obtain crystallites containing not more than 10^3 atoms when competitive growth is not desirable. Possibly for this reason, the normal growth regime is used to fabricate experimental samples rather than commercial glass.

To analyze the mean size and the size-distribution function of crystallites embedded in a glass matrix, electron microscopy, Raman scattering, or X-ray scattering may be used. A comprehensive analysis of various techniques is given by Champagnon et al. (1993). A number of groups found the Gaussian size distribution at the earlier stage and an asymmetric distribution afterwards (Lui and Risbud 1990; Woggon and Gaponenko 1995 and references therein).

A diffusion-controlled process in a glass matrix is successfully used to fabricate nanocrystals of II-VI (cadmium halcogenides) and I-VII (copper and silver halides) semiconductors. One specific point inherent in I-VII crystallites should be highlighted. While bulk II-VI compounds have a melting temperature of about 10^3°C, I-VII crystals have T_{melt} near 400°C. With decreasing size, T_{melt} falls to about 250°C for $a = 2.0$ nm. At the same time, growth temperatures are controlled to a large extent by the matrix properties and are similar to those used for II-VI nanocrystals, that is, normally $T_{\text{growth}} = 550\ldots600$°C. Therefore,

I-VII crystallites experience growth in a liquid state. Furthermore, because of the relation between the thermal expansion coefficients like $\alpha_{glass} < \alpha_{cryst}$, each crystallite appears to be inside a pore that is larger than the crystallite's volume after it cools down from T_{growth} to room temperature. This fact has been confirmed by studies of the hydrostatic pressure effect on the exciton lines position. Unlike II-VI crystallites, CuCl and CuBr exciton bands were found to exhibit no shift under pressure applied to the matrix (Kulinkin et al. 1988; Vasiliev et al. 1991).

Unlike I-VII compounds, for II-VI crystallites $\alpha_{glass} > \alpha_{cryst}$ holds. Therefore, crystallites are expected to experience a noticeable pressure from the matrix when being cooled from T_{growth} to room temperature, and they actually do (Klingshirn and Gaponenko 1992; Scamarcio, Lugara, and Manno 1992; Tomassi et al. 1992). Interestingly, at the earlier stage of semiconductor-doped glasses studies, a blue shift of absorption spectra in II-VI doped glasses as compared with the bulk crystals was tentatively attributed to the pressure effects alone until the quantum-confinement approach was advanced.

The relation $T_{melt} < T_{growth}$ in the case of I-VII crystallites provides a secondary heat treatment of crystallites at temperatures near T_{melt} to be performed. A prolonged annealing at T_{anneal} close to but less than T_{melt} was found to give rise to sharper excitonic resonances (Vasiliev et al. 1995). As T_{melt} is size-dependent, this so-called fine annealing process appears to be size-selective. Small crystallites, for which $T_{anneal} > T_{melt}$, simply undergo melting and freezing cycle during annealing and a consequent cooling to room temperature. Larger crystallites, for which $T_{anneal} < T_{melt}$, do not show any change in exciton spectra at all. Therefore, for each sample T_{anneal} should be adjusted according to the $\bar{a}$ value to give a noticeable sharpening of the exciton absorption band. The smaller the $\bar{a}$ and T_{anneal} values, the longer is the time necessary for a considerable effect. In particular, for $\bar{a} < 2.5$ nm annealing during 100 hours is desirable. The mechanism of the fine-annealing effect is not clear at present. Taking into account that the effect takes place only if $T_{anneal} \rightarrow T_{melt}$, we can propose that the large amplitude of vibrations of ions, which is close to a_L at $T_{anneal} \rightarrow T_{melt}$, assists in restoring lattice defects or in smoothing the crystallites' surface.

In connection with the structural features of the II-VI nanocrystals, it is important to note that CdS and CdSe crystallites experience a transition from the cubic (zinc blend) to hexagonal (wurtzite) lattice with increasing size from 1–2 to 5–10 nm. This behavior was revealed for the first time by Bobkova and Sinevich (1984) and later on has been confirmed by other authors (Ekimov et al. 1989; Konnikov et al. 1995). Bulk parent crystals exist usually in the wurtzite form. The zinc blend modification of CdSe and CdS bulk crystal is very unstable and tends to the wurtzite structure on heating. The dependence of the crystallite shape on growth conditions invites further investigations. Taking into account the fractal-like topology reported for other systems created by diffusion-limited aggregation (Feder 1988; Vicsek 1989) and the drastic dependence of surface structure on boundary conditions (Luczka et al. 1992; Tu, Levine, and Ridgway 1993), certain interesting results on the crystallite surface topology, depending on the growth dynamics, may be expected.

3.1.2 Nanocrystals in porous glasses

Sol-gel technology offers an alternative way of inorganic glass fabrication that does not include high temperature treatment. Porous glasses developed in this way contain nanometer-size voids that can be impregnated with dye molecules or nanocrystals. Several successful attempts have been reported of fabrication of nanocrystals embedded in a sol-gel matrix (Nogami, Nagasaka, and Takata 1990; Minti et al. 1991; Spanhel, Arpac, and Schmidt 1992; Mathieu et al. 1995). As compared with conventional glasses, porous glasses can be saturated with semiconductor materials up to very high concentrations (10 percent in the case of sol-gel matrices) thus providing an opportunity to study interactions between crystallites. Additionally, crystallites in these matrices are expected to have a smaller amount of defects because of the lower precipitation temperature.

However, the sol-gel technique in its present state encounters serious problems in providing a continuous size control and narrow size distribution. Optical spectra of these systems are usually rather broad, first of all, because of the wide size distribution. An exception is the multistep inorganic-organic processing proposed by Spanhel et al. (1992), which provides a relatively narrow size distribution of CdS nanocrystals and corresponding distinct bands in the optical absorption spectrum. Mathieu et al. have managed to get a rather large set of CdS nanocrystal sizes ranging from 2.1 to 9.2 nm, that is, covering both weak and strong confinement cases for the same crystallite matrix system (Mathieu et al. 1995).

A successful development of III-V nanocrystals embedded in porous glass has been performed. Note that these types of semiconductor nanocrystals have not yet been fabricated in the conventional glass matrices. GaAs (Justus, Tonucci, and Berry 1992) and InP (Hendershot et al. 1993) nanocrystals have been developed in porous Vycor glass by organo-metallic chemical vapor deposition without a detailed specification of the matrix structure and composition.

3.1.3 Semiconductor nanocrystals in ionic crystals

Alkali halide crystals heavily doped with copper were found to contain nanometer-size copper halide crystallites (Tsuboi 1980; Itoh, Iwabuchi, and Kataoka

1988). Beginning with NaCl, KBr, and KI monocrystals, crystallites of CuCl, CuBr, and CuI can be developed, respectively. In this procedure, for instance, CuCl is added to NaCl powder to 1 mol% concentration and the mixture is then used to grow NaCl single crystals by an ordinary technique. A subsequent annealing at a temperature near 600°C provides flat crystallites with mean size ranging from 2 to 5 nm. Unlike semiconductor-doped glasses, these structures do not exhibit a monotonic dependence of the mean size versus temperature because of a complicated interplay between diffusion and coagulation of Cu ions in a single crystal matrix. Crystallites of copper halides developed in this way were found to possess a cubic rather than spherical shape (Itoh, Iwabuchi, and Kataoka 1988; Masumoto 1996). Furthermore, crystallites of CuCl were found to show a pronounced orientation, the crystal axes of nanocrystals and of the NaCl matrix being parallel to each other (Froelich et al. 1995).

3.1.4 Nanocrystals in zeolites

Zeolites are crystalline Al–O–Si materials with regularly arranged cages of about 1-nm size. Synthesis of semiconductor clusters inside the cages provides a kind of a quantum dot array consisting of the crystallites with a very sharp size distribution and thus showing rather small inhomogeneous broadening of optical spectra. Nanocrystals of CdS (Wang and Herron 1991), PbI_2 (Tang, Nozue, and Goto 1992), and Ge (Miguez et al. 1996) have been fabricated in zeolites. Unlike the previously described techniques, zeolite matrices do not provide any possibility to vary the nanocrystal size. The latter is determined by the cage dimensions. Another problem that prevents a wide application of nanocrystals in zeolites is the very small size of the samples. Typically, zeolites are available with no more than 100 μm in all three dimensions. Therefore, conventional transmission optical spectroscopy should be replaced by the diffusive reflectance technique to study optical absorption of these species.

3.1.5 Composite semiconductor-glass films

Composite semiconductor-glass films are used to develop Si and Ge nanocrystals embedded in a SiO_2 matrix. The method is based on a planar magnetron radio frequency (rf) sputtering of Si or Ge in a hydrogen or argon atmosphere on a silicon substrate with a thin film of native silicon oxide. The latter is held on an electrode while another electrode buried by permanent magnets holds the silicon or germanium target. The size of crystallites embedded in SiO_2 films can be controlled by the substrate temperature, rf power, and the pressure of the environmental gas. The samples thus obtained are suitable for both transmission and emission optical studies as well as for X-ray and transmission electron microscopy (TEM) studies. Semiconductor-glass composite films have a concentration of nanocrystals, on the order of 10 to 30 percent, the film thickness being about few micrometers. Nanocrystals of Si (Furukawa and Miyasato 1988; Hayashi et al. 1993; Yamamoto et al. 1991) and Ge (Hayashi, Fujii, and Yamamoto 1989; Maeda 1995) have been obtained by this technique and investigated. A substrate can be heated during the sputtering deposition to provide size control. The average size obeying the $t^{1/3}$ law inherent in the diffusion-limited growth in a glass matrix has been reported for Ge crystallites (Hayashi et al. 1990).

3.1.6 Other techniques

Si nanoparticles can be produced by means of a microwave plasma decomposition of silane (SiH_4) mixed with H_2 or Ar gas (Takagi et al. 1990). It is also possible to fabricate nanometer-size Si particles from SiH_4 gas using a breakdown stimulated by nanosecond laser radiation (Kanemitsu et al. 1993). An adjustable SiH_4 pressure provides an opportunity to obtain Si nanoparticles with variable size. Silicon nanostructures containing a network of nanocrystals connected via the silicon oxide interface can be developed using porous silicon on a silicon substrate produced by electrochemical etching. These types of structures used in microelectronics were found to consist of nanometer-size silicon crystals embedded in oxide shells (Vial et al. 1992; Brus 1994; Gaponenko, Petrov et al. 1996).

High concentrations of semiconductor nanocrystals can be developed using vacuum deposited polycrystalline films (Hodes et al. 1987; Goncharova and Sinitsyn 1990; Gurevich et al. 1992). This technique, however, provides no opportunity to control the nanocrystal size and size distribution.

3.2 Inorganics in organics: semiconductor nanocrystals in organic solutions and in polymers

Nanometer-size II-VI crystallites can be developed in an organic environment using a variety of techniques based on organometallic and polymer chemistry (Rossetti, Nakahara, and Brus 1983; Weller et al. 1986; Alivisatos et al. 1988; Henglein 1989; Brus 1991; Misawa et al. 1991a; Murray, Norris, and Bawendi 1993; Woggon et al. 1993; Gaponenko, Germanenko et al. 1996). Basic features of the structures fabricated in this way can be summarized as follows. A relatively low precipitation temperature (usually not exceeding 200°C) is favorable to minimize the number of lattice defects. The ability to cap the crystallite surface by organic groups provides a way to the control of surface states. It is possible to obtain isolated clusters or to disperse them in a very thin film with a consequent structural analysis by means of X-ray or TEM facilities. Under certain conditions, an extremely narrow size distribution of clusters can be obtained.

One of these methods utilizes colloidal solutions. The main problem to be solved in this technique is to prevent a rapid agglomeration of particles due to Ostwald ripening. For this purpose, a stabilizing agent should be added to a liquid solution containing metal salts and halogen compounds. The size of crystallites thus obtained is controlled by the temperature, mixing rate of reagents, and by the concentration of stabilizer (Rossetti et al. 1984; Alivisatos et al. 1988; Eychmüller et al. 1991). To incorporate crystallites into a rigid matrix, the so-called arrested precipitation by inverse micelles has been proposed (Alivisatos et al. 1988). This procedure provides II-VI crystallites in a polymer matrix that have a controllable size and the surface capped by the phenyl groups. However, the size distributions of particles in these samples are rather broad, and individual absorption bands can be resolved under size- selective excitation only.

It is possible to use a polymer film both as stabilizer and matrix (Misawa et al. 1991a; Woggon, Bogdanov et al. 1993). Polyvinyl alcohol and polymethyl methacrylate can be used as matrix materials. Structures thus obtained exhibit sharp absorption spectra with clearly pronounced bands and an intense edge luminescence. An evident advantage of structures like semiconductor-in-an-organic-film is the ability to apply a strong electric field when studying electric field effects because the thickness of the structure, unlike glass, can be made as small as 10 μm or even less. One more important advantage of this technique as compared, for example, to the diffusion-limited growth in inorganic glass, is low defect concentration because of the low temperature of the synthesis (200–300°C) and the well-defined and controllable surface structure. CdSe (Brus 1991 and references therein), ZnS (Kortan et al. 1990), CdS (Misawa et al. 1991a; Woggon, Bogdanov et al. 1993; Gaponenko, Germanenko et al. 1996), PbS (Wang et al. 1987), and GaAs (Salata et al. 1994) nanocrystals in polymer have been developed. For II-VI nanocrystals, several studies have found that smaller crystallites show cubic (zinc blende) lattice structure that tends to hexagonal (wurtzite) structure for larger nanocrystals (Brus 1991; Bandaranayake et al. 1995). These findings are consistent with the data reported for crystallites in glass matrices (Section 3.1) and lead to a conclusion that smaller crystallites show a tendency to higher lattice symmetry.

Recently, a novel modification of the semiconductor-in-organics technology has been advanced providing an extremely narrow size-distribution and high quality of crystallites (Murray, Norris, and Bawendi 1993). At first, a rapid injection of organometallic reagents into a coordinating solvent gives rise to a homogeneous nucleation. A slow growth and annealing are performed then to obtain a uniform surface derivatization. A subsequent size-selective precipitation provides a powder of nearly monodisperse particles, which can then be dispersed in a solvent. A series of II-VI compounds was used, and a mean radius ranging from 1.2 to 11.5 nm was obtained. The samples thus developed were found to exhibit a fine structure in optical spectra that normally can be evaluated only by means of modulation or size-selective techniques (Nirmal, Murray, and Bawendi 1994). Using a similar approach, InP crystallites have been developed with a very high luminescence efficiency (Micic et al. 1996). More information on preparation and characterization of semiconductor crystallites in an organic environment can be found in the reviews (Henglein 1989; Andreas et al. 1989; Brus 1991; Wang and Herron 1991).

To develop Si nanocrystals in an organic solvent, a different approach has been proposed (Littau et al. 1993). Stable Si particles covered by an oxide shell are developed by means of a high-temperature gas-phase synthesis using disilane pyrolisis. In this procedure, nucleation, growth, annealing, and surface derivatization of free-standing crystallites take place in the aerosol state at a temperature close to 1,000°C, which is favorable to producing the most thermodynamically stable isomer. Crystallites can then be collected and embedded at a controllable concentration into ethylene glycol to obtain a stable structure with the average size of Si inclusions ranging from 3 to 8 nm.

3.3 Nanocrystals on crystal substrates: self-organized growth

The self-organized growth of nanocrystals on a crystal substrate resulting from a strain-induced 2D $\rightarrow$ 3D transition has been discovered in recent years. It provides wide opportunities in developing high-quality nanocrystals of the most important industrial semiconductors: III-V compounds, silicon, and germanium. This growth mode takes place in the case of the submonolayer hetero-epitaxy with a noticeable lattice misfit of the monocrystal substrate and the growing layer. In this case the growing monolayer exhibits the coherent growth, that is, the structure of the layer reproduces the structure of the substrate. This means that the monolayer experiences a strong pressure because of the lattice misfit. If this strain is compressive with respect to the monolayer, the latter becomes unstable and tends to a strain-relaxed arrangement. Remarkably, the strain relaxation occurs via the 2D $\rightarrow$ 3D transition, after which the strained monolayer divides into a number of hut-like crystallites. These crystallites have a well-defined pyramidal shape with a hexagonal basement. The effect was found both for the metalorganic chemical vapor deposition and for the molecular beam epitaxy.

A number of nanocrystals have been developed to date in this way: InAs and InGaAs on GaAs substrate (Nötzel et al. 1991; Fafard et al. 1994; Marzin et al. 1994; Wang et al. 1994), InP on GaInP surface (Georgsson et al. 1995), $Si_{0.5}Ge_{0.5}$ on Si (Jesson, Chen, and Pennycook 1995), (ZnCd)Se on ZnSe (Lowisch et al. 1996), GaSb in GaAs matrix (Hatami et al. 1995). In the latter case, originally developed crystallites on the GaAs surface were buried by the GaAs deposition and a number of successive cycles have been performed to produce layers of the buried GaSb nanocrystals sandwiched between layers of GaAs. The lattice misfit values for various structures are the following: 7 percent for InAs/GaAs, 4 percent for Ge/Si, InP/GaInP, and InP/GaAs. The characteristic size of nanocrystals was found to correlate with the misfit value: the larger the misfit, the smaller the crystallites. InAs pyramids on GaAs have typical sizes of 12–20 nm base diameter and heights of 3–6 nm, whereas InP nanocrystals on GaInP have base sizes of about 50 nm and heights on the order of 10–20 nm.

Although the mechanisms of self-organized growth are poorly understood at present, there is no doubt that this approach opens wide prospects for application of nanocrystals in novel optical and optoelectronic devices. Not only have high-quality crystallites with an efficient intrinsic emission been developed, but also strong evidence of the high size/shape uniformity of these structures has been reported. For example, by means of high-resolution electron microscopy, InP pyramids on a InGaAs surface were found to have well-defined {111}, {110}, and {001} facets (Georgsson et al. 1995). Furthermore, a quantum dot ensemble fabricated by successive layer-on-layer growth was found to possess a spatially regular arrangement of the dots (Priester and Lannoo 1995; Tersoff, Teichert, and Lagally 1996). Taking into account all these features, we conclude that self-assembled nanocrystals may lead to revolutionary changes in the field, providing an efficient way to the further progress in micro-, nano-, and optoelectronics.

3.4 Synopsis of nanocrystals fabricated by various techniques

In this section we provide a tabular synopsis of semiconductor nanocrystals developed to date. About 30 types of substances have been used to fabricate nanocrystals possessing the absorption edge from the near ultraviolet (e.g., CdS in the strong confinement limit and CuCl in a weak confinement regime) to the infrared region (e.g., PbS in the strong confinement regime). The data are summarized in the Tables 3.1 to 3.5 where substances are classified into groups of II-VI, I-VII, III-V, IV, and other compounds. The references provided in the tables were chosen to cite the publications containing the most comprehensive description of the synthesis and of the structural characteristics of various

Table 3.1. *II-VI nanocrystals developed by various techniques*

Nanocrystals	Environment	References
CdS	silicate glass	Ekimov, Efros, and Onushchenko 1985 Potter and Simmons 1988; Liu and Risbud 1990
	aqueous solution	Rossetti, Nakahara, and Brus 1983
	polymer	Weller et al. 1986; Misawa et al. 1991a Woggon, Bogdanov et al. 1993; Murray, Norris, and Bawendi 1993 Artemyev, Gaponenko et al. 1995
	sol-gel glass	Nogami, Nagasaka, and Takata 1990 Minti et al. 1991 Spanhel, Arpac, and Schmidt 1992; Mathieu et al. 1995
	zeolite	Wang et al. 1989
	semiconductor-glass composite film	Gurevich et al. 1992
CdSe	silicate glass	Ekimov, Efros, and Onushchenko 1985 Borrelli et al. 1987 Gaponenko, Woggon, Saleh et al. 1993
	polymer	Bawendi et al. 1989 Murray, Norris, and Bawendi 1993
	polycrystalline film	Hodes et al. 1987
CdTe	silicate glass	Potter and Simmons 1990; Liu et al. 1991
	semiconductor-glass composite film	Ochoa et al. 1996
	polymer	Murray, Norris, and Bawendi 1993 Bandaranayake et al. 1995
$CdTe_xS_{1-x}$	silicate glass	Neto et al. 1991
ZnSe	polymer	Chestnoy, Hull, and Brus 1986
	polycrystalline film	Goncharova and Sinitsyn 1990
ZnS	polymer	Kortan et al. 1990
ZnS:Mn	polymer	Bhargava et al. 1994
CdZnSe	ZnSe	Illing et al. 1995
HgS	composite film	Zylberajch, Ruaudel-Teixier, and Barraud 1989
(Zn, Cd) Se	ZnSe surface	Lowisch et al. 1996

Table 3.2. *I-VII nanocrystals developed by various techniques*

Nanocrystals	Environment	References
CuCl	silicate glass	Tsekhomskii 1978; Borrelli et al. 1987 Justus et al. 1990
	NaCl crystal	Itoh, Iwabuchi, and Kataoka 1988
	KCl crystal	Tsuboi 1980
CuBr	silicate glass	Tsekhomskii 1978; Woggon, and Henneberger 1988
	KBr crystal	Itoh, Iwabuchi, and Kirihara 1988
	sol-gel glass	Nogami, Zhu, and Nagasaka 1991
CuI	silicate glass	Gogolin et al. 1991 Vasiliev et al. 1991
	KI crystal	Tsuboi 1980; Itoh, Iwabuchi, and Kirihara 1988
AgCl	inorganic glass	Pascova, Gutzow, and Tomov 1990
AgBr	silicate glass	Tsechomskii 1978
	polymer	Marchetti, Johanson, and McLendon 1993
AgI	silicate glass	Borrelli et al. 1987
	polymer	Gurin and Grigorenko 1995
	organic solution	Schmidt, Patel, and Meisel 1988

Table 3.3. *III-V nanocrystals developed by various techniques*

Nanocrystals	Environment	References
InGaAs	GaAs surface	Fafard et al. 1994
InAs	GaAs surface	Nötzel et al. 1991; Marzin et al. 1994; Moison et al. 1994; Wang et al. 1994; Jeppesen et al. 1996
InP	polymer	Micic et al. 1996
	GaInP surface	Georgsson et al. 1995
	porous glass	Hendershot et al. 1993
GaSb	GaAs matrix	Hatami et al. 1995
GaP	organic solution	Micic et al. 1995
GaAs	polymer	Salata et al. 1994
	porous glass	Justus, Tonucci, and Berry 1992
	organic solution	Butler, Redmond, and Fitzmaurice 1993

Table 3.4. *Nanocrystals of group IV elements developed by various techniques*

Nanocrystals	Environment	References
Ge	semiconductor-glass composite film	Hayashi, Fujii, and Yamamoto 1989; Maeda 1995
	silica glass	Nogami and Abe 1994
	SiO_2	Craciun et al. 1994
Si	aerosol	Littau et al. 1993; Camata et al. 1996
	porous Si	Vial et al. 1992; Brus 1994; Gaponenko, Germanenko, Petrov et al. 1994
	Si-glass composite film	Yamamoto et al. 1991
	amorphous SiO_2	Zhu et al. 1995
$Si_{0.5}Ge_{0.5}$	Si surface	Jesson, Chen, and Pennycook 1996

Table 3.5. *Nanocrystals of other binary and ternary compounds developed by various techniques*

Nanocrystals	Environment	References
Cu_2O	silicate glass	Borrelli et al. 1987
In_2S_3	organic solution	Nosaka, Ohta, and Miyama 1990
PbI_2	zeolite	Tang, Nozue, and Goto 1992
	polymer	Artemyev, Yablonskii, and Rakovich 1995; Saito and Goto 1995
	SiO_2 film	Lifshitz et al. 1994
	porous glass	Agekyan and Serov 1996
Cu_2S	silicate glass	Klimov et al. 1995
Fe_2S	organic solution	Wilcoxon, Newcomer, and Samara 1996
TiO_2	aqueous solution	Tennakone, Jayatilake and Ketipearachchi 1991
$GaInP_2$	organic solution	Micic et al. 1995
Cd_3P_2	polymer	Kornowski et al. 1996
Ag_2S	polymer	Kagakin et al. 1995

(*continues*)

Table 3.5. (*continued*)

Nanocrystals	Environment	References
In_2S_3, In_2Se_3	organic solution	Dimitrijevic and Kamat 1988
Bi_2S_3	organic solution	Variano et al. 1987
Sb_2S_3	organic solution	Variano et al. 1987
MoS_2	organic solution	Wilcoxon and Samara 1995
HgI_2	organic solution	Micic et al. 1987
PbS	polymer	Wang et al. 1987
	silicate glass	Borrelli and Smith 1994
$CuIn_2S_2$	silicate glass	Bodnar, Molochko, and Solovey 1993
	polymer	Gurin and Sviridov 1995
$CuIn_2Se_2$	silicate glass	Bodnar, Molochko, and Solovey 1993

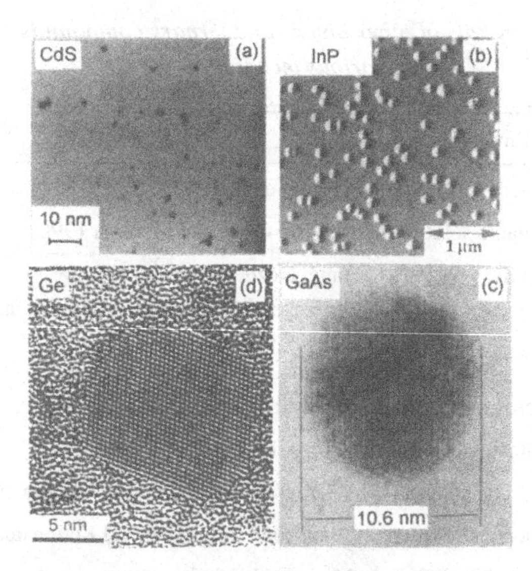

Fig. 3.3. Semiconductor nanocrystals in various environments. (a) CdS crystallites in polyvinyl pyrrolidone (Gaponenko, Germanenko et al. 1996); (b) InP islands on a GaInP surface (Carlsson et al. 1995); (c) Ge single nanocrystals in a composite semiconductor-glass film (Kanemitsu et al. 1992); (d) GaAs single nanocrystals in a polymer matrix (Micic and Nozik 1996). Note pronounced crystallographic planes in (c) and (d).

nanocrystals in different environments rather than to show the priority in the synthesis of the particular types of nanocrystals. However, an attempt has been made to cite as many groups as possible for every nanocrystal.

In Fig. 3.3 a set of nanocrystal images is presented. Clearly seen are the pronounced size dispersion [Fig. 3.3(a)] of nanocrystals in a polymer, the defined shape of nanocrystals developed by the self-organized growth [Fig. 3.3(b)], and the crystallographic planes of single nanocrystals in a composite film and in a polymer [Fig. 3.3(c),(d)]. Every real ensemble of nanocrystals dispersed in some environment possesses unavoidable inhomogeneities. For the majority of matrices this is, first of all, the size distribution. However, even if the size dispersion is minimized by some sophisticated growth procedure, crystallites remain to differ in shape, impurity and defect concentration, and in the local environment. All these factors lead to an inhomogeneous broadening of absorption and emission spectra. Therefore, optical characteristics of nanocrystals should be evaluated by means of selective spectroscopy. These experimental techniques will be the subject of Chapter 4.

4
General properties of spectrally inhomogeneous media

Because of inevitable size distribution, shape variations, different concentration of impurities and defects, and fluctuations of local environment and charge distribution, every ensemble of nanocrystals dispersed in some solid or liquid medium possesses inhomogeneously broadened absorption and emission spectra. Therefore, a number of properties inherent in molecular and atomic inhomogeneously broadened spectra can be a priori foreseen for nanocrystals. These include spectral hole-burning, fluorescence line narrowing under selective excitation, and decay time distribution. At the same time, spectrally selective techniques developed for inhomogeneously broadened molecular and atomic structures have been successfully applied to nanocrystals providing evaluation of individual parameters smeared as a result of inhomogeneous broadening. This chapter gives a brief overview of specific phenomena inherent in all spectrally inhomogeneous media and a survey of the relevant experimental techniques including nonlinear pump-and-probe spectroscopy, fluorescence excitation spectroscopy, and single molecule spectroscopy.

4.1 Population-induced optical nonlinearity and spectral hole-burning

Every real system consisting of particles with discrete energy levels exhibits absorption saturation under intense optical excitation. The only example of a nonsaturable system is an ensemble of ideal harmonic oscillators that possess an infinite number of equally spaced energy levels, optical transitions allowed only for a couple of neighboring levels, and the probability of optical transitions being proportional to the level number (Stepanov and Gribkovskii 1963). In the case of two-level particles, a system of steady-state rate equations

$$Bun_1 = n_2(Bu + A), \tag{4.1}$$

$$n_1 + n_2 = N \tag{4.2}$$

gives the following relations for intensity-dependent populations

$$n_1 = \frac{A + Bu}{A + 2Bu} N, \qquad n_2 = \frac{Bu}{A + 2Bu} N. \tag{4.3}$$

In Eqs. (4.1)–(4.3) N is the total number of particles, n_1 is population of the ground state, n_2 is population of the excited state, u is radiation density, and A and B are the Einstein coefficients for spontaneous and stimulated transitions, respectively. When $Bu \gg A$ one has

$$n_1 = n_2 = N/2. \tag{4.4}$$

In this case the absorption power takes its maximal value, that is, it saturates, whereas the absorption coefficient α equals to zero. The $\alpha(u)$ relation has the form

$$\alpha(u) = \frac{\alpha_0}{1 + au}, \tag{4.5}$$

where α_0 is the maximal value of the absorption coefficient relevant to $u = 0$, and a is the nonlinearity parameter equal to $a = 2B/A$. The relation (4.5) implies negligible radiationless transitions and transitions stimulated by equilibrium radiation. It was derived for the first time by Karplus and Schwinger (1948). Basically, it is valid for every system with discrete energy levels when the rate equations are linear (i.e., excited state decay process is monomolecular). However, in the case of a multilevel system the nonlinearity parameter for a given transition is determined by the probabilities relevant to all other transitions, that is saturation in a given absorption channel is determined by the properties of the entire system (Stepanov and Gribkovskii 1963).

The above consideration implies isotropic angular distribution of excited particles. It is valid when fixed particles experience irradiation by isotropic radiation. However, it remains relevant in the case of anisotropic or partially polarized radiation if molecules possess very fast rotational relaxation that randomizes orientation of excited and nonexcited particles. In the case of linearly polarized light, $\alpha(u)$ relation takes the form

$$\alpha(u) = \alpha_0 \left(\frac{1}{au} - \frac{1}{(3au)^{3/2}} arctg(3au)^{1/2} \right), \tag{4.6}$$

which was derived by Stepanov and Gribkovskii (1960) and later by Kramer, Tompkin, and Boyd (1986). It is relevant for molecules embedded in a rigid matrix where rotational motion is inhibited and polarization of exciting radiation becomes important (see, e.g., Gaponenko, Gribkovskii et al. 1993). In spite of the different analytical expressions, the shape of the curves corresponding to Eqs. (4.5) and (4.6) is basically similar (Fig. 4.1).

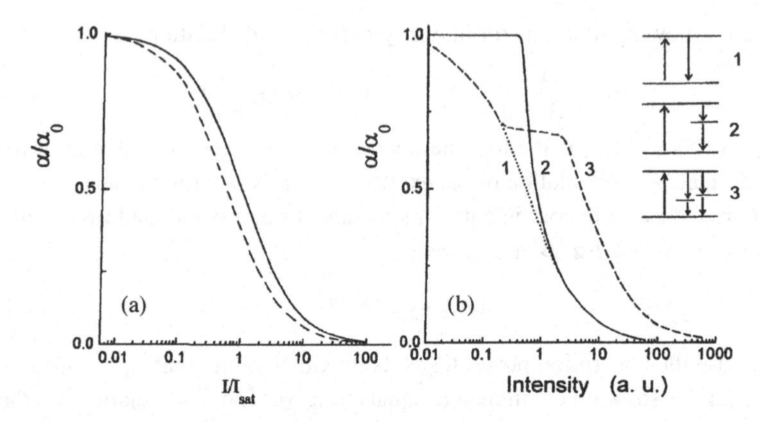

Fig. 4.1. Intensity dependence of absorption coefficient in systems with discrete energy levels. Panel (a) shows $\alpha(I)$ dependence in the case of monomolecular recombination when an ensemble of particles is irradiated by isotropic (solid line) and linearly polarized light (dashed line). The relevant analytical expressions are Eqs. (4.5) and (4.6), respectively. Panel (b) shows $\alpha(I)$ dependencies in the case of bimolecular recombination in two-, three-, and four-level systems (Gaponenko, Gribkovskii et al. 1984). In a two-level system $\alpha(I)$ function corresponds to Eq. (4.9) (curve 1). In three- and four-level systems simultaneous population and depopulation of the excited state and intermediate trap state result in more complicated saturation dynamics (curves 2 and 3). In four-level systems the initial portion of $\alpha(I)$ curve corresponds to (4.9). A pronounced step is due to redistribution of recombination between competitive paths. The final stage in all cases obeys $\alpha \propto I^{-1}$ dependence similar to Eq. (4.5).

Experimenters deal with light intensity I expressed in Watts per square centimeter, rather than with radiation density u. For this reason the relation (4.5) is commonly used in the form

$$\alpha(I) = \frac{\alpha_0}{1 + I/I_{\text{sat}}}, \tag{4.7}$$

where saturation intensity $I_{\text{sat}} = \hbar\omega N(\alpha_0 T_1)^{-1}$, T_1 is the lifetime, and ω is the resonance frequency.

In the case of bimolecular kinetics, the rates of all transitions obey a quadratic dependence

$$r_{ij} \propto n_i(N_j - n_j) \tag{4.8}$$

where n_i is the population of i-th level and $(N_j - n_j)$ is the number of free states on j-th level. The rate equations in this case are essentially nonlinear and can be solved analytically only for a few simple models. Gribkovskii (1975) derived the formula

$$\alpha(I) = \frac{\alpha_0}{1 + \frac{1}{2}aI + \sqrt{(aI/2)^2 + aI}}, \tag{4.9}$$

which was found to be valid for a two-level system, for the model of two parabolic bands, and for the model of two Gaussian bands. In all cases the specific features of the model are accounted for in the expression for nonlinearity parameter a.

Three- and four-level systems with bimolecular recombination are relevant to impurity semiconductor with one or two trap levels. These systems possess more complicated saturation dynamics because of intensity-dependent recombination rate and redistribution of recombination channels between competitive trap states (Gaponenko, Gribkovskii, et al. 1984). Two limiting cases are presented in Fig. 4.1 along with the curve corresponding to Eq. (4.9).

Spectral shape of an elementary absorption band is determined by dephasing dynamics. In the case of constant dephasing time the shape of an absorption band obeys a Lorentzian

$$A(\omega) = A_0 \frac{\Gamma^2}{(\omega - \omega_0)^2 + \Gamma^2}, \tag{4.10}$$

where ω_0 is the resonant frequency and Γ is the half-width. If spontaneous emission is the only dephasing process, then $\Gamma = 1/2T_1$ holds. In a large number of cases, population of the excited state does not affect dephasing rate. In this case absorption spectrum of a system consisting of identical particles with constant lifetime under optical excitation obeys a Lorentzian with intensity-dependent amplitude and width (see, e.g., Demtroeder 1995),

$$\alpha(\omega, I) = \frac{\alpha_0(\omega_0)}{1 + I/I_{\mathrm{sat}}} \frac{\Gamma_0^2(1 + I/I_{\mathrm{sat}})}{(\omega - \omega_0)^2 + \Gamma_0^2(1 + I/I_{\mathrm{sat}})}. \tag{4.11}$$

One can see that the linewidth in this case grows with intensity,

$$\Gamma(I) = \Gamma_0(1 + I/I_{\mathrm{sat}})^{1/2}, \tag{4.12}$$

which is shown in Fig. 4.2. Examples of absorption saturation in atomic systems with intensity-dependent dephasing dynamics (non-Markovian relaxation) are considered by Apanasevich (1977).

The above consideration implies identical absorption line-shapes inherent in every particle, that is, absorption spectrum is *homogeneously* broadened. However, in real systems broadening is often *inhomogeneous*. Atomic and molecular ensembles can experience inhomogeneous broadening as a result of velocity distribution (Doppler broadening) or environmental fluctuations. Ensembles of nanocrystals possess inhomogeneous broadening because of size and shape distribution, local environment fluctuations, and chemical inhomogeneties. In this case an inhomogeneous bandwidth can be considerably greater than a homogeneous one (Fig. 4.3). Under conditions of steady-state excitation by spectrally

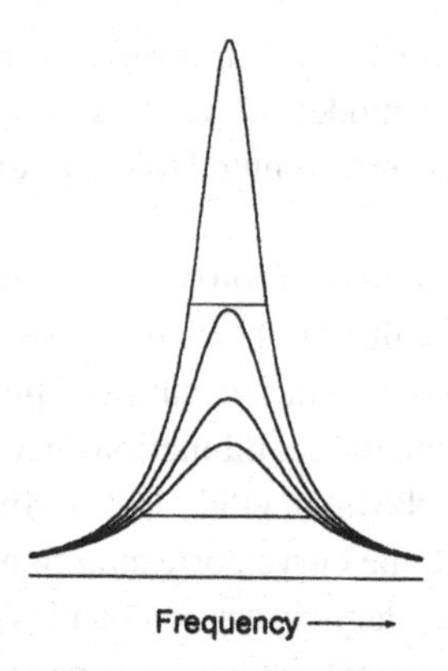

Fig. 4.2. Lorentzian-like absorption spectrum at successively growing population of the excited state. Note an increase in the bandwidth along with a decrease in the amplitude.

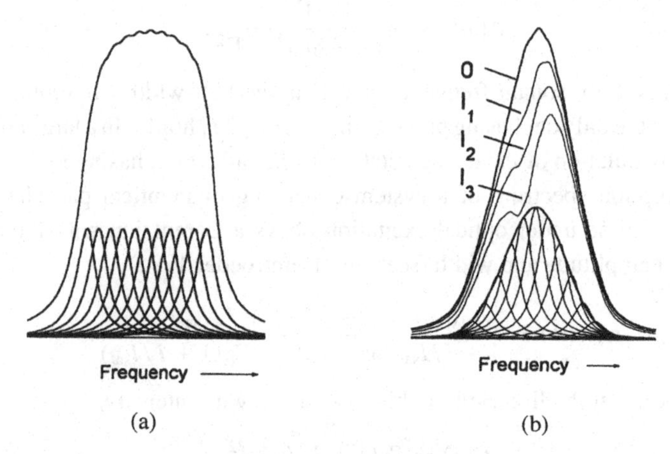

Fig. 4.3. Elementary Lorentzian-like bands and their sums in the case of inhomogeneous broadening. Case (a) corresponds to equal weights of all subbands. Case (b) corresponds to the Gaussian distribution of subbands. With narrow-band excitation, selective absorption saturation results in a dip with growing depth and width at successively growing intensity ($I_3 > I_2 > I_1$).

narrow radiation, *spectral hole-burning* takes place because of selective absorption saturation of particles resonant to exciting radiation [Fig. 4.3(b)]. As pump intensity increases, the width of the hole burned in the spectrum increases. This *power broadening* of the spectral hole results from the increasing contribution of particles whose absorption wings overlap the excitation frequency.

To analyze the dynamics of absorption saturation, consideration of the problem should include a strong monochromatic pump beam and a weak probe radiation. Tunable or wide-band probe radiation does not change the population, providing only a read-out of the population created by the pump beam. In the simplest case of uniformly distributed Lorentzians (Fig. 4.3a) the shape of the burned hole can be described by the relation (Apanasevich 1977; Letokhov and Chebotaev 1977)

$$\alpha(\omega, I) = \alpha_0(\omega)\left[1 - \left(1 - \frac{1}{\sqrt{1 + I/I_{\text{sat}}}}\right)\frac{\Gamma^2(I)}{(\omega - \omega_0)^2 + \Gamma^2(I)}\right], \quad (4.13)$$

where $\alpha_0(\omega)$ is the initial absorption spectrum of an ensemble and

$$\Gamma(I) = \Gamma_0(1 + \sqrt{1 + I/I_{\text{sat}}}) \quad (4.14)$$

is the half-width of the burned hole. The shape of the hole corresponds to a Lorentzian with an intensity-dependent width. The latter is the sum of the width of population function

$$\Gamma_{\text{pop}}(I) = \Gamma_0\sqrt{1 + I/I_{\text{sat}}} \quad (4.15)$$

and the homogeneous width Γ_0. Therefore, in a pump-and-probe experiment the width of the burned hole even when limited by small intensity is larger than the intrinsic homogeneous width. In the specific case under consideration Eq. (4.14) at $I/I_{\text{sat}} \ll 1$ gives $\Gamma(I) \rightarrow 2\Gamma_0$. At the center of the band ($\omega = \omega_0$) the absorption coefficient falls with the intensity according to the law

$$\alpha(\omega_0, I) = \frac{\alpha_0(\omega_0)}{\sqrt{1 + I/I_{\text{sat}}}}, \quad (4.16)$$

which differs from Eq. (4.7) relevant to a homogeneously broadened line.

A schematic of a pump-and-probe experiment is presented in Fig. 4.4. A tunable laser is typically used as a pump source. It can be pumped by a nitrogen laser or second and third harmonic of a Nd:YAG-laser (337, 532, and 355 nm, respectively). Low-power, wide-band radiation, obtained either by excitation of the luminescence of a dye or by the scattering of an ultrashort laser pulse ($\sim 10^{-11}$ s) in a cell with D_2O, is used for probing. It is important that both pump and probe pulses are generated using a single primary N_2- or Nd:YAG-laser, thus providing a possibility to synchronize two optical pulses in a nano- or picosecond range. A delay channel provides a control of a probe pulse with respect to the pump. A spectrograph and a multichannel analyzer based on a linear charge-coupled-device array (typically 1,000 channels) and an analog-to-digital converter (up to 18 binary digits) are used to record the transmission spectrum of the excited and unexcited samples. A personal computer is used to drive the laser power supply (typically 10–50 Hz) and to collect the data. The use of pico- and femtosecond pulses and adjustable delay of the probe pulse with respect to the exciting pulse makes it possible to analyze nonlinear

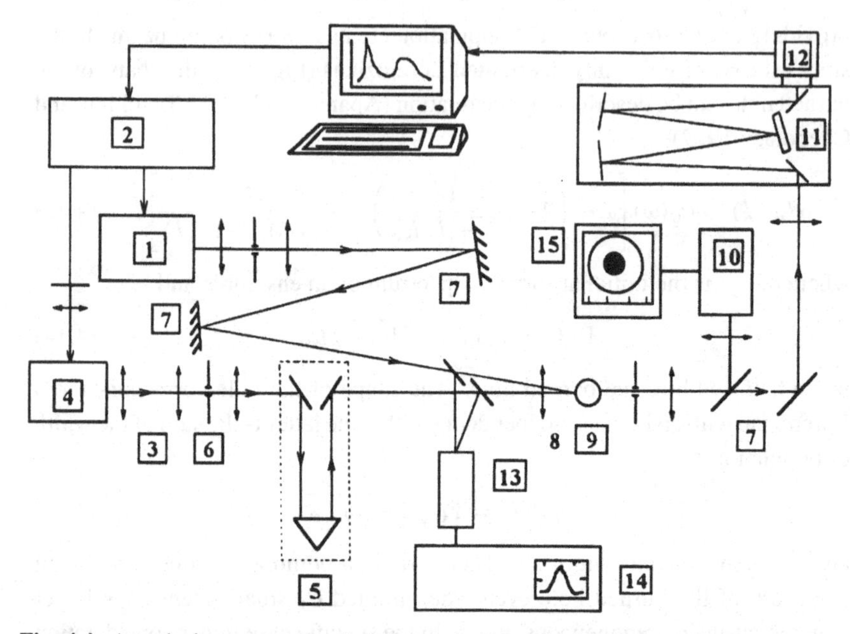

Fig. 4.4. A typical pump-and-probe set-up. 1 – tunable dye-laser or Al_2O_3:Ti-laser; 2 – YAG:Nd solid state laser or a N_2-gas laser; 3 and 8 – lenses; 4 – dye cell providing broad-band probe radiation; 5 – delay line; 6 – pinholes; 7 – mirrors; 9 – sample; 10 – microscope with CCD camera; 11 – grating monochromator; 12 – CCD array; 13 – photocell or photomultiplier; 14 – oscilloscope; 15 – monitor. A personal computer provides synchro-pulses for laser power supply and data aquisition, storage, and processing recorded by the CCD array. The delay line, the photocell, and the oscilloscope control temporal overlap of the pump and the probe pulses, whereas the microscope with the CCD camera monitor spatial overlap of the pump and the probe beams at the sample.

phenomena under conditions of nonstationary excitation and to obtain information about the energy relaxation processes on a scale shorter than the excited state lifetime.

4.2 Persistent spectral hole-burning in heterogeneous media

Under certain conditions resonant excitation of a spectrally inhomogeneous medium may result in persistent spectral hole-burning with recovery time much longer than the excited state lifetime. Persistent hole-burning can even be irreversible. The phenomenon was observed for the first time in 1974 (Gorokhovskii, Kaarli, and Rebane 1974; Kharlamov, Personov, and Bykovskaya 1974) for organic molecules in a Shpol′skii matrix and in a glassy ethanol matrix at cryogenic temperatures. It has since been established for a large variety of heterogeneous guest-host structures. It is used as an efficient technique in studies of inhomogeneously broadened spectra and is considered in the context of numerous applications for optical data storage (see Moerner 1988 and references therein).

The evident necessary conditions for persistent hole-burning are large inhomogeneous broadening and narrow linewidth of exciting light. Additionally, there must exist more than one ground state configuration of the total system (guest + host). Furthermore, an optical excitation pathway must exist that connects the ground state configurations. Finally, the relaxation among the ground states must be slower than the lifetime of any excited state. The specific mechanisms of persistent hole-burning are different for different guest-host structures. For example, color centers and rare-earth ions in glasses exhibit hole-burning as a result of selective photoionization and trapping. Organic compounds in crystals and polymers show persistent hole-burning because of photoinduced decomposition. Several complexes in alkali halides experience photoinduced effects because of reorientation in the matrix. In principle, all these processes may occur in the case of nanocrystals embedded in a solid matrix.

The persistent hole-burning phenomenon has a number of general features. Typically spectral hole-burning takes place at rather low temperature because recovery as a result of relaxation between different ground states is thermally activated (e.g., deionization of color centers). Along with the resonant hole at the excitation wavelength, additional absorption develops in the other spectral range. In the case of photochemical decomposition, induced absorption results from the product of the chemical reaction and may be shifted well from the excitation wavelength. In the case of nonphotochemical processes, a change in the local environment (nanoenvironment) results in an induced absorption immediately adjacent to the hole. Both cases are relevant to semiconductor nanocrystals and will be discussed in Chapter 7.

It is important to note that in the ideal case the hole width is twice the homogeneous width. This is because every absorbing center reacts to optical excitation in the range corresponding to homogeneous width. Every center removed from the absorption line carries with it an absorption one homogeneous width wide, so that read-out of the remaining centers yields a spectral hole with width $2\Gamma_{hom}$. In many real systems the actual hole width may be even larger, for example, because of spectral diffusion in the course of "recording" and "playback."

4.3 Luminescent properties

Under conditions of monochromatic excitation at a wavelength corresponding to the absorption long-wave wing of a wide inhomogeneously broadened band a number of selective features in emission spectrum appear. These provide evaluation of elementary absorption and emission parameters relevant to

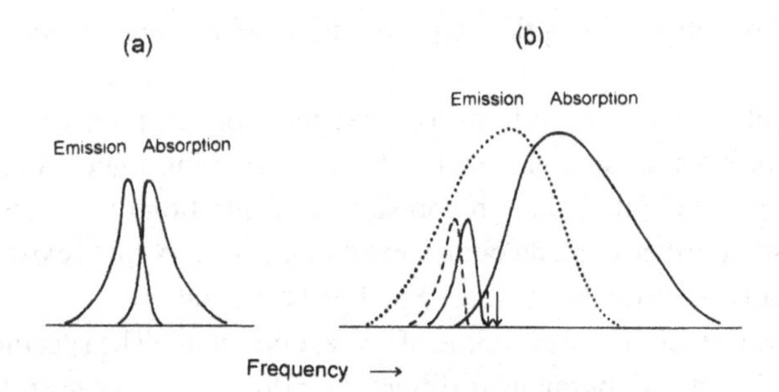

Fig. 4.5. Fluorescence line narrowing under condition of red-edge excitation. (a) Elementary absorption and emission spectra; (b) integral absorption and emission spectra and two emission spectra relevant to different excitation frequencies indicated by arrows.

individual emitters (atoms, ions, molecules, or nanocrystals). The primary effect is *fluorescence line narrowing* (Fig. 4.5). Unlike nonlinear pump-and-probe spectroscopy based on spectral hole-burning, fluorescence line narrowing does not need intense laser excitation and can be performed by means of traditional lamp sources. Tunable excitation at the low-frequency wing of the absorption band results in a narrower emission spectrum correlating with excitation wavelength. These experiments provide an estimate of the first excited state energy and dephasing rate (i.e., homogeneous linewidth) as well as the individual Stokes shift. These parameters can by no means be obtained from integral absorption.

Likewise, under conditions of narrow-band recording, the *excitation spectrum* provides information on the individual absorption spectrum including higher excited states. Basically, the excitation spectrum correlates with the absorption spectrum. However, two aspects should be mentioned. First, because of specific relaxation dynamics, some of the excited states that are pronounced in the absorption spectrum may not be efficient in luminescence excitation. For example, direct radiative decay to the ground state or competitive radiationless transitions may result in an inhibited contribution of the higher excited states to the fluorescence excitation spectrum. Therefore, excitation spectrum is determined not only by wavelength-dependent absorption, but by the wavelength-dependent *quantum yield* as well. Second, one should bear in mind that luminescence intensity (under conditions of wavelength-independent quantum yield and negligible reabsorption effect) is proportional to the number of absorbed quanta, that is, to the absorption intensity

$$A = 1 - T = 1 - e^{-\alpha d}, \tag{4.17}$$

where $T = \exp(-\alpha d)$ is transmission coefficient and d is sample thickness.

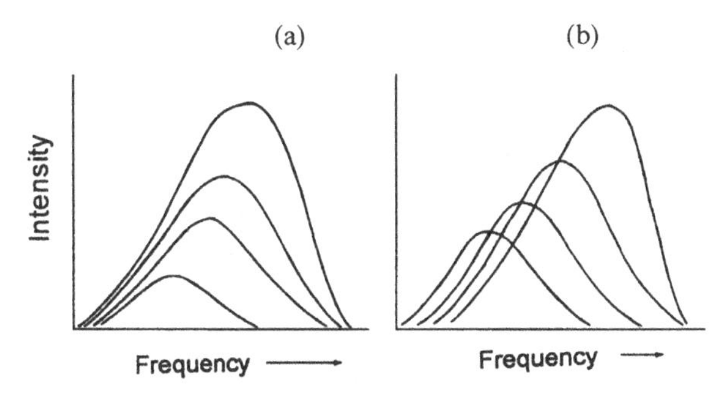

Fig. 4.6. Gated emission spectra of an inhomogeneously broadened band in the case of (a) absence and (b) presence of spectral diffusion.

Therefore, an evaluation of absorption features based on excitation spectrum is meaningful as long as $\alpha d < 1$. At $\alpha d \gg 1$ luminescence intensity saturates and does not depend on excitation wavelength.

A distribution of resonant absorption and emission frequencies within an inhomogeneously broadened band may occur along with the distribution of decay times. In this case, recording of gated or time-resolved emission spectra can provide additional information. If no energy transfer occurs between different particles within a given ensemble, then distribution of decay times results in a monotonous shift of emission maximum with time with no overlap of gated spectra recorded at sequential instances. However, in a number of cases such a transfer takes place, resulting in burning-up of luminescence at one side of the band, with simultaneous quenching at the opposite side. Typical time-resolved spectra relevant to these cases are sketched in Fig. 4.6.

Evaluation of the *decay time distribution* relevant to an inhomogeneous band by means of time-resolved studies is a typical ill-posed inverse problem. In recent years a number of numerical approaches have been proposed to examine dispersive decay laws inhomogeneously broadened emission bands. They imply some model distribution of lifetimes that is physically meaningful with a subsequent regularization procedure (Siemiarczuk and Ware 1989; Gakamsky et al. 1992; Gaponenko, Germanenko, Petrov et al. 1994). In this way, a single- or double-exponential decay relevant to a large number of homogeneous emission bands is replaced by a distribution of lifetimes (Fig. 4.7).

4.4 Single molecule spectroscopy

Discussion so far has been related to evaluation of individual spectral features in spectrally inhomogeneous systems by means of *spectrally* selective techniques.

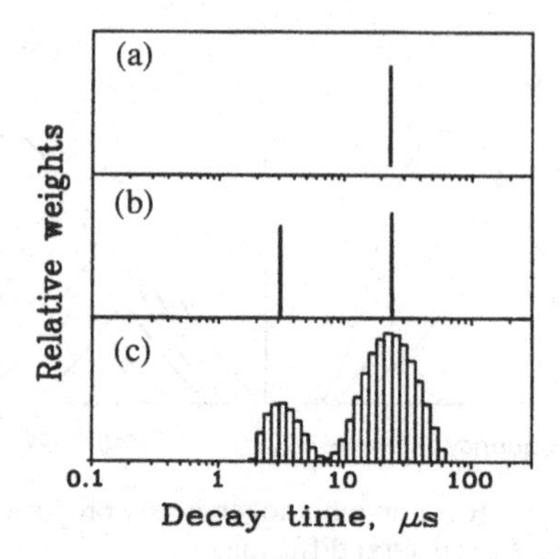

Fig. 4.7. Decay time distributions. (a) Monoexponential decay; (b) biexponential decay; (c) decay time distribution found for aged porous Si (Gaponenko, Germanenko, Petrov et al. 1994).

Along with spectrally selective performance, laser sources allow *spatial* selection of molecules, ions, or nanocrystals to be performed. Single-mode laser radiation with Gaussian beam cross section can be focused to a spot of a diameter

$$d \approx 2\lambda f/d_0 \tag{4.18}$$

where λ is the laser wavelength and f/d_0 is the focus-to-aperture ratio of a focusing system applied. Using high-quality microscope objectives with $f/d \approx 1$ one can get for $\lambda \approx 500$ nm a focal spot of $d \approx 1$ μm. In the case of a rather dilute solution this corresponds to a few molecules within an excitation spot. Further selectivity can be enhanced by means of a red-edge excitation regime. In this way a number of challenging experiments on single molecule fluorescence have been performed, providing intriguing data on dynamic broadening, time-dependent spectral diffusion, and microenvironment fluctuations (Moerner 1994b; Trautman et al. 1994 and references therein).

Possible versions of single-molecule fluorescent measurements are sketched in Fig. 4.8. In one case, the same microscope objective is used for excitation, focusing, and collection of emitted light by means of a semitransparent or beam-splitting mirror. The collected light is then dispersed by a spectrograph and recorded with a cooled intensified CCD-array. Under continuous-wave excitation the number of emitted photons can be on the order of 10^4–10^6 per second, whereas detector sensitivity can be about one count per photon. Because of the low temperature of a detector, data aquisition during a period of minutes to an hour is possible, thus providing a time-averaged single-molecule signal instead of statistically averaged instant spectra, as in the case of conventional spectroscopic techniques.

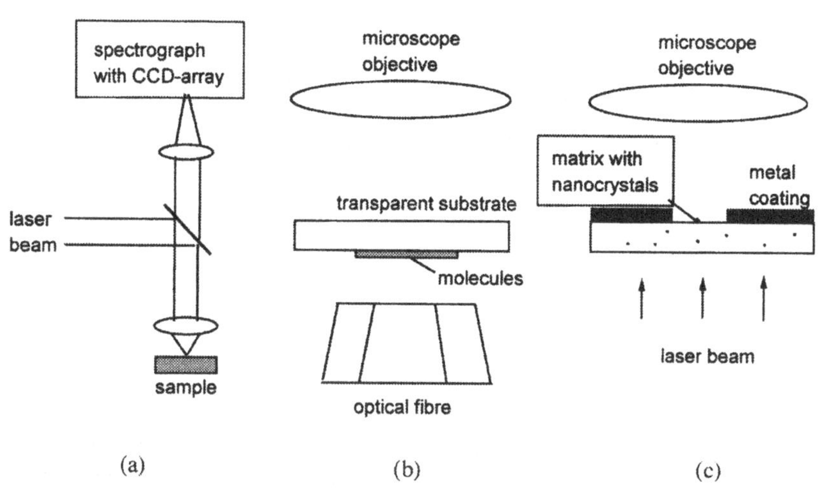

Fig. 4.8. Schematics of single-molecule fluorescence techniques. (a) Excitation and detection are performed through the same lens; (b) excitation is performed with a tapered optical fiber of about 100 nm diameter (near-field excitation) whereas emitted photons are collected by an objective through a transparent substrate; (c) excitation is performed with a wide laser beam, whereas luminescence is detected through a hole of about 1 μm diameter made in a metal film covering a solid matrix with nanocrystals.

In the another version, a tapered optical fiber provides *near-field* excitation [Fig. 4.8(b)] of a single molecule, whereas emitted light is collected through a transparent substrate. A combination of a solid transparent matrix and photolithography has been successfully performed in the case of single nanocrystal spectroscopy (Samuelson et al. 1995; Gammon et al. 1996a and 1996b). In this case, presented in Fig. 4.8(c), a wide laser beam is used for excitation, whereas the spatial selection is performed by means of submicron holes etched in a thin metal film covering the sample.

5

Absorption and emission of light by semiconductor nanocrystals

In this chapter we consider the optical processes in nanocrystals that can be interpreted in terms of creation and annihilation of a single electron-hole pair within a crystallite. Size-dependent absorption and emission spectra and their fine structures, as well as size-dependent radiative lifetimes, will be discussed for nanocrystals of II-VI, I-VII compounds and, where possible, for nanocrystals of III-V compounds and of group IV elements. Nontrivial aspects of exciton-phonon interactions that manifest themselves in homogeneous linewidths and/or intraband relaxation processes will be outlined. Challenging experiments providing the optical information relevant to a single nanocrystal will be discussed as well. Most of these results have become possible because of a number of the spectrally and spatially selective techniques described in Chapter 4. An influence of a microcavity on spontaneous emission of nanocrystals, the competitive recombination mechanisms, and the electric field induced effects will be analyzed as well.

5.1 Size-dependent absorption spectra. Inhomogeneous broadening and homogeneous linewidths

5.1.1 Experimental evidence for quantum-size effects in real nanocrystals

In the early 1980s A. I. Ekimov and A. A. Onushchenko (Ekimov and Onushchenko 1982; Ekimov and Onushchenko 1984) and L. Brus with co-workers (Brus 1983; Rossetti, Nakahara, and Brus 1983) published pioneering articles in which size-dependent absorption spectra of semiconductor nanocrystals resulting from quantum confinement were demonstrated for the first time. During the same period S. V. Gaponenko et al. reported on inhomogeneous broadening of the optical absorption spectra of glasses doped with semiconductor nanocrystals (Gaponenko, Zimin, and Nikeenko 1984). These findings were followed by an extensive exploration of quantum-size effects on the absorption spectra of nanocrystals and evaluation of the individual bands smeared by inhomogeneous broadening (see, e.g., reviews by Henglein 1989; Brus 1991; Ekimov 1991; Woggon and Gaponenko 1995; Gaponenko 1996). In this section we provide a description of the basic features revealed in these studies using a few representative examples based on I-VII and II-VI nanocrystals as the model objects.

Copper halide nanocrystals provide an opportunity to trace optical properties of quantum dots within the framework of the weak confinement limit (see Section 2.1). Particularly, for a CuCl crystal the exciton Bohr radius $a_B = 0.7$ nm is close to the lattice constant $a_L = 0.5$ nm (see Table 1.2). Therefore, one has a condition

$$a_B \ll a \ll \lambda \tag{5.1}$$

in the wide size range. The large value of the exciton Rydberg energy ($Ry^* = 200$ meV for CuCl) provides optical detection of exciton resonance even at room temperature. Bulk CuCl nanocrystal possesses two exciton bands, Z_3 and Z_{12}, relevant to the different valence subbands (branches 1 and 2 are degenerate at the $k = 0$ point and are split off branch 3). Larger CuCl nanocrystals show nearly the same absorption features with the absorption bands coinciding with that of the bulk monocrystal [Fig. 5.1(a)]. As the mean nanocrystal size $\bar{a}$ decreases to a few lattice constants, the exciton bands experience a monotonic shift towards higher photon energy because of the spatial confinement of excitons. At the same time, the width of the bands is getting larger, first of all, because of the size distribution under the conditions of the quantum confinement. These features are clearly seen in Fig. 5.1(b). Size dependence of the absorption peak is close to the $E \propto \bar{a}^{-2}$ function predicted for a particle-in-a-box problem (Fig. 5.2). However, in the case of the finite size distribution, the experimental peak position differs by a constant coefficient from the energy shift derived for a single particle [Eq. (2.1)]. Specifically, for the case of the Lifshitz-Slyozov function [Eq. (3.4)] relevant to the asymptotic distribution for diffusion-limited growth, an expression has been obtained (Ekimov, Efros, and Onushchenko 1985)

$$E_{1S1s} = E_g - Ry^* + 0.67\gamma_1 \frac{\pi^2\hbar^2}{2M\bar{a}^2}. \tag{5.2}$$

In Eq. (5.2) $\gamma_1 = 0.53$ is the Luttinger parameter; other notations are the same as in Eq. (2.1). This formula was found to fit the experimental size dependence of the Z_3 exciton band adequately. Therefore, size-dependent absorption spectra

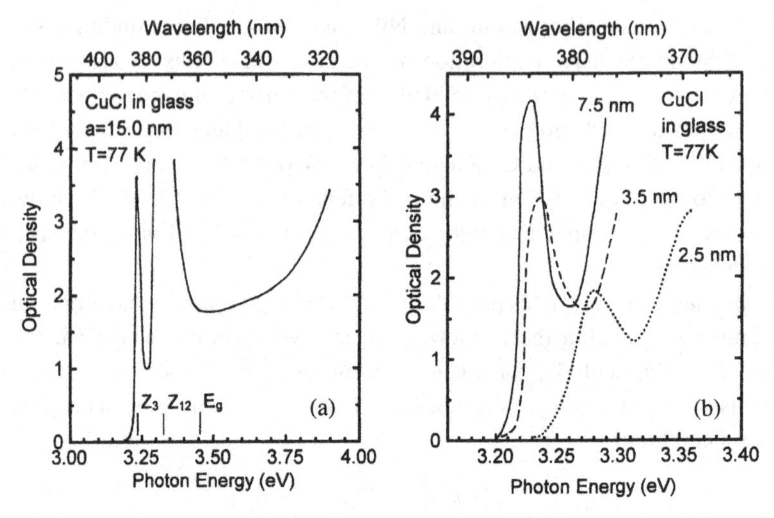

Fig. 5.1. Quantum size effect on exciton absorption of CuCl nanocrystals in a glass matrix (adapted from Zimin, Gaponenko et al. 1990). Panel (a) shows the absorption spectrum of larger nanocrystals ($a = 15$ nm) containing two exciton bands (Z_3 and Z_{12}) and a flat interband absorption for $\hbar\omega \geq E_g$. Large nanocrystals exhibit basically the same features as bulk crystals. The sharp increase of absorption at $\lambda < 330$ nm is due to matrix absorption. With decreasing size the exciton band shows a monotonic blue shift [panel (b)]. As the absolute value of the shift remains considerably smaller than exciton Rydberg energy, $Ry^* = 200$ meV, this is a typical example of the quantum-size effect in the weak confinement limit.

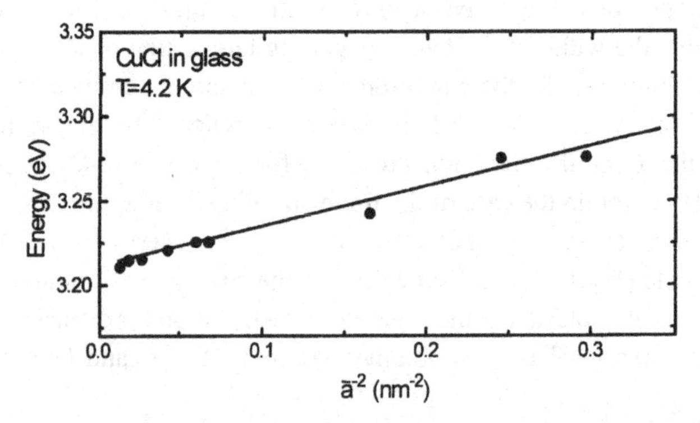

Fig. 5.2. Size dependence of the Z_3 exciton band for CuCl nanocrystals in a glass matrix (Ekimov, Onushchenko et al. 1985). The experimental dependence (dots) features $\bar{a}^{-2}$ behavior typical of a particle-in-a-box model. Size distribution of nanocrystals results in a modified $E(\bar{a})$ dependence differing from the exact analytical solution for a single particle in a box by a numerical coefficient A specific for every distribution function. In the case of asymptotic Lifshitz-Slyozov distribution relevant to the sample under consideration $A = 0.67$ holds.

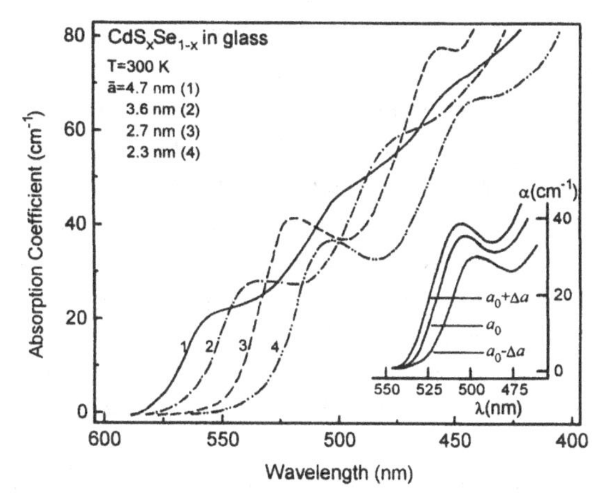

Fig. 5.3. Absorption spectra of CdS_xSe_{1-x} nanocrystals in a glass matrix at room temperature (Gaponenko, Woggon et al. 1993). Insert shows absorption spectra for small deviations of the mean size around $\bar{a}_0 = 2.3$ nm. The spectra show a clear blue shift with decreasing mean radius $\bar{a}$ due to the quantum confinement effect. The absolute value of the shift is one order of magnitude larger than the exciton Rydberg energy, which indicates strong confinement regime. Electron-hole states in this case cannot be separated into bound and free e-h pairs. Accordingly, the bulklike exciton lines and continuum absorption inherent in larger crystallites are replaced by a set of similar bands. For smaller $\bar{a}$ values the shift of the first band with growing $\bar{a}$ systematically results in an increase in the total absorption (see data for $\bar{a} = 2.3$, 2.7 and the insert), which indicates the normal growth stage. At this stage, mean size growth takes place due to increase in the total volume of the semiconductor phase in the matrix. Therefore, absorption spectra at this stage do not overlap.

of semiconductor nanocrystals actually can be described in terms of textbook quantum mechanics if one considers the weak confinement regime.

A behavior of the Z_{12} exciton band nearly obeys the $E \propto \bar{a}^{-2}$ dependence as well. However, an adequate description has been developed by means of a more complicated expression as compared with (5.2) to account for a deviation of the real Z_{12} exciton from an ideal particle with the scalar mass (Ekimov, Onushchenko, et al. 1985). Quantum confinement effect on exciton spectra has also been reported for other I-VII compounds, CuBr and CuI (Ekimov, Efros, and Onushchenko 1985; Itoh, Iwabuchi, and Kirihara 1988; Woggon and Henneberger 1988; Gogolin et al. 1991; Vasiliev et al. 1991), AgI (Borelli et al. 1987), and AgBr (Marchetti, Johansson, and McLendon 1993).

Many examples demonstrating size-dependent absorption spectra of nanocrystals in the case when $a < a_B$ can be found among the numerous studies of II-VI nanocrystals dispersed in inorganic glass matrix or in organic polymers and various solvents. In Fig. 5.3 are plotted the absorption spectra of a number of glasses doped with CdS_xSe_{1-x} crystallites of the same stoichiometry but of different sizes. These samples have been developed strictly following the commercial recipe and processing. Evidently, every absorption spectrum is drastically different from that of monocrystals. At least two wide bands can be resolved in these spectra even at room temperature, whereas a room-temperature monocrystal spectrum is always flat and structureless (see, e.g., Fig. 1.9(b) for comparison). The first absorption peak rapidly shifts towards higher energy for smaller size. The absolute value of the first peak energy, E_1, grows from 2.2 eV (560 nm) for $\bar{a} = 5$ nm to 2.5 eV (500 nm) for $\bar{a} = 2$ nm showing an increase by 0.3 eV, while the Rydberg energies are 0.016 eV and 0.030 eV for CdSe and CdS, respectively. Therefore, II-VI nanocrystals actually do provide an opportunity to examine the quantum confinement effect in the strong confinement limit, when the kinetic energies of the confined electron and hole are much greater than the exciton Rydberg energy.

Two aspects are noteworthy in connection with II-VI nanocrystals in glasses. First, spectra similar to those presented in Fig. 5.3 have been known for commercial glasses designed for cut-off filters for the visible range at least since the early 1960s (Buzhinskii and Bobrova 1962). Moreover, empirically adjusted commercial recipes systematically imply a possibility to obtain variable color for the same matrix/ingredients combination by means of adjustable heat treatments, that is, by means of the quantum confinement effect. Second, color versus size dependence in the strong confinement range can be distinguished by the naked eye. For example, glasses in Fig. 5.3 show distinct light-red, orange, and yellow colors. Therefore, we conclude that these species illustrate a macroscopic manifestation of the Heisenberg's uncertainty relation as soon as the latter provides an estimate of the particle-in-a-box zero energy (see Section 1.1).

Larger II-VI crystallites with size on the order of 10 nm and more show bulklike exciton bands featuring a small shift with size in accordance with the weak confinement model. Thus, II-VI compounds provide evidence for the smooth evolution of the fundamental absorption spectrum from the crystallike to the clusterlike properties via weak, intermediate, and strong confinement ranges. An example is given in Fig. 5.4, where optical absorption spectra of CdS nanocrystals in glass are presented with size ranging from 9.2 down to 2.1 nm. These data show a nice qualitative agreement with the predicted tendency (see Fig. 2.4 for comparison). A spectrum consisting of a number of close and narrow lines related to the hydrogenlike exciton followed by an interband absorption onset corresponds to the larger $\bar{a} = 9.2$ nm. This spectrum evolves to a number of wide bands shifted with respect to the bulk band gap energy, $E_g^0 = 2.58$ eV, by about 0.4 eV (the curve corresponding to $\bar{a} = 2.1$ nm

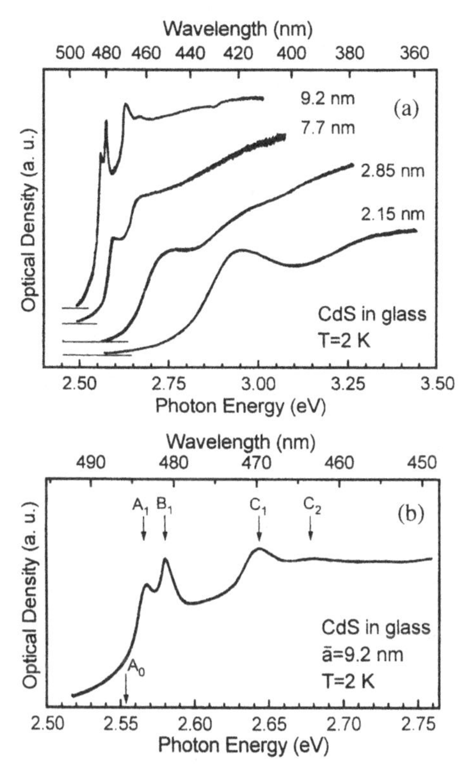

Fig. 5.4. Evolution of absorption spectra of CdS nanocrystals in a glass matrix from crystal-like to cluster-like behavior (Mathieu et al. 1995). Larger crystallites ($\bar{a}$ = 9.2 nm) feature distinctive exciton bands and structureless interband absorption. Smaller crystallites ($\bar{a}$ = 2.1 nm) possess a number of similar wide bands due to the discrete electron-hole states.

in Fig. 5.4). Size-dependent optical absorption spectra have been reported also for other nanocrystals of II-VI compounds such as CdTe (Potter and Simmons 1990; Murray et al. 1993), PbS (Wang and Herron 1991; Borrelli and Smith 1994), as well as for a number of III-V compounds: GaAs (Salata et al. 1994), InP (Micic et al. 1996), GaSb (Hatami et al. 1995), GaP (Ekimov 1991; Micic et al. 1995), Si (Brus 1994), Ge (Maeda 1995) nanocrystals, and for nanocrystals of Fe_2S (Wilcoxon, Newcomer, and Samara 1996), Cu_2S (Klimov et al. 1995), MoS_2 (Wilcoxon and Samara 1995), In_2S_3 and In_2Se_3 (Dimitrijevic and Kamat 1988; Nosaka, Ohta, and Miyama 1990), Cd_3P_2 (Kornowski et al. 1996), $CuInS_2$, $CuInSe_2$ (Bodnar, Molochko, and Solovey 1993), $GaInP_2$ (Micic et al. 1995), HgI_2 (Micic et al. 1987) and PbI_2 (Lifshitz et al. 1994; Saito and Goto 1995; Agekyan and Serov 1996).

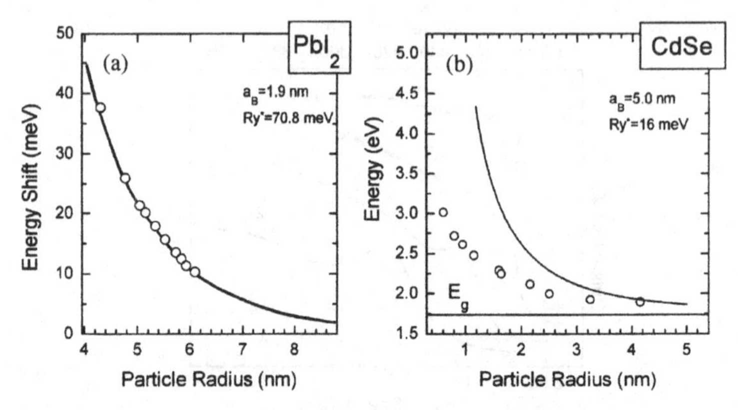

Fig. 5.5. Size dependence of the first absorption maximum in the case of intermediate (a) and strong (b) confinement ranges. Panel (a) shows experimental data on exciton band shift obtained for PbI_2 nanocrystals in a SiO_2 film (circles) along with the calculated dependence (solid line) derived for the model of acceptorlike exciton (adapted from Lifshitz et al. 1994). In this model the heavier electron is localized near the center of a crystallite, whereas the lighter hole experiences confined motion. Panel (b) presents numerical data reported for CdSe particles in a polymer film (after Norris et al. 1994) versus analytical function derived for the model of strong confinement (Eq. 2.15). The discrepancy observed for the smaller size can be reduced by correctly considering the finite potential height at the boundaries and for the dielectric confinement effect, as well as for the specific size distribution.

The latter case related to PbI_2 nanocrystals is an interesting example of the so-called acceptor-like exciton, which is characterized by an electron localized at the center of a spherical crystallite and a hole experiencing confined motion. This is possible if a crystallite has the size relevant to the intermediate confinement range ($a_B < a < 3a_B$) provided that the electron effective mass is much larger than that of a hole ($m_e \gg m_h$). This is just the case for PbI_2 nanocrystals. The localization of the heavier electron occurs because of the potential of the hole, whose motion is confined by the boundaries of a crystallite. This localization is further enhanced by the difference in dielectric permitivity between the PbI_2 particle and the SiO_2 matrix. Experimental findings have been compared with the theory developed within the framework of the effective mass approach for the acceptor-like exciton model (Ekimov et al. 1989), which is shown in Fig. 5.5(a). A nice correspondence of experimental and calculated data is evident.

A number of authors have compared the experimentally observed energy of the first absorption band in the strong confinement limit with the numerical results provided within the framework of a particle-in-a-box model. The effective mass model was found to overestimate the energy of the first transition for smaller crystallites (Weller et al. 1986; Wang and Herron 1991; Murray et al. 1993). Better agreement has been obtained in cases where the finite

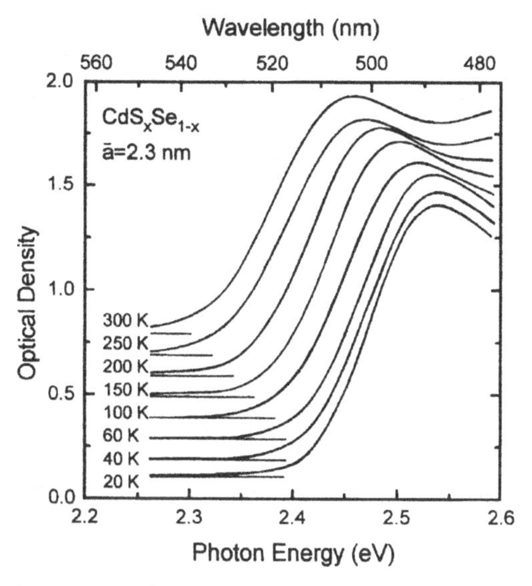

Fig. 5.6. Absorption spectra of CdS_xSe_{1-x} nanocrystals in a glass matrix at various temperatures (Woggon, Gaponenko et al. 1994). Mean nanocrystal radius is $\bar{a} = 2.3$ nm. Note that, unlike monocrystals, the shape and the width of the main absorption band are nearly independent of temperature (see Fig. 1.9 for a comparison).

potential barrier is taken into account (Tran Thoai et al. 1990; Nosaka 1991). A typical example of such a dependence is given in Fig. 5.5(b) using colloidal particles of cadmium selenide as a model object. In this connection, one should bear in mind that a comparison of the experiment and theory is not yet perfect. It is still difficult to get the monodisperse nanocrystal ensemble or to derive an explicit size distribution function. The latter should be taken into account to calculate a correction to the dependencies evaluated for a single semiconductor particle towards an actual inhomogeneously broadened spectrum. Eventually, for smaller particles the effective mass approximation is not justified and should be replaced by an explicit quantum-chemical consideration.

Remarkably, optical spectra of organic and inorganic matrices containing nanocrystals of a mean size about 3 nm and smaller remain nearly the same when being cooled from room to liquid helium temperature (Fig. 5.6). This is evidence of the fact that the observed bandwidths result from temperature-independent inhomogeneous broadening. In the case of semiconductor nanocrystals, inhomogeneous broadening arises, first of all, from the finite size distribution. Further physical mechanisms resulting in inhomogeneous broadening are variations in stoichiometry, shape, surface structure, defect concentration, charge, local environment, and so on. To evaluate contributions of individual crystallites to the absorption spectrum of an ensemble, spectrally selective techniques should be applied.

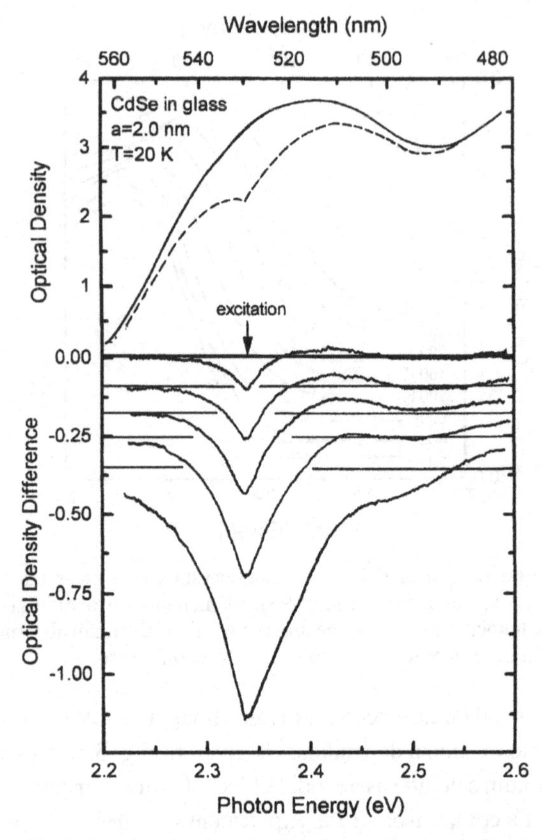

Fig. 5.7. Spectral hole-burning due to selective absorption saturation in an ensemble of CdSe nanocrystals in a glass matrix (Gaponenko, Woggon et al. 1994). Mean nanocrystal radius is $\bar{a} = 2.0$ nm. Temperature is 20 K. Upper panel shows the initial absorption spectrum of the unexcited sample (solid curve) and the spectrum recorded under excitation by a nanosecond laser pulse (photon energy 2.34 eV, duration is 5 ns, intensity is 7 MW/cm^2) and simultaneous probing by a wide-band weak radiation (dashed curve). Lower panel contains a set of differential absorption spectra for successively growing excitation intensity from 90 kW/cm^2 to 7 MW/cm^2. Negative change in optical density corresponds to bleaching. Note a change in the scale for differential optical density in the lower panel as compared with the optical density in the upper panel.

5.1.2 Selective absorption spectroscopy: spectral hole-burning

Inhomogeneous broadening of the absorption spectrum of II-VI nanocrystals in glasses has been examined by means of the selective absorption saturation resulting in a power-dependent spectral hole-burning peaking at the excitation wavelength (Fig. 5.7). In agreement with the theory of absorption saturation in inhomogeneously broadened systems (see Section 4.1) the width of the burned

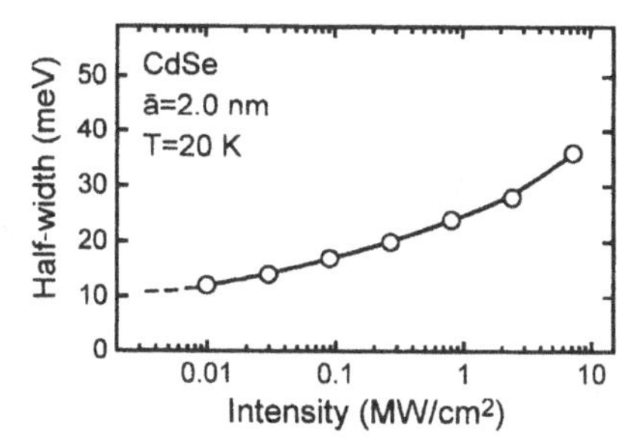

Fig. 5.8. Intensity dependence of the half-width of the spectral hole due to selective absorption saturation in an ensemble of CdSe nanocrystals in glass (Langbein et al. 1992). Mean nanocrystal radius is $\bar{a} = 2.0$ nm, temperature is 20 K, excitation wavelength is 530 nm. In the limit of negligible intensity the half-width of the burned hole tends to the true homogeneous width $2\Gamma_0$ if the individual line-shape is described by a Lorentzian.

hole monotonically grows with rising excitation intensity (Figs. 5.7 and 5.8). The spectral shape of the burned hole can be fitted by a Lorentzian. Therefore, in the limit of negligible intensity one has

$$2\Gamma(I) \xrightarrow{I \to 0} 4\Gamma_0, \tag{5.3}$$

where $2\Gamma_0$ is the homogeneous FWHM (full width at half maximum) value and $2\Gamma(I)$ is the FWHM value of the burned hole. This relation provides the upper estimate of the true homogeneous width. For the data presented in Fig. 5.7 the homogeneous linewidth was found to be $2\Gamma_0 = 11$ meV at $T = 10$ K, whereas the inhomogeneous FWHM value is more than one order of the magnitude larger. Similar results indicative of the narrow homogeneous linewidth as compared with the inhomogeneous bandwidth have been reported for II-VI nanocrystals in organic environments (Alivisatos et al. 1988; Norris and Bawendi 1995).

Spectral hole-burning experiments for CuCl nanocrystals dispersed in a glass matrix of a size corresponding to the weak confinement regime have been performed by Kippelen et al. (1991) and by Wamura et al. (Wamura, Masumoto, and Kawamura 1991). These studies revealed homogeneous linewidths on the order of $\Gamma_0 = 1$ meV and less. Small Γ_0 values for I-VII nanocrystals have been confirmed also by means of the persistent hole-burning technique (Masumoto et al. 1995).

5.1.3 Selective emission spectroscopy

Considerable quantum yield of intrinsic luminescence makes it possible to distinguish the substructure in inhomogeneous absorption spectra by means of photoluminescence excitation (PLE) spectroscopy. In what follows we discuss

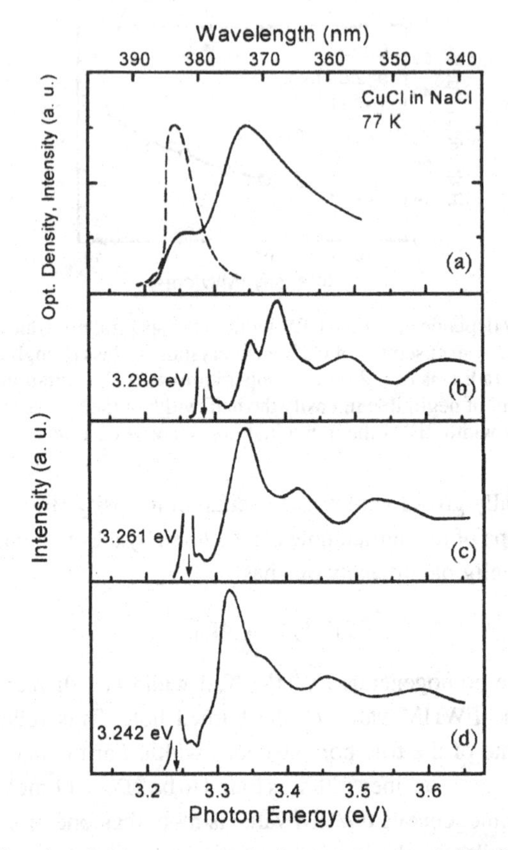

Fig. 5.9. Selective photoluminescence excitation spectroscopy of CuCl nanocrystals in NaCl (as adapted from Itoh, Iwabuchi, and Kirihara 1988). Panel (a) shows absorption spectrum (solid curve) and emission spectrum (dashed curve) recorded under condition of a broad-band UV-excitation. Absorption spectrum contains inhomogeneously broadened Z_3 and Z_{12} exciton bands. Emission spectrum contains inhomogeneously broadened band due to annihilation of Z_3 exciton. Because of the inhomogeneous broadening, excitation spectra are different for different emission wavelengths [panels (b)–(d)]. Arrows indicate emission wavelengths. Mean radius of crystallites $\bar{a} = 3.3$ nm.

two representative examples related to CuCl and CdSe nanocrystals as model objects.

In CuCl nanocrystals, broad-band Z_3 exciton luminescence recorded under spectrally wide and nonresonant excitation [Fig. 5.9(a)] was found to change drastically under conditions of excitation by monochromatic light tunable within Z_3 and Z_{12} exciton absorption. Accordingly, excitation spectrum is strongly dependent on the emission wavelength [Fig. 5.9(b–d)].

Extensive PLE experiments for the Z_3 exciton emission band of CuCl nanocrystals have revealed a complicated structure of absorption spectra and made

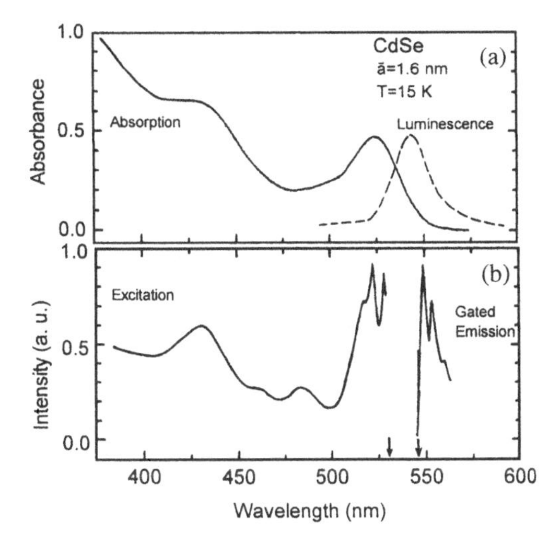

Fig. 5.10. Selective photoluminescence spectroscopy of CdSe nanocrystals in organic matrix (Bawendi, Wilson et al. 1990). Mean nanocrystal radius is $\bar{a} = 1.6$ nm, temperature is 15 K. Panel (a) shows absorption and emission spectra. Panel (b) presents excitation spectrum recorded for the emission wavelength 530 nm and the gated emission spectrum under pulse excitation at 545 nm (excitation pulse duration is 10 ns, registration time gate is 20 ns, a delay between excitation and registration is 30 ns).

it possible to identify a number of excited states of Z_3 and Z_{12} excitons (Itoh, Iwabuchi, and Kirihara 1988). Identification of subbands in the excitation spectrum and analysis of their dependence on the emission wavelength have led to a conclusion that CuCl crystallites in NaCl crystal have cubic, rather than spherical, shape. The conclusion has been confirmed by parallel X-ray scattering experiments (Itoh, Iwabuchi, and Kataoka 1988). Cubic shape has also been reported for the fine structure revealed in persistent hole-burning studies for CuCl nanocrystals in NaCl (Masumoto 1996).

Similar effects have been demonstrated for CdSe nanocrystals of sizes smaller than the exciton Bohr radius (Bawendi, Wilson et al. 1990; Ekimov 1991). A comparison of the data presented in Fig. 5.10(a) and (b) shows that wide emission and absorption spectra possess fine structure that manifests itself under conditions of the monochromatic red-edge excitation, narrow-band, and time-resolved detection.

5.1.4 Other manifestations of inhomogeneous broadening

Additional indications of small homogeneous widths as compared with inhomogeneous widths have been obtained for a number of nanocrystals by means of other experimental techniques. For example, a number of structures have been found to exhibit persistent reversible spectral hole-burning in the absorption spectra resulting from modification of the local environment (Masumoto et al. 1995). Smaller semiconductor nanocrystals in organic matrixes show permanent, irreversible photochemical hole-burning in the absorption spectra due to selective photodestruction (Gaponenko, Germanenko et al. 1996). Several studies have demonstrated evaluation of individual emission spectra by means of near-field spectroscopy, providing *spatial selection* of emitting species. These experiments will be discussed in Section 5.5. One more technique that is based on size-dependent melting temperature T_{melt} inherent in small crystallites has been proposed for CuCl nanocrystals possessing T_{melt} in the range 200–400°C (Vasiliev et al. 1995). In the case of a noticeable size distribution, annealing at the temperature specific for each size results in a spectrally selective modification (see Section 7.2).

5.1.5 Correlation of optical properties with the precipitation stages for nanocrystals in a glass matrix

Narrow-band spectral hole-burning presented in Fig. 5.7 is not typical for II-VI nanocrystals embedded in a glass matrix. Typically, CdSe and CdS_xSe_{1-x} crystallites of the mean size $\bar{a}$ less than the exciton Bohr radius a_B possess homogeneous linewidth of the order of the inhomogeneous one. In this case, monochromatic excitation results in a broad-band bleach. The position of the bleach peak correlates with the excitation wavelength, but generally the bleaching spectrum resembles an inverted linear absorption spectrum. A typical example is given in Fig. 5.11. Similar behavior has been reported by several groups (Gaponenko, Zimin et al. 1984; Peyghambarian et al. 1989; Roussignol et al. 1989; Spiegelberg, Henneberger, and Puls 1991; Uhrig et al. 1991; Tokizaki et al. 1992; Woggon et al. 1992). However, similar crystallites in an organic environment were found to possess $\Gamma_0 < 10$ meV, which is one order of magnitude less than the typical inhomogeneous bandwidth for II-VI crystallites of size $\bar{a} < a_B$ (Alivisatos et al. 1988).

The reason for the discrepancy between the values of Γ_0 for crystallites in glass and in polymers was established in experiments on glasses containing II-VI crystallites that experienced different growth dynamics (Gaponenko, Woggon et al. 1993). It was shown that the values of Γ_0 can differ by almost an order of magnitude for crystallites in glass depending on the growth regime. The large values of Γ_0 were found to be typical for crystallites formed in the process of competitive growth. However, crystallites prepared in glass using nucleation and normal growth stages possess narrow homogeneous width on the order of 10 meV. The correlation of homogeneous linewidth with the precipitation stage is related to the additional electron-hole dephasing mechanisms in the case of the crystallites that experienced the competitive growth. The higher

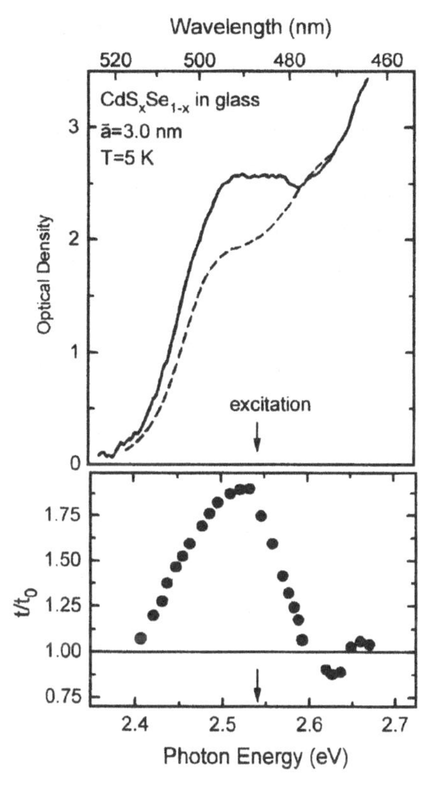

Fig. 5.11. Typical absorption saturation behavior of CdS_xSe_{1-x} nanocrystals that have experienced the competitive growth regime in a glass matrix (Klingshirn and Gaponenko 1992). Mean size of crystallites is 3.0 nm, temperature is 4.2 K. Upper panel shows the linear absorption spectrum (solid line) and the absorption spectrum under excitation at $\hbar\omega_{\text{pump}} = 2.54$ eV (dashed line). Lower panel presents relative transmission change. The shape of the bleach is rather wide and correlates with the linear absorption spectrum rather than with a position of the excitation photon energy.

dephasing rate may be a result of the imperfection of the lattice structure, surface defects, or uncompensated surface charge. Therefore, thc revealed correlation may be interpreted in terms of the dependence of crystallite's topology and/or morphology on its growth dynamics. This conclusion is supported in several instances in which the complicated shape and surface of the particles experienced diffusion-limited growth under conditions of a restricted source of atoms were reported (Feder 1988; Vicsek 1989; Luczka, Gadomski, and Grzywna 1992; Tu et al. 1993).

To summarize, size-dependent optical absorption spectra resulting from the quantum confinement effect have been demonstrated to date for a number of nanocrystals in various environments. Size dependence qualitatively agrees with the theory described in Chapter 2. A quantitative agreement with the analytical relation based on the effective mass approximation has been obtained for the case of the weak confinement range. For smaller crystallites the effective mass approach provides a reasonable agreement with the experiments under condition of the correct account for the electron-hole Coulomb interaction and the finite potential barrier, which can be performed numerically. A number of selective techniques show unambiguously a considerable inhomogeneous broadening of the absorption and emission spectra, the homogeneous linewidths being typically one order of magnitude less than the inhomogeneous ones. Minimal absolute values of homogeneous widths on the order of 0.1 meV (1 cm^{-1}) have been reported for nanocrystals corresponding to the weak confinement range, whereas the same parameter for crystallites smaller that the exciton Bohr radius is about 10 meV.

5.2 Valence band mixing

In this section we consider the fine structure of the absorption spectra of real nanocrystals, which is determined by the deviation of hole properties from those of an ideal particle with scalar effective mass.

In Section 2.1 we discussed predictions based on the effective mass approximation with the substructure of the valence band taken into account. A multitude of the otherwise forbidden transitions involving a variety of hole states gives rise to rather complicated absorption spectrum when nanocrystal size is comparable to or less than the exciton Bohr radius (see Fig. 2.5). The typical absorption spectrum of nanocrystals dispersed in some environment like that presented in Fig. 5.4 not only contains the individual absorption lines condensed in wide inhomogeneous bands but also hides the fine structure of every individual absorption spectrum. The complicated structure of the first absorption feature in the case of II-VI nanocrystals has been revealed, and the assignment of the lowest optical transitions has been performed by means of nonlinear pump-and-probe spectroscopy (Woggon, Gaponenko et al. 1993), on the basis of the second-derivative analysis of the absorption spectrum (Ekimov et al. 1993), and using highly monodisperse ensembles of perfect nanocrystals developed in an organic matrix (Norris et al. 1994). In what follows these experiments are discussed in more detail.

In the case of the nonlinear pump-and-probe technique, application of a tunable excitation laser provides elucidation of the fine substructure of the inhomogeneous band. The spectral width of the hole burned at the excitation wavelength, which is determined by both the homogeneous width and the applied power, must be less than the separation of the subbands under investigation. To satisfy this condition one needs to use samples with a small $\bar{a}$ value (i.e., large energy sublevels separation), high-quality nanocrystals

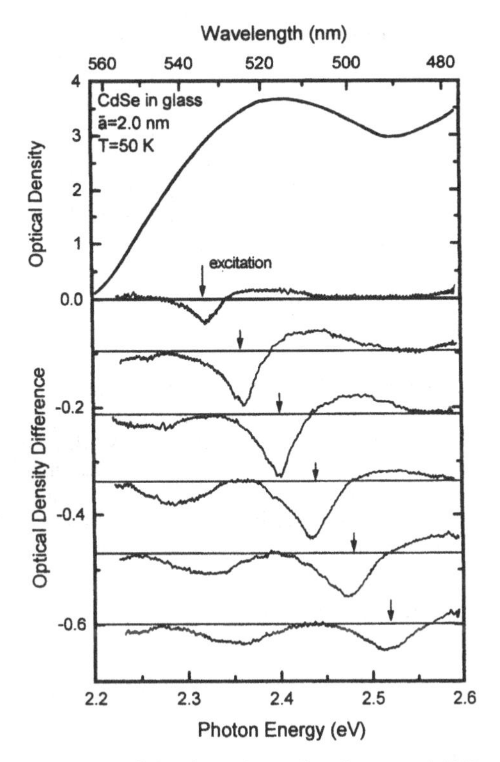

Fig. 5.12. Doublet structure of the first absorption feature of CdSe nanocrystals elucidated by means of nonlinear pump-and-probe technique (Woggon, Gaponenko et al. 1993). Mean crystallite radius is $\bar{a} = 2.0$ nm, temperature is 50 K. Upper panel presents the linear absorption spectrum of a glass sample containing CdSe crystallites. Lower panel presents a set of the differential absorption spectra recorded under monochromatic excitation by a tunable laser (pulse duration is 5 ns, intensity is about 100 kW/cm^2) with simultaneous read-out by a weak broad-band probe beam. Negative differential optical density corresponds to bleaching. Two-band bleaching is due to simultaneous saturation of optical absorption relevant to creation of excitons in the first excited state and in the ground state. Note a different scale for the differential spectra as compared with the linear spectrum.

(narrow homogeneous bandwidth), and low excitation power (small power broadening). A representative example demonstrating a clear doublet structure of the first absorption feature of CdSe nanocrystals is given in Fig. 5.12 (Woggon, Gaponenko et al. 1993). A comparison with the theory (Sweeny and Xu 1989; Xia 1989; Sercel and Vahala 1990) provides an interpretation of subbands in terms of the ground and the first excited exciton states.[1] The ground state of the exciton corresponds to an electron in $1s$-state and a hole in $1S_{3/2}$-state of the mixed (s, d)-symmetry. The first excited state corresponds to an

[1] Note that the ground state of the exciton is relevant to the first excited state of a nanocrystal.

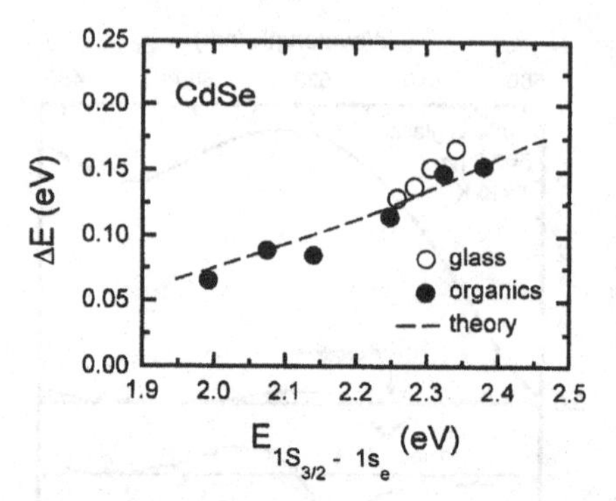

Fig. 5.13. Energy separation ΔE of the exciton first excited and the ground state versus the energy of the ground state for CdSe nanocrystals in an inorganic glass matrix (circles), in organic matrix (dots), and calculated within the framework of the effective mass approximation with the electron-hole Coulomb interaction and the multiple valence band structure taken into account (Woggon and Gaponenko 1995). Data for glass matrix are taken from Fig. 5.11, data for organic matrix are given after Norris et al. (Norris et al. 1994). Theoretical results are taken from Ekimov et al. (Ekimov et al. 1993).

electron in $1s$-state and a hole in $2S_{3/2}$-state. The common electron state and the fast relaxation of an exciton from the first excited state to the ground state with time similar to recombination time is the reason for the simultaneous bleaches at two frequencies. In this case the pump frequency is resonant to $1s2S_{3/2}$-state for a given size, whereas the wide-band probe beam reads out the exciton population both in the excited and in the ground $1s1S_{3/2}$-state. Therefore, tuning of the pump frequency within the band provides the size dependence of the energy separation of the first excited and the ground exciton state

$$\Delta E = E_{2S_{3/2}1s_e} - E_{1S_{3/2}1s_e}$$

to be evaluated. In Fig. 5.13 are plotted the values of energy separation ΔE evaluated by means of nonlinear spectroscopy for CdSe crystallites in an inorganic glass matrix and in an organic environment, along with the numerical data calculated by Ekimov et al. (1993). A good agreement of experiment and theory is evident. Therefore, one can see that the effective mass approach in terms of the particle-in-a-box model not only provides an explanation of the size-dependent absorption spectra of nanocrystals but also predicts the actual fine substructure of the spectrum as well.

Noteworthy is a small, systematic deviation of the data reported for crystallites in glass from that observed for the same crystallites in an organic matrix. The energy splitting in the case of the glass matrix seems to obey the sharper increase with decreasing size. The reason might be connected with the strain effect on the hole energy as crystallites in glass are known to experience a noticeable pressure from the matrix (see discussion in Section 3.1).

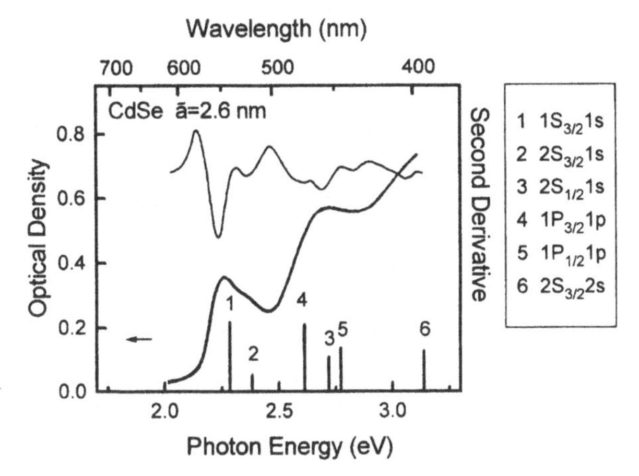

Fig. 5.14. Manifestation of the fine structure of absorption spectrum of CdSe nanocrystals in the second-derivative spectrum (adapted from Ricard et al. 1995). Vertical lines show resonances calculated by means of the effective mass approximation. The insert shows the assignment of the optical transitions. Mean crystallite radius is $\bar{a} = 2.6$ nm.

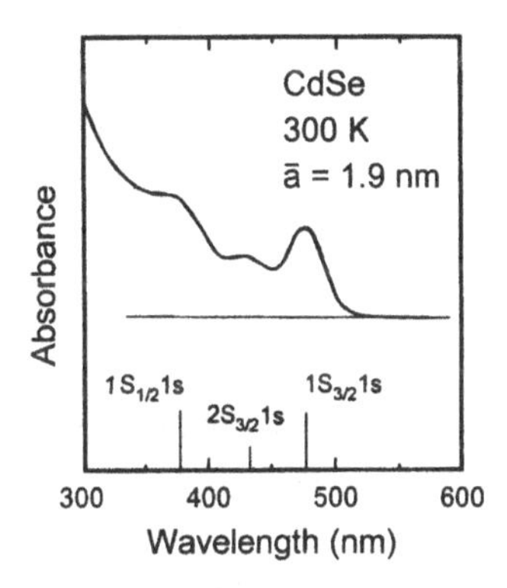

Fig. 5.15. Linear absorption spectrum of the perfect CdSe nanocrystals dispersed in an organic matrix with the reduced size distribution (after Norris et al. 1994) and the assignment of the resolved optical transitions. Mean radius of crystallites is 1.9 nm, temperature is 300 K.

The substructure of the absorption spectrum can be elucidated in the second-derivative spectra (Fig. 5.14). This approach, however, does not provide size-selective information and yields the data relevant to the mean $\bar{a}$ size only. Therefore, to examine the size dependence of energy spectrum on crystallite size one needs to analyze a set of samples with different $\bar{a}$ values. In the case of the reduced size distribution the fine structure of the absorption spectrum can be observed in conventional absorption data without any further processing or impact. Such a monodisperse ensemble of nanometer-size CdSe crystallites has been developed by Norris et al. (1994) (Fig. 5.15).

Consideration of complicated valence band structure is just the first step from the simple particle-in-a-box problem towards real nanocrystals. Further analysis includes the effects of the hexagonal lattice, nonspherical shape, and electron-hole exchange interaction. When these aspects are included, the eight-fold degenerate $1S_{3/2}1s_e$ ground exciton state splits into five sublevels (Norris et al. 1996; Norris and Bawendi 1996; Efros et al. 1996). Influence of shape and exchange interaction will be discussed in more detail in Section 5.4.

5.3 Exciton-phonon interactions

In Section 1.2, in the discussion of the crystal Hamiltonian, we divided a crystal into two subsystems related to the electrons of the outer shells and to the remaining atomic cores consisting of the nuclei and the electrons of the inner shells. This adiabatic approximation has provided an opportunity to consider various events in the electron subsystem in terms of electrons, holes, and excitons. However, elementary excitations in the electron subsystem do interact with the lattice, and we are going to consider optical manifestations of these interactions in this section. Note that lattice properties of nanocrystals do not differ so much from those of the bulk crystals as electron properties do because in the majority of cases nanocrystal size is considerably larger than the lattice period.

5.3.1 Lattice oscillations

Atomic cores in a crystal experience continuous oscillations near the equilibrium positions coinciding with the lattice sites. Amplitudes of these oscillations increase with increasing temperature but remain much smaller than the lattice period. Otherwise, crystal structure is destroyed to produce the liquid phase, that is, the melting process develops.[2] Therefore, the "snapshots" of a real solid-state lattice are different for every instant and never coincide with the ideal geometric lattice.

In a crystal consisting of N elementary cells with n atoms in each, ($3nN$-6) elementary oscillations exist, which are called *normal modes*. This number is nothing but a sum of all degrees of freedom of all particles, $3nN$ minus six degrees of freedom inherent in a solid body, or three translational and three rotational ones of the whole crystal. Elementary oscillations are independent of each other and may coexist. Every crystal state can be described in terms of a superposition of normal modes. As the N value is rather large in a bulk

[2]In the bulk solids, crystal and liquid phases can be easily distinguished by means of several macroscopic parameters like hardness, reflectance, density, etc. However, in the case of smaller nanocrystals and clusters the ratio of the oscillation amplitude and interatomic distance is probably the only criterion of the solid and liquid state.

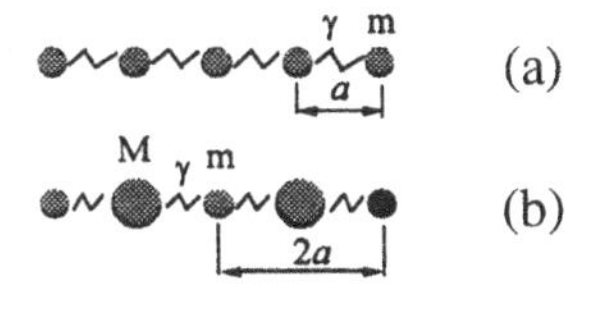

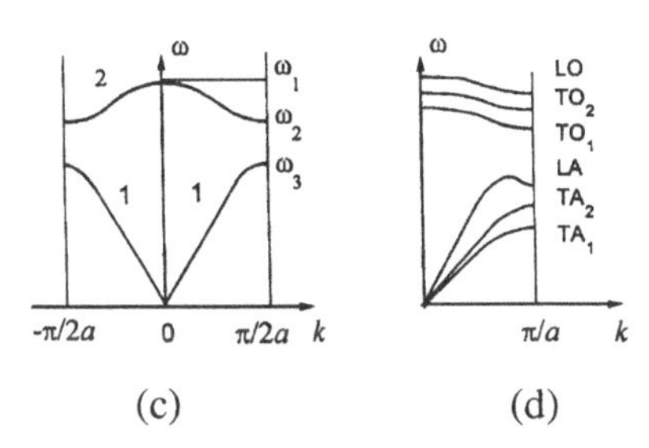

Fig. 5.16. Model linear (a) monoatomic and (b) diatomic chains; (c) dispersion laws for acoustic (curve 1) and optical (curve 2) branches of oscillation modes in a diatomic chain within the first Brilloin zone; (d) sketch of the dispersion relation of lattice vibrations for a three-dimensional anisotropic lattice with partly ionic binding (adapted from Blakemore 1985). LO, TO, LA, and TA labels correspond to longitudinal optical, transverse optical, longitudinal acoustic, and transverse acoustic modes, respectively.

crystal, we can omit six in the above sum. This is justified for nanocrystals as well, since we agreed that a nanocrystal is still large as compared with the lattice period.

All $3nN$ normal modes can be classified in terms of $3n$ different groups (branches) containing N oscillations each. Three of them are called *acoustic branches* whereas the remaining ($3n$-3) modes are referred to as *optical branches*.

To show the difference between acoustic and optical modes, consider a model atomic lattice in the form of a linear chain of solid particles connected by elastic springs. If all particles are identical [monoatomic chain, Fig. 5.16(a)], then the elementary cell contains a single particle. It possesses only one degree of freedom. Accordingly, the only branch of normal oscillations exists with the dispersion law

$$\omega(k) = 2\sqrt{\gamma/m}\left|\sin\frac{ka}{2}\right|, \tag{5.4}$$

where γ is a force constant representing the harmonic character of oscillations in the model under consideration and m is the particle mass. Note that $\omega = 0$ for $k = 0$.

In the diatomic chain [Fig. 5.16(b)] the elementary cell contains two atoms ($n = 2$) with different masses, m and M. It has two branches of oscillation with the more complicated dispersion laws (see, e.g., Gribkovskii 1975; Klingshirn 1995)

$$\omega_{\text{opt}}^2 = \gamma\left(\frac{1}{m} + \frac{1}{M}\right) + \gamma\left[\left(\frac{1}{m} + \frac{1}{M}\right)^2 - \frac{4}{mM}\sin^2 ka\right]^{1/2} \tag{5.5a}$$

and

$$\omega_{ac}^2 = \gamma\left(\frac{1}{m} + \frac{1}{M}\right) - \gamma\left[\left(\frac{1}{m} + \frac{1}{M}\right)^2 - \frac{4}{mM}\sin^2 ka\right]^{1/2}, \tag{5.5b}$$

which are plotted in Fig. 5.16(c) for the first Brillouin zone

$$\left(-\frac{\pi}{2a} < k < \frac{\pi}{2a}\right).$$

Eq. (5.5a) describes the optical branch, whereas Eq. (5.5b) corresponds to the acoustic branch. The principal difference of the two branches is connected with their features at $k \to 0$. The optical branch at $k = 0$ takes the finite frequency

$$\omega_1 = \sqrt{2\gamma\left(\frac{1}{m} + \frac{1}{M}\right)}, \tag{5.6}$$

whereas the acoustic branch possesses $\omega = 0$. Nonzero energy with $k = 0$ becomes possible because in the case of an optical mode different atoms are displaced in antiphase. At the boundary of the first Brillouin zone ($k = \pm\pi/2a$) one has

$$\omega_2 = \sqrt{2\gamma/m}, \tag{5.7}$$

that is, only the light atoms oscillate. In the case of acoustic modes at the boundary of the first Brillouin zone only the heavy atoms oscillate, resulting in

$$\omega_3 = \sqrt{2\gamma/M}. \tag{5.8}$$

If two particles in the elementary cell are charged (i.e., the binding is at least partly ionic), then their oscillations can be described as oscillating electric dipoles. As we have eigenmodes for negligible k, these modes can be generated by photons via resonant absorption. This fact provides a justification of the term "optical" branch.

In a three-dimensional lattice with more than one atom in the primitive cell there are always three types of acoustic modes and three types of optical modes, namely one longitudinal and two transversal. The latter may be degenerate in isotropic structures, whereas in an anisotropic lattice the degeneracy is lifted [Fig. 5.16(d)].

5.3.2 Concept of phonons

In terms of quantum mechanics within the framework of harmonic oscillations, the Hamiltonian of an atomic lattice can be written in the form

$$H = \sum_i \left(\frac{P_i^2}{2M} + \frac{M\omega_i^2}{2} x_i^2 \right). \tag{5.9}$$

Therefore, the general Schroedinger equation

$$H\Psi = E\Psi \tag{5.10}$$

reduces to the set of equations

$$\left(\frac{P_i^2}{2M} + \frac{M\omega_i^2}{2} x_i^2 \right) \varphi(x_i) = E_i \varphi(x_i) \tag{5.11}$$

where

$$\Psi = \prod_i \varphi(x_i), \qquad E = \sum_i E_i. \tag{5.12}$$

Equation (5.11) is the standard harmonic oscillator equation, which results in the energy spectrum

$$E_i = \hbar\omega_i (n_i + 1/2), \quad n = 0, 1, 2, \ldots \tag{5.13}$$

The state of a given oscillator corresponding to a normal mode of a given branch is determined by the quantum number n_i. A set of n_i numbers for all modes of all branches provides an identification of the lattice state.

The total energy of the lattice then reads

$$E = \frac{1}{2} \sum_i \hbar\omega_i + \sum_i n_i \hbar\omega_i. \tag{5.14}$$

The first term in Eq. (5.14) is the zero-energy, which is a finite constant value, whereas the second term is a variable describing the excitation of the system. Various excited states are characterized by different sets of n_i numbers. In other words, every vibrational state of a lattice can be described in terms of the state of an ideal gas of noninteracting quasi-particles called *phonons*.[3] Phonon number in every state can take any value, that is, phonon gas can be described by the Bose-Einstein statistic. The total number of phonons in crystals does not conserve because it is simply the sum of all n_i numbers. Thus, an ensemble of

[3] The concept of phonons was introduced by I. E. Tamm in 1930 (i.e., just after appearance of the term "photon" for quanta of electromagnetic radiation which was proposed by G. N. Louis in 1929).

Table 5.1. *Longitudinal optical phonon energies for several semiconductors*

Crystal	Energy (meV)
CdS	38
CdSe	26
CuCl	20
InP	45
GaAs	35

Source: Landolt-Boernstein 1982.

phonons has the zero chemical potential, and the phonon distribution function reduces to Plank's formula

$$n_i = \left(\exp \frac{\hbar \omega_i}{kT} - 1 \right)^{-1} \tag{5.15}$$

known for photons. Similarly to photons, probability of processes with phonon absorption has the coefficient n_i, whereas the process with phonon emission is characterized by the coefficient $(n_i + 1)$. Equations (5.14) and (5.15) combined with the dispersion laws provide calculations of all thermodynamic properties of the lattice in terms of the ideal gas of bosons. In the case of nonharmonic oscillations, a gas of phonons should be considered as nonideal gas. Phonon characteristic energies for several semiconductors are listed in Table 5.1.

Exciton-phonon interaction is of considerable importance in semiconductor optics. In semiconductor monocrystals exciton-phonon interaction determines the shape of absorption and emission spectra. At low temperatures ($kT \ll Ry^*$) in perfect crystals, exciton-photon coupling dominates over exciton-phonon scattering. In this case coupled exciton-photon excitation (polariton) propagates throughout a crystal and the concept of the absorption coefficient fails to describe the specific features of light-crystal interaction (Haug and Koch 1990; Klingshirn 1995). Both the integral exciton absorption and the line-shape are controlled by exciton-phonon dynamics to give rise to the temperature-dependent oscillator strength (i.e., failure of absorption coefficient concept) and to the non-Lorentzian line-shape (i.e., non-Markovian phase relaxation). At high temperature ($kT > Ry^*$) exciton-phonon interaction results in noticeable absorption at photon energies lower than the band gap energy and the $1S$-exciton resonance [see, e.g., Fig. 1.9(b)]. The exponential absorption tail (Urbach rule) observed for a large number of perfect single crystals is a direct consequence of the exciton-phonon interaction.

The role of exciton-phonon interaction in nanocrystals is of great importance as well. Contrary to bulk monocrystals it does not determine the integral exciton absorption because of negligible exciton-phonon coupling when $a \ll \lambda$ (see Section 2.3). However, phonons remain to control the intrinsic dephasing and energy relaxation processes and therefore determine the absorption and emission linewidths and the luminescence Stokes shift. Manifold aspects of exciton-phonon interactions are still the subject of extensive research. In what follows we restrict ourselves to a number of principal effects related to experimental results. For more detail the reader is referred to the theoretical works cited below.

5.3.3 Vibrational modes in small particles

In nanocrystals acoustic phonon modes become discrete because of finite size. Acoustic properties of a spherical crystallite can be described in terms of elastic vibration of homogeneous particles. In the case of an elastically isotropic sphere the two kinds of eigenmodes exist, namely *torsional modes* and *spheriodal modes* (Takagahara 1996). These effects are to be considered in the problem of exciton-phonon interactions.

Optical modes in nanocrystals experience some modification as well. Along with conventional optical modes inherent in the proper infinite lattice, a number of *surface modes* appear, satisfying a condition (Hayashi 1984)

$$\varepsilon(\omega) = -\frac{l+1}{l}\varepsilon_m, \quad l = 1, 2, 3, \ldots, \tag{5.16}$$

where $\varepsilon(\omega)$ is the dielectric function of nanocrystal material and ε_m is that of the surrounding medium. The multitude of surface optical modes covers a range within the interval $\omega_{TO} < \omega < \omega_{LO}$. The most effective mode corresponds to $l = 1$ when polarization is uniform over a spherical particle (*Froelich mode*). Its frequency, ω_F can be derived from the expression (Froelich 1958)

$$\left(\frac{\omega_F}{\omega_{TO}}\right)^2 = \frac{\varepsilon_0 + 2\varepsilon_m}{\varepsilon_\infty + 2\varepsilon_m}, \tag{5.17}$$

where ε_0 and ε_∞ are the static and the optical permittivities of crystallite material, respectively. One can see that Froelich frequency depends on the medium in which nanocrystals are embedded. Its relative weight is expected to increase in proportion to the surface-to-volume ratio. Experimentally, surface phonons in nanocrystals have been tested by means of resonant Raman spectroscopy using III-V and II-VI nanocrystals in various liquids (Hayashi 1984; Pan et al. 1990). These studies revealed an additional peak in the Raman spectra corresponding to the Froelich mode.

5.3.4 "Phonon bottleneck"; selective population of the higher states

Consider the simplified scheme of the optical transitions including the ground and the two lowest excited states of a nanocrystal. In the ground state, $|0\rangle$, no excitons exist. The first excited state, $|1\rangle$ corresponds to one exciton in its ground state, that is, $1s$-state in a weak confinement case and $1s_e 1S_{3/2}$-state in the strong confinement limit. The second state, $|2\rangle$ is then the $2s$-state or $1s_e 2S_{3/2}$-state, respectively. The energy difference between the two excited states, ΔE_{12}, grows rapidly with decreasing size (see, e.g., Fig. 5.12). For CdSe nanocrystals when $a = 2$ nm, one has $\Delta E_{12} \approx 200$ meV (1600 cm^{-1}). This is a very large value as compared with LO-phonon energy, which is for CdSe equal to $E_{LO} = 26$ meV. The latter is close to the kT value corresponding to room temperature. Therefore, for rather large size and temperature ranges a relation $\Delta E_{12} \gg E_{LO}$, kT holds. In this case relaxation between two excited states $2 \rightarrow 1$ is possible via multiple emission of optical and acoustical phonons. In bulk semiconductors exciton energy relaxation results mainly from cascade LO-phonons emission, whereas interaction with acoustic phonons is less important. In nanocrystals of size less than 10 nm when the discreteness of electron-hole states becomes essential, optical phonons cannot provide an efficient relaxation channel for the following reasons. First, the dispersion curve of optical phonons is nearly independent of wave number, and LO-phonon energies can only weakly deviate from the E_{LO} value [see Fig. 5.16(c)]. Therefore, the relaxation via multiple LO-phonon emission requires a condition

$$\Delta E_{12}(a) = nE_{LO}, \quad n = 1, 2, 3, \ldots \qquad (5.18)$$

to be fulfilled. Second, contrary to bulk crystals possessing a continuous spectrum for quasi-particles, in nanocrystals no intermediate state exists between the E_1 and E_2 states. Therefore, even if the condition (5.18) is satisfied, multiple phonon emission should occur via virtual intermediate states. Probability of this process is very low and the corresponding relaxation rate is lower then the typical electron-hole recombination rates. Thus, the relaxation between excited states in nanocrystals is essentially inhibited. This effect is often referred to as the "phonon bottleneck." The above qualitative consideration has been confirmed by quantum-mechanical calculations (Bockelman and Bastard 1990; Benisty, Sotomayor-Torres, and Weisbuch 1991; Inoshita and Sakaki 1992; Bockelman 1993). In the case

$$\Delta E_{12} = E_{LO} \pm \delta E \qquad (5.19)$$

two-phonon LO $\pm$ LA processes were shown to be important if δE is about $0.1E_{LO}$ (Inoshita and Sakaki 1992). Probably, the enhanced relaxation effect from the matching energy condition (5.19) has been observed in CdSe nanocrystals (Norris and Bawendi 1995). Within the lowest absorption feature two bands have been elucidated by means of nonlinear size-selective absorption. The lowest band is narrow, with the FWHM of about a few meV, whereas the higher band has the FWHM as large as 30 meV. This wider band, however, has not been detected in fluorescence line narrowing experiments. Remarkably, that energy difference between two bands fits LO-phonon energy. Therefore, it is reasonable to suppose that the large width of the higher band results from the fast relaxation by emission of LO- and LO $\pm$ TA phonons.

Because of the rigid matching conditions for optical phonons, for smaller crystallites multiple emission of acoustic phonons appears to be the main relaxation mechanism. Acoustic phonons have a continuous energy spectrum and a condition similar to (5.18) is not necessary. However, at low temperature, size quantization of acoustic modes should manifest itself. Elastic modes cannot exist in a crystallite if

$$\lambda/2 > a \tag{5.20}$$

holds, where λ is the phonon wavelength. At lower temperatures long-wave (low-energy) modes should predominate in agreement with the Bose-Einstein distribution function. As these modes cannot develop in smaller crystallites, for every a value inequality (5.20) leads to some threshold temperature, lower than which relaxation via acoustic phonons is inhibited as well. Thus, at low temperature and for small size the energy relaxation time may be comparable with, or even larger than, the exciton annihilation time. If this is the case, quasi-steady-state population of a higher excited state becomes possible.

A manifestation of this effect was observed for CdSe nanocrystals in a glass matrix (Woggon, Gaponenko et al. 1994). At lower temperature resonant excitation to the $1s_e2S_{3/2}$ exciton state resulted in a clear absorption saturation in channel $|0\rangle \rightarrow |2\rangle$, whereas at higher temperatures fast relaxation from $1s_e2S_{3/2}$ to $1s_e1S_{3/2}$ state gave rise to predominating bleaching at lower photon energies (Fig. 5.17). Note that just for these reasons the doublet structure due to valence band mixing (Fig. 5.12) has been established at intermediate temperatures $T \approx 50$ K, when relative weights of nanocrystals in the first and in the second states are approximately the same.

One more manifestation of quantization of acoustic phonons has been reported by T. Kobayashi with co-workers (Misawa et al. 1991a, 1991b). Temperature dependence of the radiative lifetime τ in CdS nanocrystals corresponding to the intermediate confinement ($a \approx a_B$) shows a linear $\tau(T)$ increase for $T > 50$ K, whereas below 50 K the lifetime is temperature-independent. The explanation of this behavior implies that the temperature-independent decay results from the superradiant emission, which is possible as long as the radiative rate dominates over dephasing. The latter is inhibited at low temperature

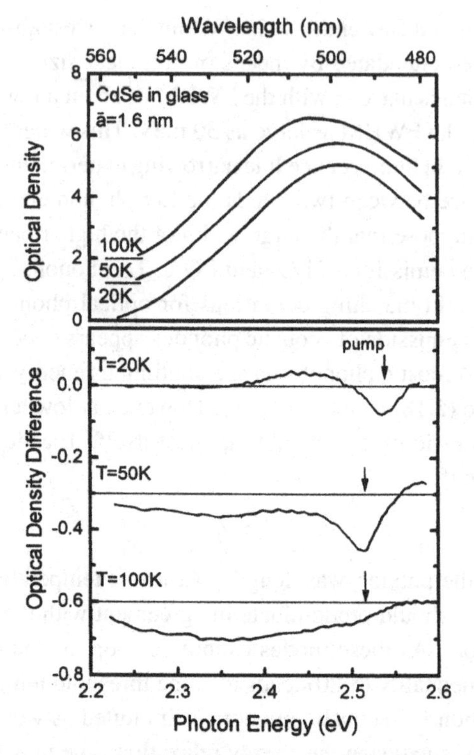

Fig. 5.17. Selective population of the higher excited states in nanocrystals due to phonon bottleneck monitored by pump-and-probe spectroscopy in the nanosecond time range (Woggon, Gaponenko et al. 1994). Upper panel shows linear absorption spectrum of CdSe nanocrystals dispersed in a glass matrix at three different temperatures. Lower panel presents differential absorption spectra recorded under selective excitation. Excitation photon energy is marked by the arrows. At lower temperature ($T = 20$ K) a pronounced bleach at the excitation wavelength is indicative of inhibited relaxation from the higher excited state of a nanocrystal corresponding to $1s2S_{3/2}$ exciton. At higher temperatures ($T = 100$ K) fast relaxation of excitons occurs from the resonant $1s2S_{3/2}$ state to lower $1s1S_{3/2}$ state by means of acoustic phonon emission. This results in the pronounced bleach at lower energy with respect to the excitation quantum. At the intermediate temperature ($T = 50$ K) both exciton states are populated. The excitation photon energy in this experiment has been tuned with temperature to account for the temperature dependence of the linear absorption spectrum.

because of quantization of acoustic modes. At higher temperatures exciton scattering by acoustic phonons destroys coherence and the lifetime grows proportionally to the temperature. Assuming a nanocrystal to be an elastic sphere, the frequency of a radial mode can be estimated as

$$\omega = \frac{\pi}{a} v, \tag{5.21}$$

where v is the longitudinal sound velocity. For a crystallite with $a = 2.5$ nm, the phonon cut-off frequency is 30.0 cm^{-1}, which corresponds to 3.7 meV threshold phonon energy and is equivalent to 43 K. The latter agrees well with the observed threshold temperature.

5.3.5 Line-shapes and linewidths

Spectral line-shapes of an ideal atomic system are determined by dynamic processes that destroy the phase coherence of the excited state. In many cases the dephasing dynamics can be described in terms of the constant dephasing rate, Γ. In spectroscopy this case is usually referred to as the Markovian relaxation. In the case of the constant dephasing rate for a given excited state, the corresponding line-shape obeys a Lorentzian

$$I(\omega) = A\frac{\Gamma^2}{(\omega - \omega_0)^2 + \Gamma^2}, \tag{5.22}$$

where ω_0 is the resonant frequency, Γ is the dephasing rate that gives the half-width of the line, and A is the frequency independent coefficient of the proper dimensionality and value providing the true integral absorption of the band. Otherwise the dephasing process corresponds to non-Markovian relaxation and the line-shape cannot be described by a Lorentzian (Apanasevich 1977).

Semiconductor nanocrystals are characterized by the discrete energy spectrum, the energy distance between neighboring excited states being much larger than the spectral linewidth. This is proved by successful elucidation of the discrete transitions in spectral hole-burning studies (see, e.g., Fig. 5.12). Moreover, spectral hole-burning experiments have shown that the lowest absorption band in nanocrystals can be satisfactorily fitted by a Lorentzian (Fig. 5.18) (Gaponenko, Woggon et al. 1994). Therefore, it is reasonable to discuss the problem of exciton line-shapes and linewidths in nanocrystals in terms of the Lorentzian shape and frequency-independent dephasing rate for a given state at a given temperature.

Basically, the phase relaxation rate can be described in terms of a number of independent processes characterized by the respective time constants, that is,

$$\Gamma = \frac{1}{T_2} = \frac{1}{2T_1} + \sum_i \frac{1}{T_i}, \tag{5.23}$$

where T_2 is the total dephasing time (so-called transversal relaxation time), T_1 is the pure population lifetime, and T_i characterizes the purely dephasing processes without energy deviation.

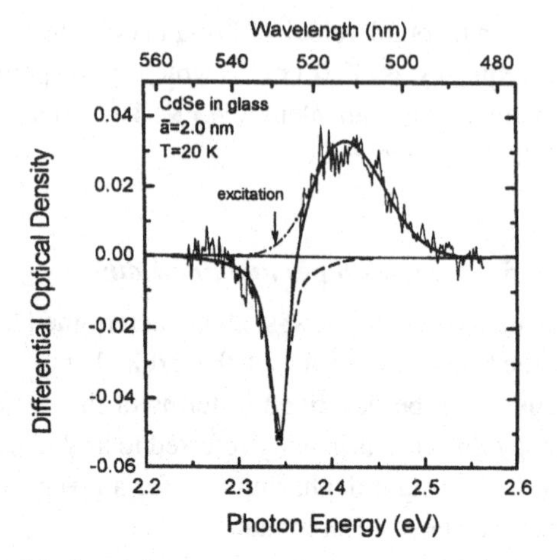

Fig. 5.18. Differential absorption spectrum recorded for CdSe nanocrystals in pump-and-probe spectroscopy at smaller excitation densities (adapted from Woggon and Gaponenko 1995). Mean radius of nanocrystals is 2 nm, temperature is 20 K, excitation intensity is 10^5 W/cm^2, excitation photon energy is 2.34 eV (marked by the arrow). The negative part of the differential spectrum relevant to the saturation of the exciton resonance can be satisfactorily fitted by a Lorentzian (dashed line), whereas the positive part due to excited-state absorption obeys a Gaussian (dashed-dotted line). Solid line is the sum of the Lorentzian and Gaussian.

5.3.6 Lifetime broadening

Let us discuss first the contribution from the lifetime broadening that results from the radiative annihilation rate and from the scattering rate providing a transition from a given exciton state to others. Within the framework of the perturbation approach, the scattering rate for the i-th exciton state can be expressed as follows:

$$\Gamma_i(T) = \Gamma_i^0 + \sum_{j<i} \gamma_{ij}[n(E_{ij}, T) + 1] + \sum_{j>i} \gamma_{ij} n(E_{ij}, T), \tag{5.24}$$

where Γ_i^0 is the radiative annihilation rate, $\gamma_{ij} = \gamma_{ji}$ determine phonon-assisted scattering from i-th to j-th state, and $n(E, T)$ is the Bose-Einstein distribution function describing the number of acoustic phonons. We can neglect scattering by emission/absorption of optical phonons because of the exact matching conditions resulting in a phonon bottleneck.

Consider contributions from different terms in Eq. (5.24). The first term is not related to phonons and is temperature-independent. The second term describes scattering to lower states because of phonon emission. The third term corresponds to scattering to higher states because of phonon absorption. For the ground state the second term equals to zero for all temperatures. Generally, at $T \to 0$ the third term tends to zero because $n(E, T) = 0$ at $T = 0$.

Therefore, at $T \to 0$ the ground state is expected to possess the purely radiative lifetime broadening, which is the basic linewidth limit. For higher temperature the ground state experiences monotonic broadening due to the phonon-assisted scattering to the excited states. Every excited state should possess an extra broadening due to scattering to lower states. If one takes the typical radiative lifetime of an exciton on the order of 10^{-8} s, then in the limit $T \to 0$ the lifetime broadening is on the order of 10^{-5} eV (or 10^{-1} cm^{-1}). These linewidth values have been observed for larger GaAs nanocrystals by means of the single dot spectroscopy (Gammon et al. 1996a and 1996b). Other nanocrystals show considerably larger linewidths that by no means can be interpreted in terms of the lifetime broadening.

The above estimate testifies that purely dephasing processes rather than the lifetime broadening should be taken into consideration. The main mechanisms of dephasing are the exciton-phonon interaction, carrier-carrier (or exciton-exciton) interactions, and scattering by defects and other extrinsic imperfections. Carrier-carrier interactions are absent under conditions of the low excitation when no more than one electron-hole pair within a crystallite is created. The influence of defects can be neglected in perfect nanocrystals. Therefore, basic limits and temperature dependence of the homogeneous linewidth are controlled by exciton-phonon interactions. These interactions are also responsible for the luminescence Stokes shift.

5.3.7 Dephasing due to exciton-phonon interactions

Lattice vibrations give rise to stochastic fluctuations of the exciton energy, which in turn manifest themselves in the exciton line broadening. Study of exciton-phonon interactions in small nanocrystals ($a \ll a_B$) is of special interest because smaller crystallites exhibiting strong confinement are expected to be novel artificial species with controllable discrete spectra (so-called artificial atoms). Various aspects of exciton-phonon interactions in the case of the strong confinement limit have been analyzed both experimentally and theoretically. Experiments based on spectral hole-burning (Alivisatos et al. 1988; Gaponenko, Woggon et al. 1993; Norris et al. 1994), fluorescence line narrowing due to the red-edge excitation (Nirmal et al. 1994), nondegenerate four-wave mixing (Woggon and Portune 1995), and photon echo techniques (Mittleman et al. 1994) provide data on absolute values of homogeneous linewidths and dephasing rates and their temperature dependencies. Additional data on exciton interaction with optical phonons can be obtained by means of resonant Raman scattering (Silvestri and Schroeder 1994; Scamarcio et al. 1996).

In Figs. 5.19 and 5.20 the data reported for II-VI nanocrystals are shown as the representative example. At liquid helium temperatures the full width

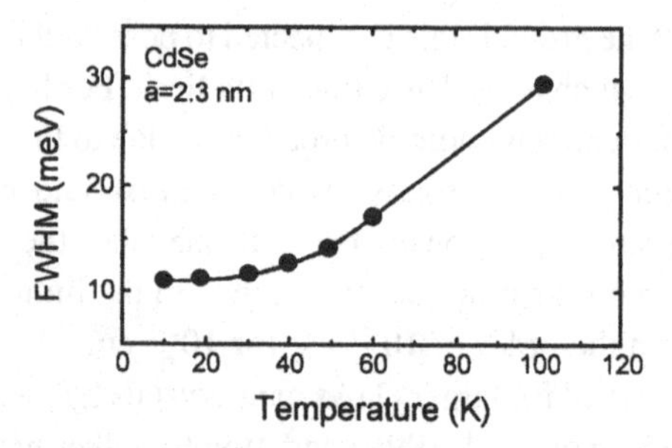

Fig. 5.19. Temperature dependence of the full homogeneous linewidth of CdSe nanocrystals in a glass matrix derived from spectral hole-burning studies (adapted from Gaponenko, Woggon et al. 1994). Mean size of nanocrystals is 2.3 nm.

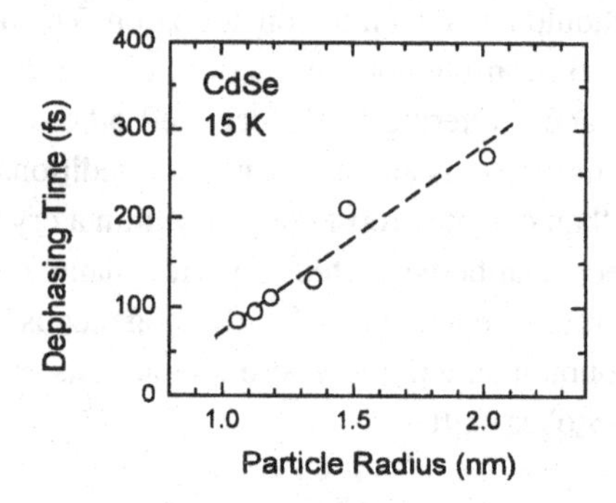

Fig. 5.20. Size dependence of the dephasing time of excitons in CdSe nanocrystals evaluated by means of the photon-echo technique. Points are plotted according to Mittleman et al. (1994), and the line is the guide for the eye.

at half-maximum is about 10 meV and exhibits a monotonic growth with increasing temperature. The size dependence of the linewidth features steady increase with decreasing size. A good agreement of data reported by various groups and close absolute values for the parameters evaluated for crystallites in organic and inorganic matrices provide a background for comparison with the theory. According to the estimate (Mittleman et al. 1994) the purely lifetime contribution to the observed linewidths is on the order of 10%. To explain the remaining contributions, exciton-phonon interactions should be examined.

Exciton-phonon coupling in an ideal nanocrystal may be considered in terms of the two contributions resulting from interactions with optical phonons and acoustic phonons. The interaction with optical phonons can be described as polar Froelich coupling of the electric field created by vibrations of ionic cores with the Coulomb field of the electron-hole pair. Interaction with acoustic phonons results in a coupling to longitudinal acoustic modes via the deformation potential and to the transverse acoustic modes by piezoelectric interaction.

Exciton coupling to LO phonons mediated by the Froelich interaction is usually described in terms of the Huang-Rhys factor S, which determines the linewidth and the luminescence Stokes shift,

$$2\Gamma_0 = \hbar\omega_{LO} 2S^{1/2}, \qquad \Delta_{\text{Stokes}} = 2S\hbar\omega_{LO}. \tag{5.25}$$

For data presented in Fig. 5.19 one has $S = 0.04$. To evaluate the contribution of the exciton-phonon coupling the experimental S value should be compared with the calculated one.

Theoretical studies of the exciton-LO-phonon interaction in the strong confinement limit have been performed using several approximations. In the earlier work Schmitt-Rink, Miller, and Chemla predicted vanishing coupling to LO phonons for smaller crystallites (Schmitt-Rink, Miller, and Chemla 1987). Their model implies that the electron and hole wave functions are identical. Later, Klein et al. (1990) used a donorlike exciton model with the heavier hole localized near the center but neglecting the electron-hole correlation to predict a finite but size-independent coupling. More recently the consideration has been extended taking into account electron-hole correlation, valence-band degeneracy, nonparabolicity of the conduction band, and the proper confined phonon wave functions (Efros 1992b; Nomura and Kobayashi 1992; Marini, Stebe and Kartheuser 1993; Takagahara 1996). In this case a systematic increase of the Huang-Rhys factor with decreasing size has been obtained, providing a reasonable agreement with the experimental findings. A more detailed comparison is difficult because the calculated exciton-LO-phonon coupling is very sensitive to the spatial electron-hole charge distribution. Particularly, the fixed position of the hole in the center of a spherical nanocrystal gives S growing from 0.25 for $a = 4.0$ nm to $S \approx 1$ for $a \approx 1$ nm for CdSe nanocrystals. This is larger than the values observed in perfect samples. Experimentally, the charge separation may be enhanced due to surface polarization effects (Banyai et al. 1992), which are sensitive to the barrier height at the crystallite-matrix interface.

Size-dependent coupling to acoustic phonons has been considered for deformational and piezoelectric coupling (Nomura and Kobayashi 1992; Takagahara 1993b; Takagahara 1996). In the case of the deformation-potential coupling the size dependence of dephasing rate scales as a^{-2}. In the case of piezoelectric coupling the size dependence obeys a^{-1} behavior. Absolute values of dephasing rate due to the exciton coupling to acoustic phonons have been evaluated using wurtzite-type CdSe as a model. Interaction with the deformation potential provides an explanation of the temperature-dependent linewidth.

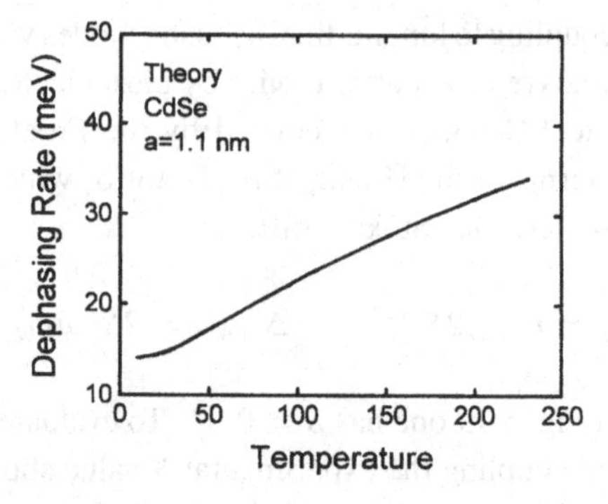

Fig. 5.21. Temperature dependence of the homogeneous linewidth due to exciton interaction with acoustic phonons calculated for CdSe nanocrystals (adapted from Takagahara 1996).

The calculated $\Gamma(T)$ dependence as presented in Fig. 5.21 corresponds well to the data reported for CdSe and CdS_xSe_{1-x} nanocrystals (Mittleman et al. 1994; Gaponenko, Woggon et al. 1994).

Interaction with optical phonons of hydrogenlike excitons experiencing weak confinement in the larger nanocrystals seems to be diminished. For example, several groups outlined a strong resonant zero-phonon exciton emission band for CuCl nanocrystals (Fig. 5.22), whereas in the bulk crystals LO phonon replicas predominate (Itoh, Iwabuchi, and Kirihara 1988; Ekimov 1991; Gaponenko, Germanenko et al. 1993). Accordingly, spectral hole-burning performed for CuCl nanocrystals showed linewidths less than 1 meV (Wamura et al. 1991; Masumoto et al. 1995). The size dependence of the homogeneous width was found to obey a^{-2} dependence, which is indicative of the main contribution from coupling to deformational potential (Takagahara 1996). Further peculiarities of exciton-phonon interactions in the weak confinement case have been reported recently by Itoh and co-workers (Itoh et al. 1995). The fine structure in the luminescence spectra has been observed, which exhibits strong resonance behavior when the distance between the $1P$ and $1S$ states of the confined exciton equals the energy of one or two optical phonons. The effect is explained by the formation of an exciton-phonon complex. Under this condition, the forbidden optical transition to the $1P$ exciton state is allowed in a nanocrystal as a result of the formation of a hybrid exciton-phonon complex with the angular momentum equal to 1.

In general, the problem of exciton-phonon interactions is being extensively investigated. Recently the interest in this problem has been promoted by challenging experiments on the single dot spectroscopy, which will be reviewed in Section 5.5.

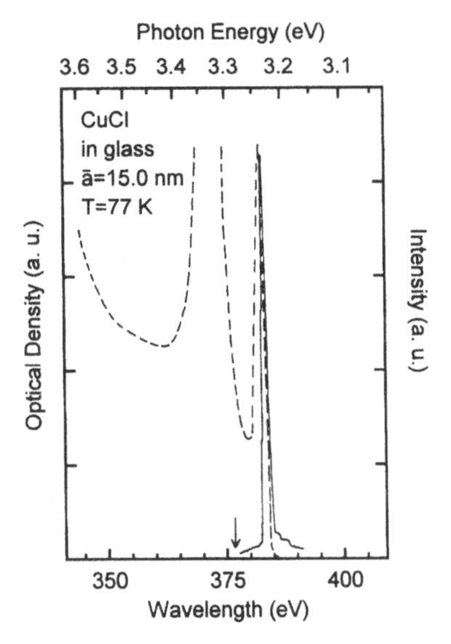

Fig. 5.22. Absorption (dashed line) and emission (solid line) spectrum of CuCl nanocrystals in a glass matrix (adapted from Gaponenko, Germanenko et al. 1993). Mean radius of nanocrystals is 15 nm, temperature is 77 K.

5.4 Size-dependent radiative decay of excitons

5.4.1 Superradiant decay of excitons in larger crystallites

In Section 2.1 the mesoscopic enhancement of excitonic oscillator strength in larger nanocrystals was discussed. Within the size range

$$a_B \ll a \ll \lambda_{\mathrm{exc}} \tag{5.26}$$

where a_B is the exciton Bohr radius and λ_{exc} is the wavelength relevant to the excitonic resonance, a subrange exists in which the oscillator strength of an exciton transition is proportional to the number of atoms, that is, to the volume of the crystallite. This proportionality results from the fact that the transition dipole moment of every atom is added coherently throughout the crystallite (Takagahara 1987; Hanamura 1988a; Kayanuma 1988). In this case the radiative decay rate, which is proportional to the oscillator strength, should obey the $\tau_{\mathrm{rad}} \propto a^{-3}$ dependence as well. The fact that the probability of a radiative transition is proportional to the number of atoms participating in the transition makes it possible to classify the photon emission process in this case as *superradiance* by analogy with the similar phenomenon occurring in atomic and molecular ensembles.

The phenomenon of the size-dependent superradiant decay of excitons confined in a crystallite has been observed in CuCl nanocrystals. These species embedded in a glass or ionic matrix show strong photoluminescence resulting from annihilation of the Z_3 exciton (Fig. 5.22). Because of the negligible Stokes shift the photoluminescence kinetics is strongly influenced by reabsorption effects. Multiple absorption-emission in the sample resulted in a slowing down of luminescence kinetics recorded in the experiments. Fine experiments on CuCl luminescence aimed at evaluating the intrinsic size-dependent radiative lifetime have been performed by Itoh and co-workers (Itoh, Ikehara, and Iwabuchi 1990; Itoh, Furumiya et al. 1990). In these experiments, CuCl nanocrystals embedded in NaCl were found to possess a very high internal quantum yield of photoluminescence, η. The η value reaches 90% and decreases continuously to 50% as size decreases from 10 to 2 nm. Under this condition the reabsorption effect should be explicitly considered when radiative parameters are evaluated from experimental data. Itoh et al. used the following approach. The transient decay time τ_{tr} just after the short excitation pulse is considered as the luminescence decay time with no reabsorption effect. After excitation the decay time continuously increases and reaches the stationary value τ_{stat} long after the excitation. The two times are related by an equation

$$\tau_{\text{stat}} = \frac{\tau_{tr}}{1 - \eta\beta\delta}, \tag{5.27}$$

where η is the internal luminescence quantum yield, β is the generation rate of the excitons with the same energy as that of the absorbed photons, and δ is the reabsorption rate. For a thin platelet sample δ can be expressed as

$$\delta = 1 - \frac{(1 - \cos\theta)(1 - \exp(-\alpha d))}{\alpha d}, \tag{5.28}$$

where α is the absorption coefficient, θ is the critical angle of the total reflection at the surface of the sample, and d is the sample thickness. Using the measured τ_{tr} and τ_{stat} values, the η, β, and radiative lifetime τ_{rad} can be evaluated as follows. Exciton densities that yield resonant and nonresonant luminescence, n and n^*, respectively, satisfy the rate equations

$$\frac{dn}{dt} = -\frac{n}{\eta\tau_{\text{rad}}} + \delta\beta\frac{n}{\tau_{\text{rad}}}, \tag{5.29a}$$

$$\frac{dn^*}{dt} = -\frac{n^*}{\eta^*\tau^*_{\text{rad}}} + \delta^*\beta^*\frac{n^*}{\tau^*_{\text{rad}}} + \delta(1 - \beta)\frac{n}{\tau_{\text{rad}}}, \tag{5.29b}$$

where asterisks denote the quantities relevant to nonresonant contribution.

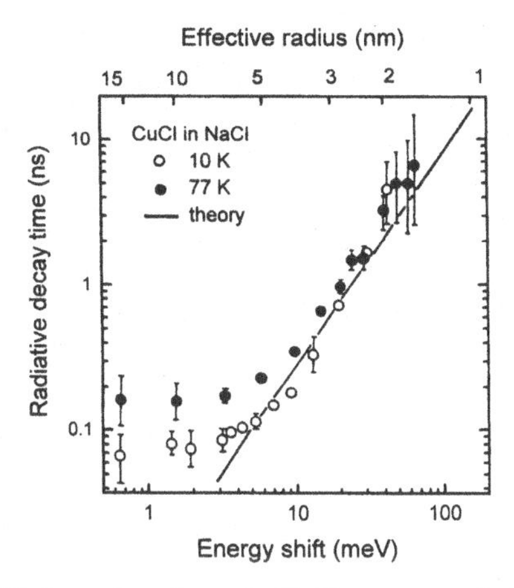

Fig. 5.23. Radiative lifetime of excitons in CuCl nanocrystals in NaCl matrix versus mean radius (Itoh, Furumiya et al. 1990). Experimental data correspond to 10 K (open circles) and 77 K (closed circles). A solid line represents the theoretical dependence.

Using some further assumptions, quantum yield can be expressed as

$$\eta = \left(1 - \frac{\tau_{tr}}{\tau_{\text{stat}}}\right)\frac{1}{\beta\delta}, \tag{5.30}$$

whence $\tau_{\text{rad}} = \tau_{tr}/\eta$.

Radiative lifetime versus size derived in this way is shown in Fig. 5.23 (Itoh, Furumiya et al. 1990). In accordance with the theory, there exists a size range in which $\tau_{\text{rad}} \propto a^{-3}$ holds. For $a > 5$ nm the probability of annihilation of a Z_3 exciton no longer depends on the crystallite size.

There are two possible mechanisms that could be responsible for the saturation of the decay rate. First is the so-called retardation effect resulting from the finiteness of the photon wavelength. It leads to the breakdown of the coherent addition of atomic transition dipoles throughout a nanocrystal. The other mechanism is the breakdown of mesoscopic enhancement brought on by homogeneous broadening, which induces the overlap of the excited excitonic states within the homogeneous linewidth of the lowest exciton state. A nonlocal theory of the radiative decay rate of excitons in semiconductor quantum dots was developed by Takagahara (1993d), and the observed size and temperature dependence for CuCl microcrystallites has been explained successfully. The retardation effect was found to be irrelevant for explaining the experimental findings. Instead, the homogeneous broadening effect was shown to be responsible for the saturation of the mesoscopic enhancement of the excitonic radiative decay rate. In other words, the homogeneous broadening rather than the light wavelength was found to determine the effective exciton coherence length.

5.4.2 Radiative transitions in nanocrystals of indirect-gap materials

Indirect-gap bulk crystals possess the valence-band maximum and conduction-band minimum at different values of wavenumber k (see, e.g., Fig. 1.6). This means that the "electrical" band gap measured as the minimal energy required to create one electron-hole pair is less than the "optical" band gap that determines the onset of vertical interband transitions. The typical examples of indirect-gap semiconductor materials are silicon, germanium, and silver bromide. Interestingly, these semiconductors are of great practical importance. Silicon and germanium are the main electronic materials, whereas silver bromide microcrystals are the essential component of photographic emulsions. Optical absorption in indirect-gap crystals takes place because of phonon-assisted processes. The absorption coefficient obeys the frequency dependence (Gribkovskii 1975)

$$\alpha(\omega) = \text{const} \times (\hbar\omega - E_g)^{3/2}, \tag{5.31}$$

the absolute values of α being several orders of magnitude less than those of direct-gap materials because of the low probability of many-particle processes. Likewise, the radiative decay rate is extremely slow, typically on the order of $10^{-3} - 1$ s. Evidently, recombination in this case is controlled either by nonradiative processes or by radiative defect-related transitions, the intrinsic luminescence quantum yield being negligibly small.

As indirect-gap band structure in a monocrystal results from certain translational symmetry, it is reasonable to ask whether it is possible to get an enhancement of radiative transitions in nanocrystals of indirect-gap materials as a result of the translational symmetry breaking. To answer this question, extensive theoretical and experimental research has been performed in recent years. These studies have been initiated by a challenging hypothesis that the fundamental origin of visible luminescence in porous silicon films (known since 1984; Pickering et al. 1984) is related to quantum-confinement effects in nanometer size silicon wirelike and dotlike patterns (Canham 1990; Lehmann and Gösele 1991). In what follows we consider the results obtained for Si nanocrystals with a few remarks on other nanocrystals of indirect-gap materials, namely Ge, AgBr, and indirect-gap modification of CdSe. Calculations made within the framework of the effective mass approximation with the multiple valence-band structure taken into account show that size restriction results in the simultaneous high-energy shift of the band-gap energy and in an enhancement of the radiative decay rate (Fig. 5.24). Experiments have been performed using specially fabricated Si nanocrystals in organic environments and nanostructured porous silicon. Comprehensive studies of visible photoluminescence in porous silicon films developed by means of electrochemical etching have shown that the intense

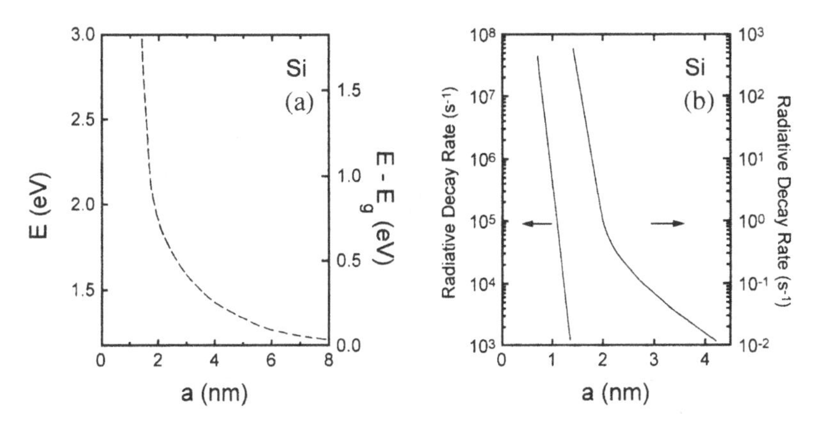

Fig. 5.24. Calculated energy band gap and radiative decay rate of Si nanocrystals versus the nanocrystal radius (adapted from Takagahara and Takeda 1992). Calculations are based on the effective mass approximation with the electron-hole Coulomb interaction and manifold valence band structure taken into account.

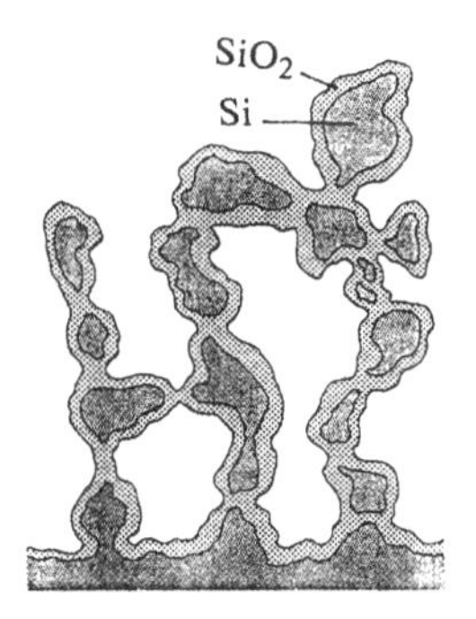

Fig. 5.25. Schematic presentation of a porous silicon film on a silicon substrate (after Vial et al. 1992). Porous silicon film obtained by electrochemical etching consists of nanometer-size dotlike or needlelike silicon grains buried in silicon oxide.

red-orange emission in these species originates from elongated nanocrystals of silicon buried in silicon oxide (Fig. 5.25), forming a random rigid network with a preferable orientation along the etching direction (Brus 1994; Lockwood 1994; John and Singh 1995; Kanemitsu 1995; Vial 1995; Gaponenko, Petrov et al. 1996). Therefore, it is reasonable to consider the spectroscopic properties of purposefully fabricated silicon nanocrystals along with those of porous silicon. The main features of luminescence can be summarized as follows.

The emission spectrum shifts significantly to a higher photon energy as compared to the bulk Si. Typically, luminescence peak is in the range 600–800 nm (1.55–2.15 eV), whereas the bulk band gap energy $E_g = 1.17$ eV. The emission spectrum is rather wide (about 0.3 eV) exhibiting the similar width even at liquid helium temperatures [Figs. 5.26(b) and 5.27(b)].

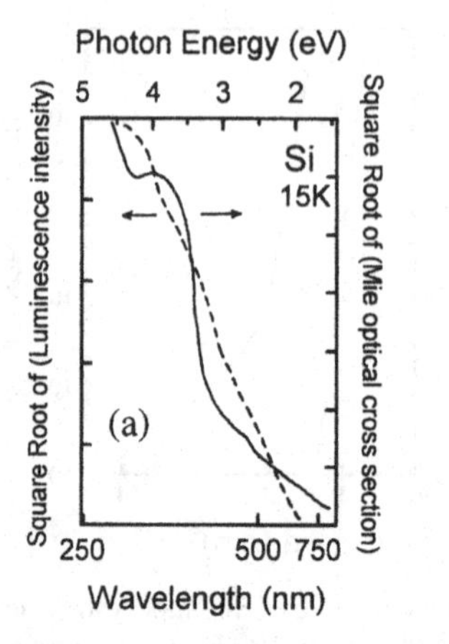

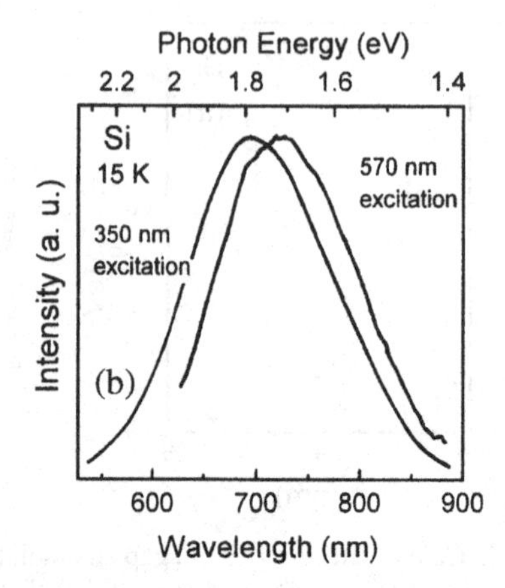

Fig. 5.26. Steady-state spectroscopic properties of silicon nanocrystals in ethylene glycol (adapted from Brus et al. 1995). Panel (a) represents the excitation spectrum relevant to the emission wavelength 600 nm (solid line) along with the calculated bulk silicon absorption spectrum (dashed line). Panel (b) contains emission spectra recorded at two excitation wavelengths. All data were taken at temperature 15 K.

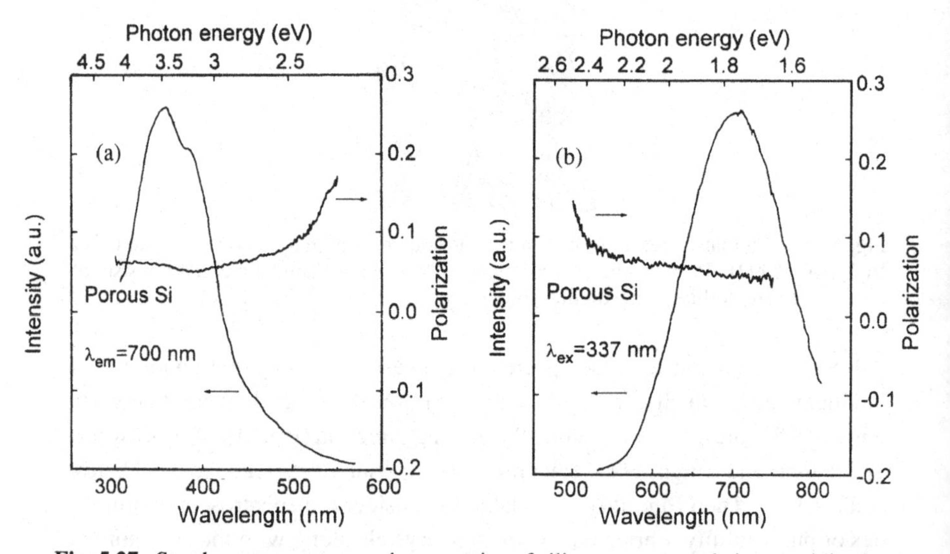

Fig. 5.27. Steady-state spectroscopic properties of silicon nanocrystals in an oxidized porous silicon film (Gaponenko, Petrov et al. 1996). Panel (a) shows the excitation spectrum relevant to the emission wavelength 700 nm. Panel (b) presents the emission spectrum recorded under excitation at 337 nm. All data are taken at room temperature. Luminescence is characterized by a noticeable positive polarization, which manifests itself both in polarization-resolved (a) excitation and (b) emission spectra.

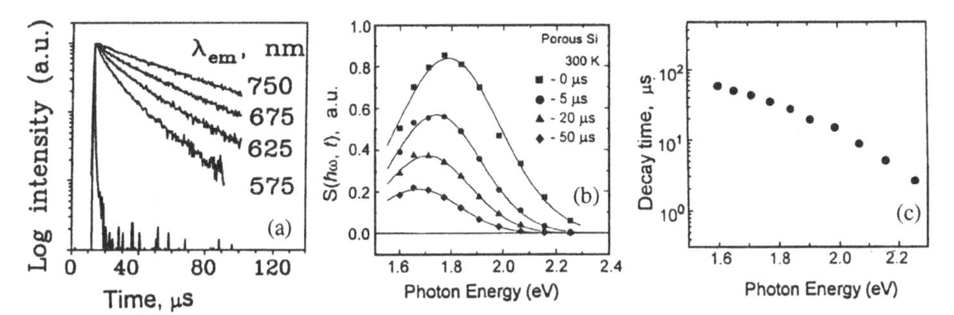

Fig. 5.28. Kinetic properties of luminescencing silicon nanocrystals in oxidized porous silicon film (Gaponenko, Germanenko, Petrov et al. 1994; Gaponenko, Petrov et al. 1996). (a) Decay curves in a semilogarithmic scale for several emission wavelengths along with the excitation pulse; (b) recovered time-resolved emission spectra along with their Gaussian approximations for various times: 0 μs (squares); 5 μs (full circles); 20 μs (triangles); 50 μs (diamonds); (c) wavelength dependence of the mean lifetime.

The excitation spectrum is very smooth and without substructure in the range of 2–5 eV [Fig. 5.26(a)]. A pronounced shoulder or maximum in the excitation spectrum of porous silicon [Fig. 5.27(a)] is attributed either to absorption by a substrate or absorption by larger crystallites possessing bulk-like spectrum with the subsequent energy transfer to smaller emitting nanocrystals. The spectral position of the maximum or shoulder correlates with the direct $\Gamma'_{25} - \Gamma'_2$ optical transition in the bulk silicon, which has an onset at 3.4 eV.

The emission spectrum shows rather weak dependence on the excitation wavelength [see, e.g., Fig. 5.26(b)], which implies that the homogeneous broadening dominates over the inhomogeneous broadening. The quantum yield is rather high, growing from a few percent at room temperature to nearly 100 percent at low temperatures.

The typical mean decay time ranges from 1 to 100 μs at room temperature slowing down to the millisecond range at low temperatures. Typically, the decay is nonexponential, and the mean lifetime shows considerable dependence on the wavelength. Therefore, luminescence decay takes place along with the shift of the emission band maximum towards longer wavelengths (Fig. 5.28).

The above features have been reported by various groups whose original publications are analyzed in reviews (Brus 1994, Kanemitsu 1995; Fauchet 1996; Gaponenko, Petrov et al. 1996). Additionally, at liquid helium temperature and with red-edge excitation, porous silicon was found to exhibit a clear step-like substructure correlating with the phonon energy (Calcott et al. 1993; Suemoto et al. 1993; Brus et al. 1995).

These findings have led to a conclusion that silicon nanocrystals, in spite of a lack of translational symmetry, still exhibit phonon-assisted absorption and emission of photons inherent in the parent indirect-gap monocrystal.

Nevertheless, the size-dependent short-wave shift of the emission band with respect to the monocrystal is very significant. Absolute values of the decay time and of the spectral position of the emission band have been successfully explained in terms of quantum-confinement effect and phonon-assisted transitions (Hybertsen 1994; Takagahara and Takeda 1996). Extremely small silicon clusters have been synthesized in a condensed krypton matrix, and their optical properties were examined (Honea, Kraus, and Bower 1993). Clusters Si_{25}, Si_{35}, Si_{45}, and Si_{55} of the serial magic numbers provide a possibility to trace the build-up of optical absorption bands. Remarkably, absorption spectra of these clusters are smooth and structureless with a shoulder in the range of 3–4 eV, which replicates the relevant direct-gap transition in the bulk Si at 3.4 eV. Thus, the properties of smaller Si crystallites and clusters are drastically different from those of direct-gap nanocrystals such as CdS or CdSe.

At large, the high luminescence quantum yield in silicon nanocrystals originates not from the strongly enhanced radiative decay rate but from the inhibited nonradiative electron-hole recombination. The latter is connected with the statistical distribution of luminescence quenchers over the nanoparticle ensemble and with the reduced transport of excitations between different sites (Brus 1994, Gaponenko, Germanenko, Petrov et al. 1994; Gaponenko, Petrov 1996). For example, in the case of nanocrystals in porous silicon structures the mean number of traps per electron-hole pair was found to be less than one (Gaponenko, Germanenko, Petrov et al. 1994). Therefore, the statistical deficiency of defects with respect to confined electron-hole pairs provides high emission efficiency despite a rather slow radiation decay rate.

Comparing II-VI nanocrystals as a typical example of quantum confined direct-gap structure with silicon nanoparticles, one can see a difference in size restriction effects on optical properties. While the main effect in II-VI nanocrystals is the concentration of oscillator strength in a finite number of discrete bands, in Si nanoparticles the main effect is kinetic, resulting in a dramatic increase in luminescence efficiency even though the emission spectrum and absorption spectrum remain smooth and structureless, indicative of indirect optical transitions.

Several authors have reported on the blue-green emission band in Si nanocrystals and in porous Si (Andrianov et al. 1993; Zhao et al. 1994), which was originally suggested as a manifestation of strong confinement effect in extremely small crystallites. However, further studies have provided strong evidence in favor of the model in which the short-lived blue-green emission is related to silicon oxide rather than silicon nanocrystals (Kontkiewicz et al. 1994; Loni et al. 1995; Gaponenko, Petrov et al. 1996 and references therein).

Basically similar results are expected for Ge nanocrystals. However, studies performed for Ge nanoparticles are not yet exhaustive. Germanium nanocrystals

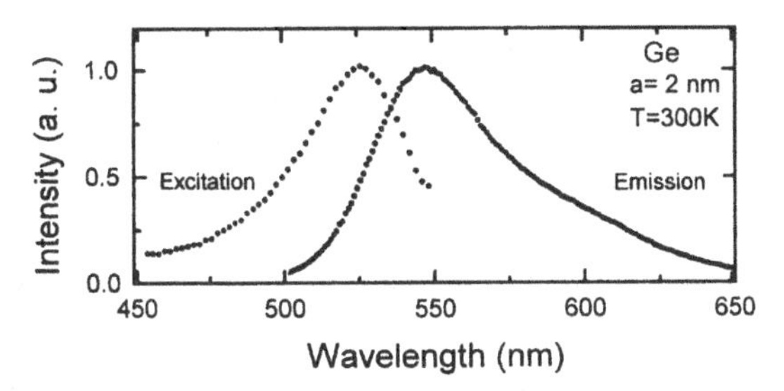

Fig. 5.29. Photoluminescence of Ge nanoparticles developed in SiO_2 film (Kanemitsu et al. 1992). The mean radius of crystallites is 2 nm. Emission spectrum (right-hand curve) was recorded under excitation at 488 nm. Excitation spectrum (left-hand curve) corresponds to the emission wavelength 550 nm. All data were taken at room temperature.

were found to possess emission bands in the range 500–600 nm (Fig. 5.29), whereas the bulk band gap corresponds to 1.7 μm (Kanemitsu et al. 1992; Hayashi et al. 1993; Maeda 1995). Unlike silicon crystallites, Ge nanocrystals exhibit decay time of about 1 ns. Therefore, it was suggested that they possess direct optical transitions. However, the observed decay may be due to non-radiative pathways. This explanation is favored by the low quantum yield (less than 1 percent). To come to an unambiguous conclusion, further studies are necessary. Moreover, the crystallographic structure of germanium nanocrystals in SiO_2 films was found to differ from the diamond structure inherent in bulk germanium. Thus, the observed energy shift and enhanced decay rate of germanium nanocrystals as compared with the monocrystal might be related not only to the quantum-confinement effects due to the size restriction but to the lattice modification as well.

Silver bromide is a typical indirect-gap semiconductor that can be studied in the weak confinement range. AgBr crystallites of submicron size are incorporated in commercial photographic emulsions. By means of precipitation in the inverse micellar solution, nanocrystals of a few nanometers in size can be developed, whereas the exciton Bohr radius $a_B = 3.0$ nm. Bulk AgBr crystals exhibit a rather weak free-exciton line at 460 nm, whereas a bound exciton band peaking at 495 nm dominates (Marchetti et al. 1993). Several groups have reported on a dramatic enhancement of the free exciton emission as compared with the bound exciton band with size restriction (Johansson, McLendon, and Marchetti 1991; Kanzaki and Tadakuma 1991; Masumoto et al. 1992). When the particle diameter decreases from 25 to 8 nm, intensity of free-exciton emission increases more than 10^3 times! The quantum yield is about 15 percent in the best samples. Such an enhancement does not necessarily imply a rapid radiative decay, which might be expected from the quantum confinement effects on the exciton annihilation rate. To evaluate the origin of the strong enhancement of the excitonic luminescence in the context of quantum confinement, effects of impurities and defects should be examined thoroughly. Nanometer-size particles are expected to contain a definite number of impurity atoms distributed according to Poisson statistic. A certain fraction of crystallites may simply be free of defects. Isolation of impurities will result in the inhibition of impurity-related recombination pathways, which in turn will be manifested as an increase in the intrinsic quantum yield. Therefore, to elucidate the mechanism of free-exciton luminescence, a correct evaluation of the size-dependent radiative decay time is necessary. Because of the considerable discrepancies in the lifetimes reported by different authors (Masumoto 1992; Marchetti et al. 1993; Chen et al. 1994) further studies are necessary to come to an unambiguous conclusion.

Direct-gap CdSe monocrystals possessing either wurtzite or zinc blende crystallographic structure are known to experience a structural phase transition to the rock salt lattice under conditions of strong hydrostatic pressure. This is accompanied by a transition from the direct-gap to the indirect-gap band structure. Therefore, it is possible to investigate whether the indirect-gap energy structure transforms into a direct-gap energy structure under conditions including extreme size restriction, use of well-defined, perfect CdSe nanocrystals in a polymer host, and application of hydrostatic pressure. Such experiments were performed by Tolbert et al. (1994). The clear wurtzite-rock salt structural transition has been observed at a hydrostatic pressure of about 9 GPa. A pronounced size dependence of the absorption onset has been revealed for the indirect-gap rock salt modification. However, the absorption spectrum remains very flat and structureless, which indicates indirect-gap forbidden transitions (Fig. 5.30). The indirect-gap feature was found to remain even when the crystallite radius was less than 1 nm.

In view of these results, the transition from the indirect-gap to the direct-gap energy structure in small crystallites because of a lack of translational symmetry appears to be rather questionable. Surprisingly, even smaller crystallites that possess no translational symmetry still feature indirect optical transitions inherent in the parent bulk crystal. One may conclude that, in the case of the above discussed materials, the stoichiometry and the short-range order rather than the translational symmetry determine the probability of optical transitions.

5.4.3 Polarization of luminescence

In the case of optical anisotropy of individual nanocrystals an ensemble of randomly oriented nanocrystals will emit radiation possessing a finite polarization under conditions of excitation by the linearly polarized light. A similar effect is known in molecular spectroscopy and is often referred to as *photoselection*. Polarization of luminescence is usually described by a quantity of the degree

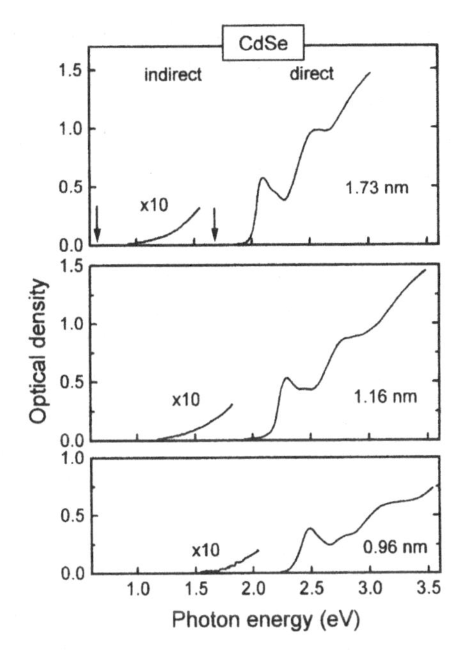

Fig. 5.30. Electronic absorption of 17.3, 11.6, and 9.6 Å CdSe nanocrystals in the wurtzite (atmospheric pressure) and rock salt phases (9.3 ± 5 GPa) (after Tolbert et al. 1994). The spectra have been normalized so that the concentration and path length in the direct- and indirect-gap phases are the same. Additionally, spectra have been scaled so that the integrated areas under the first two absorption features in the direct-gap phase are the same. Indirect-gap spectra have been multiplied by 10 to improve clarity. The bulk CdSe rock salt band gap at 9 GPa and the bulk wurtzite band gap at atmospheric pressure are indicated with arrows.

of polarization defined as

$$P = \frac{I_{\parallel} - I_{\perp}}{I_{\parallel} + I_{\perp}} \tag{5.32}$$

with $I_{\parallel}$ ($I_{\perp}$) being the components of emitted radiation with parallel (orthogonal) polarization with respect to the polarization of the excitation radiation. In a simple oscillator model, the polarization degree satisfies a relation (see, e.g., Stepanov and Gribkovskii 1963)

$$P = \frac{3\cos^2\alpha - 1}{\cos^2\alpha + 3}, \tag{5.33}$$

where α is the angle between the absorbing and emitting oscillators. Therefore, the maximal P value corresponds to the fully anisotropic oscillators ($\alpha = 0$), $P^{\max} = 1/2$, whereas the minimal value occurs at $\alpha = 90°$ and corresponds to $P^{\min} = -1/3$. Note that the ensemble under consideration is macroscopically isotropic because of the random orientation of particles.

Nanocrystals of a number of II-VI compounds possessing the wurtzite structure are anisotropic, exhibiting different optical properties for linearly polarized light with respect to the hexagonal c-axis.

The transition from the A-valence subband to the conduction band is allowed for light polarization vector perpendicular to the c-axis, whereas the transition from the B-valence subband is allowed for both polarizations, parallel and perpendicular to the c-axis, but with different probabilities. Therefore, an ensemble of randomly oriented CdS or CdSe nanocrystals of hexagonal structure should exhibit linearly polarized luminescence under conditions of excitation by linearly polarized radiation (Efros 1992). The finite positive polarization degree has been observed experimentally in the edge emission of larger mixed CdS_xSe_{1-x} crystallites with $a = 30$ nm at low temperatures (Chamarro, Gourdon, and Lavallard 1992). The maximal P value was found to be close to the theoretical limit, which for the case under consideration was evaluated to be 1/7.

The finite polarization degree may result also from an asymmetric shape of randomly distributed nanocrystals. This seems to be the case for elongated silicon crystallites in porous silicon films (Fig. 5.27). The latter was found to exhibit a noticeable positive polarization of emission (Starukhin et al. 1992; Andrianov et al. 1993; Gaponenko et al. 1995) whereas silicon nanocrystals of a sphere-like shape were found to emit unpolarized radiation (Brus 1994). An explanation has been proposed in terms of dielectric confinement effect under condition of ellipsoidal crystallite shape (Kovalev et al. 1995; Lavallard and Suris 1995). The anisotropy of the emitted radiation results from the anisotropy of the lightwave electric field distribution inside a dielectric ellipsoid embedded in some medium with dielectric constant differing from that of an ellipsoid.

Finally, a finite polarization of luminescence may result from the anisotropy of the defect states if the latter are responsible for the emission. Brus and coworkers have reported on the negative polarization degree of CdSe nanocrystals in an organic solvent (Bawendi et al. 1992). The negative luminescence degree of polarization indicates unambiguously that the absorbing and the emitting electronic states are different and efficient energy transfer from absorbing to emitting centers occurs. In the case under consideration the absorption occurs because an electron-hole pair is created corresponding to the volume state, whereas the emission process involves the surface state, which is nearly resonant with the volume state.

5.4.4 Exchange interaction and Stokes shift

The simple model that implies an electron-hole pair with the scalar effective mass in a spherical quantum box predicts a monotonic size-dependent radiative decay rate not only for larger crystallites (with respect to the exciton Bohr radius) but for the smaller ones as well, although the lifetime versus size does not obey a a^{-3}-dependence (Kayanuma 1988). However, experiments on size-dependent radiative decay for smaller nanocrystals were hindered for a long period by the low quantum yield of intrinsic luminescence in nanocrystals of II-VI and III-V compounds that are suitable for optical studies relevant to the strong confinement of electron-hole pairs. The recent advances in fabrication of high-quality II-VI and III-V nanocrystals exhibiting a quantum yield of about 50 percent and higher (Murray et al. 1993; Micic et al. 1996) allow evaluation of the exciton radiative decay rate and its size dependence. Interestingly, time-resolved studies of edge luminescence at low temperatures have revealed luminescence decay in the microsecond range both for CdSe (Nirmal et al. 1994) and InP (Micic et al. 1997) crystallites, whereas radiative decay of excitons in the relevant bulk crystals features lifetimes on the order of 10^{-9} s (Klingshirn 1995). At the same time, theoretical calculations predict a forbidden optical transition from the lowest excited to the ground state for smaller CdSe (Efros 1992a) and CdS (Kuzmitskii et al. 1996) particles. The problem of the low radiative rate in II-VI and III-V nanocrystals has been successfully solved by means of the next step from the simplified models towards real nanocrystals.

In Section 5.2 we saw that accounting for the multiple valence-band structure of the relevant bulk crystals provided an adequate description of the substructure in the optical absorption spectrum of smaller II-VI nanocrystals. The reasonable step towards further understanding of real quantum-dot structures is to introduce electron and hole *spin* into consideration. Spin effects result in the singlet-triplet splitting of the exciton state because of the electron-hole exchange interaction. However, this splitting, ΔE_{ST}, is rather small in the bulk semiconductors as compared with the kT value even for helium temperatures. For example, for typical crystals, CdSe and Si, these values are $\Delta E_{ST}^{\mathrm{CdSe}} = 0.13$ meV and $\Delta E_{ST}^{Si} = 0.15$ meV. However, quantum confinement was found to increase the ΔE_{ST} value by several orders of magnitude (Calcott et al. 1993; Martin et al. 1994; Romestain and Fishman 1994; Nirmal et al. 1995; Takagahara and Takeda 1996). Enhancement of the electron-hole exchange interaction is promoted by the spatial overlap between an electron and a hole which is favored by quantum confinement. In the bulk crystals under conditions of $\Delta E_{ST} \ll kT$, the optically passive triplet state does not manifest itself because of the effective thermal population of the neighboring singlet state. This is not the case for nanocrystals. Significant enhancement of the exchange interaction results in a slowing down of exciton decay because of higher ΔE_{ST} values and the fast singlet-triplet relaxation. In experiments, the optically active singlet state provides an efficient absorption and excitation channel, whereas the optically passive triplet state controls the exciton decay at low temperatures.

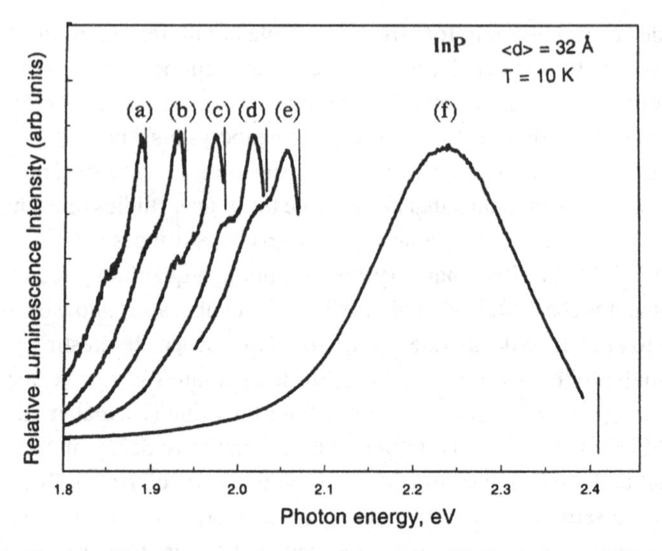

Fig. 5.31. Photoluminescence spectra of an ensemble of InP nanocrystals for different excitation energies (Micic et al. 1997). The mean radius is 1.6 nm, temperature is 10 K. Curves (a)–(e) result from excitation (1.895–2.070 eV) in the red tail of the onset region of the absorption spectrum. The excitation photon energy is indicated by vertical bars near every curve. Curve (f) is a global emission spectrum recorded at the excitation at 2.41 eV, i.e., well to the blue of the first absorption peak.

Systematic experiments on high-quality CdSe nanocrystals in a polymer exhibiting the quantum yield of about 90 percent at liquid helium temperatures have shown that luminescence decay exhibits a dependence on the external magnetic field in agreement with theory (Nirmal et al. 1995). Singlet-triplet splitting was found to contribute significantly to the observed luminescence Stokes shift of the zero-phonon line. The latter was found to grow monotonically from 1 to 20 meV as the crystallite radius decreases from 5.0 to 1.2 nm in agreement with the calculated $\Delta E_{ST}(a)$ dependence. Therefore, the Stokes shift of the exciton luminescence in smaller II-VI quantum dots is controlled not only by the exciton-phonon interaction (Section 5.3) but by the singlet-triplet splitting as well, that is,

$$\Delta_{\text{Stokes}} = 2S\hbar\omega + \Delta E_{ST},$$

where S is the Huang-Rhys factor and $\hbar\omega$ is the characteristic phonon energy.

The stokes shift of the intrinsic luminescence under narrow-band edge excitation has also been studied for InP nanocrystals (Micic et al. 1997). The value of the Stokes shift ranges from 5 to 15 meV (Fig. 5.31) and has been successfully explained in terms of the size-dependent singlet-triplet splitting. The observed luminescence decay shows a nonexponential behavior with the mean

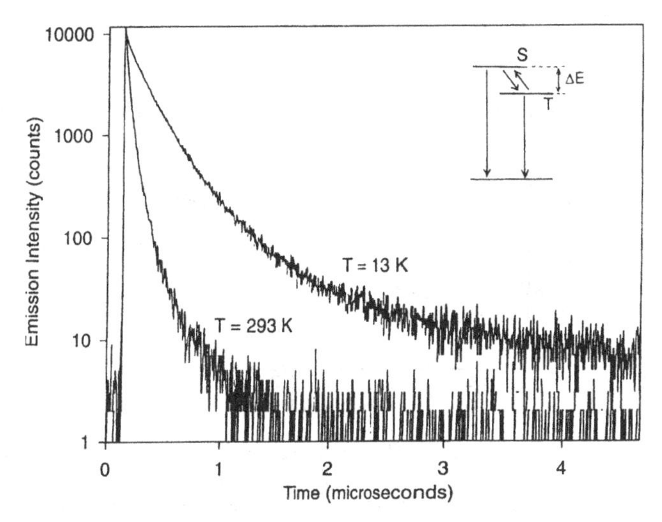

Fig. 5.32. Photoluminescence decay for InP nanocrystals at 293 K and 13 K (Micic et al. 1997). The insert shows a three-level scheme including the crystallite ground states and the two excitonic state, singlet and triplet, separated by ΔE_{ST} due to the electron-hole exchange interaction.

time of about 0.2 μs at low temperatures ($kT < \Delta E_{ST}$). At room temperature ($kT \geq \Delta E_{ST}$) a noticeable enhancement of the decay rate occurs because of electron-hole annihilation from the thermally populated singlet state (Fig. 5.32). The temperature dependence of the radiative exciton lifetime resulting from the thermal balance between the occupation of the singlet and triplet states obeys the relation (Calcott et al. 1993)

$$\tau_{\text{rad}} = \tau_{\text{tripl}} \left[\frac{1 + (1/3)\exp(-\Delta E_{ST}/kT)}{1 + (1/3)(\tau_{\text{tripl}}/\tau_{\text{sing}})\exp(-\Delta E_x/kT)} \right],$$

where τ_{sing} is the radiative lifetime of the singlet state and τ_{tripl} is the radiative lifetime of the triplet state.

The singlet-triplet splitting in silicon nanocrystals has been the subject of extensive research (Calcott et al. 1993, Martin et al. 1994; Takagahara and Takeda 1996). The absolute ΔE_{ST} values for crystallite size corresponding to the visible range were found to be 5 . . . 10 meV, exhibiting monotonic growth with decreasing size (Fig. 5.33). The interplay between S-T states is responsible for the peculiar temperature dependence evaluated for silicon nanostructures. Typically, silicon nanocrystals show a monotonic increase in the visible emission when being cooled to 50–70 K. However, further cooling to liquid helium temperature results in a decrease in the visible emission and in the slowing down of the radiative decay rate. This behavior can be explained reasonably

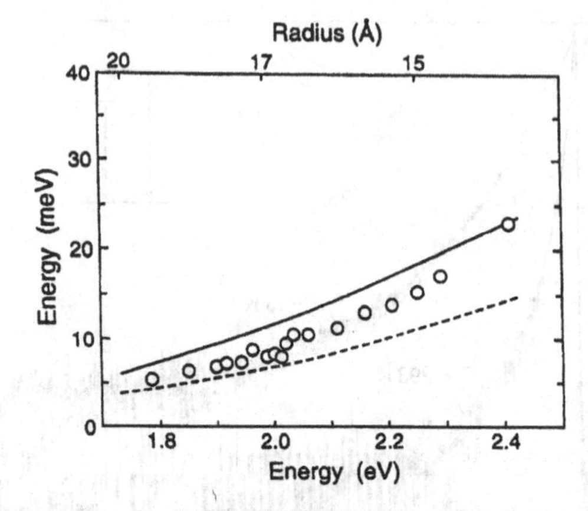

Fig. 5.33. Photoluminescence Stokes shift and excitonic exchange splitting in silicon nanocrystals (Takagahara and Takeda 1996). The theoretical excitonic exchange splitting and the result including the Stokes shift due to acoustic phonon modes are plotted by a dashed line and a solid line, respectively, as a function of the excitonic transition energy. Circles represent the experimental data. On the upper abscissa, the nanocrystal radius is shown corresponding to the exciton energy of the lower abscissa.

in terms of a simple three-level structure (Fig. 5.32) in which at low temperatures ($kT < \Delta E_{ST}$) the radiative recombination occurs from the triplet state with the inhibited rate whereas the competitive nonradiative channels predominate. At $kT \gg \Delta E_{ST}$ the recombination from the singlet state dominates, resulting in an enhanced quantum yield. Noteworthy is the excellent agreement of the experiment and theory. The latter predicts for the exciton energy $E_{\text{exc}} = 1.8 \ldots 2.0$ eV corresponding to the red-orange spectral range the splitting value $\Delta E_{ST} = 5 \ldots 7$ meV (Fig. 5.33), which gives the temperature interval of interest approximately 60–80 K. Similar to CdSe nanocrystals, the luminescence Stokes shift in silicon crystallites can be described in terms of the comparable contributions of singlet-triplet splitting and exciton-phonon interactions (Fig. 5.33). In this case because of the phonon-assisted recombination the zero-phonon line is not pronounced and the Stokes shift is measured as the distance between the excitation energy and the emission onset.

In the previous sections we have seen that an ensemble of nanocrystals embedded in some matrix or solution in essence has a lot of common features with molecules in matrices and solutions. The above interplay of singlet-triplet states is also typical for organic molecules in solid hosts. In this case an efficient S-T conversion results in longer decay times and noticeable population of the triplet state (see, e.g., Gaponenko, Germanenko, Stupak et al. 1994 and references therein).

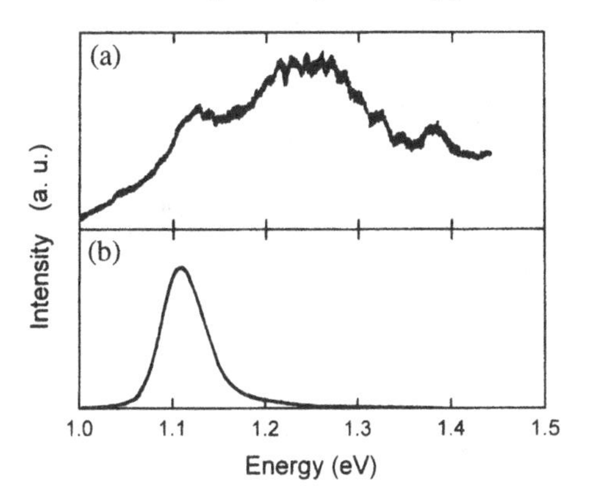

Fig. 5.34. Absorption (a) and emission (b) spectra of InAs nanocrystals developed on GaAs surface (Grundmann et al. 1995).

5.5 Single dot spectroscopy

Several groups have reported on challenging experiments providing luminescent data inherent in a single nanocrystal. Because of the small Stokes shift between resonant absorption and emission bands, luminescence of a single nanocrystal can be collected by means of spatial selection using near-field optics or micrometer size diaphragms rather than by means of the spectral selection as performed in the case of molecular ensembles (Moerner 1994b). Single dot luminescence has been studied for a number of III-V nanocrystals developed on crystal substrates (Grundmann et al. 1995; Nagamune et al. 1995; Samuelson et al. 1995; Gammon et al. 1996a; Bockelmann et al. 1996) and for II-VI colloidal nanocrystals in organic films (Empedocles, Norris, and Bawendi 1996; Nirmal et al. 1996; Blanton, Hines, and Guyot-Sionnest 1996). In the case of self-assembled III-V nanocrystals on a crystal substrate, the size of crystallites is rather large corresponding to the weak confinement limit, whereas CdSe crystallites examined correspond to the strong confinement regime. For this reason we discuss data on III-V and II-VI nanocrystals separately.

The absorption spectrum of InAs nanocrystals on GaAs monocrystal surface features a number of broad bands overlapping each other [Fig. 5.34(a)]. The photoluminescence spectrum is usually a single broad band (FWHM on the order of 50 meV) resonant to the first absorption feature [Fig. 5.34(b)]. However, high resolution cathodoluminescence experiments have revealed a number of very narrow reproducible spikes [Fig. 5.35(a)] that can be assigned to the different dots by means of the combined spectral/spatial high-resolution performance. Linewidths were beyond the spectral band-pass, which was as low as 0.09 nm in the experiments under consideration. To this accuracy FWHM

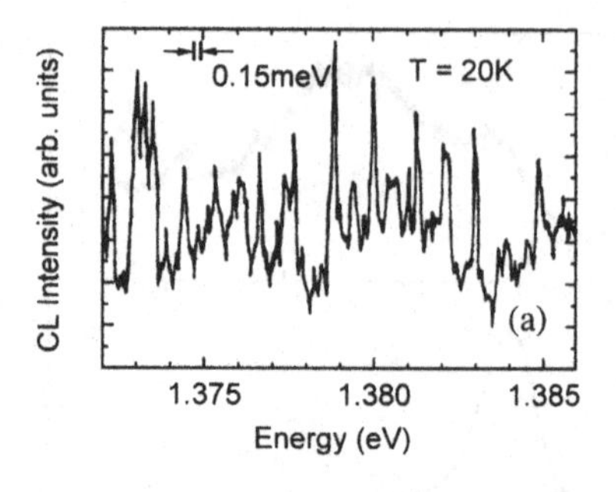

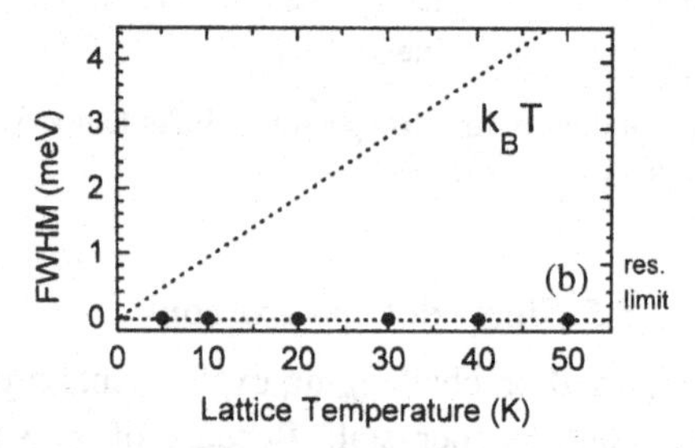

Fig. 5.35. High spatial resolution cathodoluminescence spectrum of InAs nanocrystals on GaAs surface (Grundmann et al. 1995). Temperature is 20 K. Discrete peaks are well reproducible and correspond to the emission of individual crystallites. The width of emission lines is limited by the instrumental spectral resolution up to $T = 50$ K, which is shown by dots in panel (b). Dashed line shows kT value.

value is independent of temperature [Fig. 5.35(b)] within a rather wide temperature range. The higher spectral resolution realized in the experiments on GaAs crystallites buried in a $Al_xGa_{1-x}As$ layer provides homogeneous exciton linewidth as small as 50 μeV (0.40 cm^{-1}) at $T = 5$ K increasing to 150 μeV at $T = 50$ K (Gammon et al. 1996a). The relevant luminescence excitation spectrum consists of a set of discrete and extremely narrow peaks (Fig. 5.36). Note that submillielectronvolt linewidths found for III-V nanocrystals in a weak confinement regime are of the same order of magnitude as compared with large copper halide nanocrystals whose linewidths were evaluated by means of persistent hole-burning and nonlinear pump-and-probe spectroscopy (Masumoto et al. 1995). To summarize, nanocrystals in a weak confinement limit were found to possess discrete atomlike absorption and emission lines with widths two orders of magnitude less than the kT value.

Pioneering results on single dot spectroscopy in the strong confinement limit have been recently reported by M. Bawendi and co-workers (Empedocles et al. 1996). In these experiments perfect colloidal CdSe particles with mean sizes $\bar{a}$ ranging from 2 to 2.5 nm were examined. While spectrally selective techniques like the nonlinear pump-and-probe spectroscopy and fluorescence line

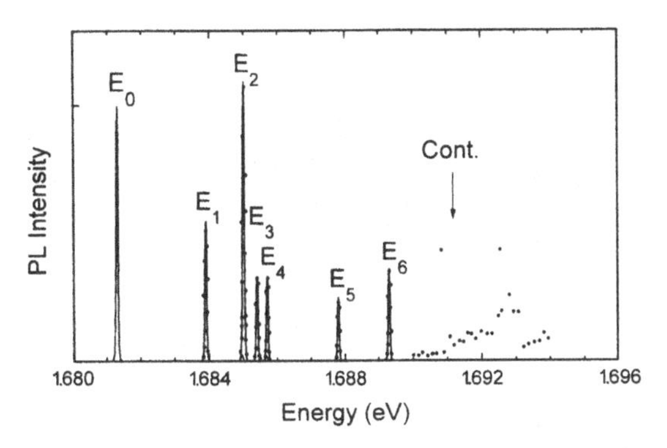

Fig. 5.36. Photoluminescence emission and excitation spectra of a single GaAs nanocrystal buried in a $Al_xGa_{1-x}As$ layer (adapted from Gammon et al. 1996b). E_0 denotes the emission resonance, and E_1 to E_6 denote excitation bands.

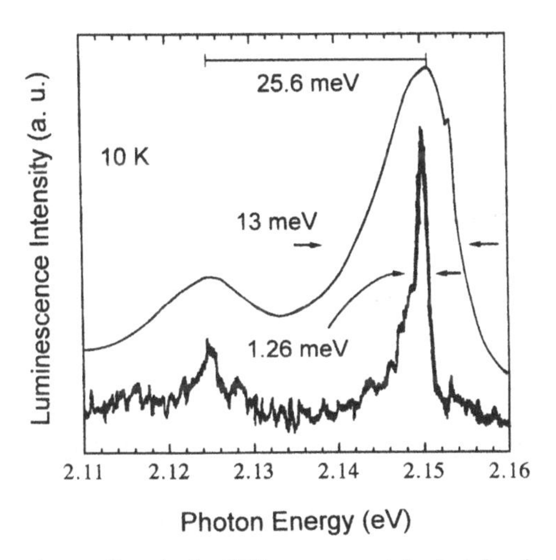

Fig. 5.37. Comparison of a single CdSe nanocrystal photoluminescence spectrum versus fluorescence spectrum obtained during presence of selective red-edge excitation (Empedocles, Norris, and Bawendi 1996). Mean size is $\bar{a} - 2.2$ nm, temperature is 10 K. Both spectra contain two bands separated by LO phonon energy equal to 26 meV.

narrowing (see Section 5.1) give FWHM values about 10 meV at liquid helium temperature, single dot spectroscopy reveals narrower widths in the submillielectronvolt range (Fig. 5.37). These studies highlighted important features of single dot spectroscopy data. The single dot emission spectrum shows a pronounced dependence on the number of photons collected during the measurements. The emission spectrum widens with increasing data acquisition time and with increasing excitation intensity (Fig. 5.38). A number of sequential spectra detected from the same dot during equal time intervals exhibit clear

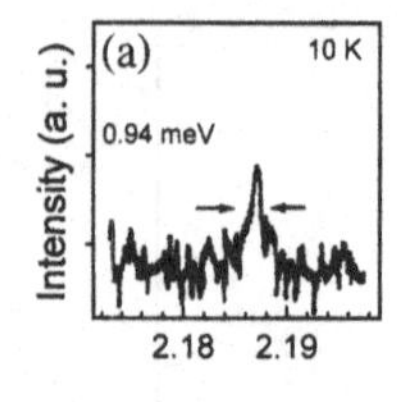

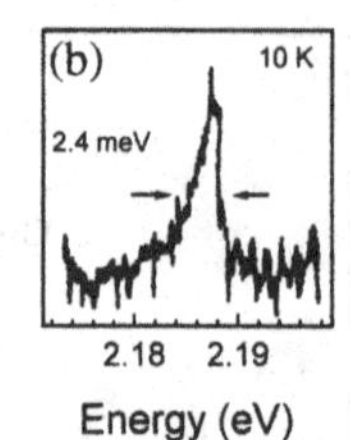

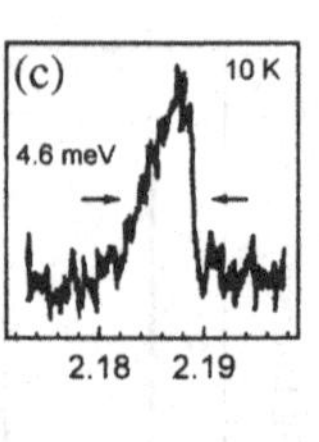

Fig. 5.38. Emission spectra collected for a single CdSe crystallite during 1 min at various excitation intensity (Empedocles, Norris, and Bawendi 1996). Crystallite size is $\bar{a} = 2.2$ nm, temperature is 10 K. Excitation power is 65 (a), 150 (b), and 314 W/cm^2 (c).

spectral diffusion resulting in the individual shape of every spectrum. The spectral diffusion observed in these fine experiments has been attributed to the photo-induced modification of the nanocrystal local environment ("nanoenvironment"), for example, from local electric field effects.

In this connection two questions arise. What is the true homogeneous width? Is it possible to interpret the parameters of emission and absorption spectra of a given nanocrystal (atom or molecule) in terms of the homogeneous width? As the discussion of these important issues is beyond the scope of this book, we restrict ourselves to a few comments. First of all, as soon as a single nanocrystal (atom or molecule) has been selected, the relevant spectroscopic data do not acquire broadening resulting from difference of the different nanocrystals. This means that the *static* broadening due to variable size, barrier height, impurity concentration, and stable charge distribution around nanocrystals is inhibited. Nevertheless, the *dynamic* broadening from environmental time-dependent fluctuations and/or ionization-deionization events still remains. These remaining broadening mechanisms are common for all nanocrystals. Therefore, it is reasonable to remember the ergodic theorem. This theorem is a background of statistical physics and can be formulated as follows: Averaging over an ensemble of similar particles is equivalent to averaging over time applied to a given particle. Thus, averaging of a single dot emission spectrum in a case of rather long data acquisition should be equivalent to a snapshot which is, however, taken not for the whole set of nanocrystals but for the subset consisting of crystallites with the same static parameters listed above.

In the context of this discussion one may expect that single-dot spectra corresponding to the large number of photons emitted either because of lengthy accumulation or high excitation power should converge with the homogeneous linewidths evaluated by means of spectral hole-burning and fluorescence line narrowing techniques. This seems to be just the case (compare Fig. 5.38 with

Figs. 5.7 and 5.8). The lowest estimate of FWHM value for CdSe nanocrystals from hole-burning spectroscopy is $2\Gamma_0 = 7$ meV for $a = 2$ nm (Gaponenko 1996), whereas the single dot spectrum features $2\Gamma_0 = 4.6$ meV for similar crystallites in an organic environment [Fig. 5.38(c)].

To summarize, it is reasonable to conclude that instant single dot spectroscopy data provide the basic homogeneous width determined by the lifetime broadening and by the exciton-phonon interaction, whereas long-time averaged data are relevant to the width driven by dynamic environmental fluctuations.

5.6 Quantum dot in a microcavity

In the previous sections we have shown that spontaneous photon emission in nanocrystals can be considerably modified by electron-hole spatial confinement. However, the spontaneous decay of electron-hole pairs is not an intrinsic property but results from interaction with the electromagnetic field. In an environment other than a vacuum (or free space), spontaneous emission of photons can be controlled to a large extent by the modified *photon density of states*. The latter can be controlled either by means of a creation of artificial media (e.g., periodic dielectric lattice of a period comparable to the photon wavelength, the so-called *photonic crystals*) or by means of a planar or spherical resonator of a size comparable with the photon wavelength, or the so-called *microcavity*. In the case of a microcavity one can discuss the photon confinement effects on the spontaneous emission rate. Furthermore, in the case of a spherical or box-like cavity, the term *photonic quantum dot* seems to be appropriate. In this section we consider the effect of photon confinement on spontaneous emission of quantum dots, which manifests itself when a quantum dot is embedded in a microcavity. The phenomenon connected with continuous media with modified photon density of states will be discussed in Chapter 8.

The cavity effect on spontaneous emission rate can be understood in terms of Fermi's golden rule. Suppose we have a single two-level atom interacting with an electromagnetic field. Consider the two states of the atom and the field. One state corresponds to an excited atom state $|e\rangle$ whereas the field contains no photons. The state of the system can be labeled as $|e, 0_k\rangle$, where 0_k indicates the lack of photons of wave number k. Another state corresponds to the ground atom state $|g\rangle$ and a single photon, which can be denoted as $|g, 1_k\rangle$. The transition rate from the first to the second state can be written as follows

$$W_{eg} = \frac{2\pi}{\hbar}\rho(\omega_k)|\langle g, 1_k|\mathrm{H}_{\mathrm{int}}|e, 0_k\rangle|^2, \tag{5.33}$$

where $\mathrm{H}_{\mathrm{int}}$ is the interaction part of the Hamiltonian that couples the atom to the field, ω_k is the frequency of the emitted photon, and ρ is the density of modes at the frequency corresponding to the wave number k.[4]

The contribution of a cavity is to provide the mode density ρ at the frequency of interest. If no mode is available, then no spontaneous emission occurs. In the case of a multilevel system the properties of a cavity can modify the decay rates via different pathways resulting in an enhancement of certain emission bands and inhibition of competitive bands. This effect has been known since 1946, when it was predicted by E. M. Purcell (1946). Controlled spontaneous emission resulting from nontrivial boundary conditions provided in a microcavity with a limited number of modes within the emission spectrum of the embedded species is of great scientific and practical importance. Not only the basic features of the matter-field interaction can be revealed in these studies but also the novel generation of optical devices in which light is spontaneously emitted at the desired wavelength with the desired rate in the desired direction. Modification of spontaneous emission of molecules in a Fabry-Perot interferometer was reported for the first time in 1970 (Rubinov and Nikolaev 1970). It has been extensively investigated both theoretically and experimentally for atoms, molecules, and semiconductor microstructures in microcavities (see, e.g., Haroche and Kleppner 1989; Berman 1994; Tanaka et al. 1995; Yamanishi 1995; Jhe and Jang 1996; Barnes et al. 1996 and references therein).

However, study of the microcavity effect on spontaneous emission of nanocrystals is still at the preliminary stage. Saunders et al. (1992) considered a pointlike quantum dot embedded in a finite dielectric sphere of size compared with the wavelength of emitted light. For the specific parameters of the problem corresponding to GaAs quantum dot embedded in a $Al_{0.4}Ga_{0.6}As$ microsphere, the theory predicts a monotonic increase in the spontaneous emission rate W by two orders of magnitude when the cavity radius a_c grows from a few nm to about 150 nm. For a larger cavity $W(a_c)$ dependence exhibits oscillations corresponding to resonance conditions between the emission wavelength and the cavity modes. A structure of this type has not been developed to date. However, recent advances in the synthesis of semiconductor nanoparticles covered by a layer of the different composition (Schooss et al. 1994; Mews et al. 1996) may lead to a breakthrough in this field in the near future.

Modification of the emission spectrum of quantum dots in a planar Fabry-Perot resonator has been examined using microstructures based on porous

[4]In the case of the free three-dimensional space, the photon density of states is $\rho(\omega) = \omega^2/\pi^2 c^3$. Note that it is one of the coefficients in Plank's formula

$$u(\omega) = \frac{\omega^2}{\pi^2 c^3} \hbar\omega \frac{1}{\exp(\hbar\omega/kT) - 1} = \hbar\omega\rho(\omega) f(\omega),$$

whereas the other coefficients are nothing but the photon energy $\hbar\omega$ and Bose-Einstein distribution function $f(E)$ with the zeroth chemical potential inherent in photon gas.

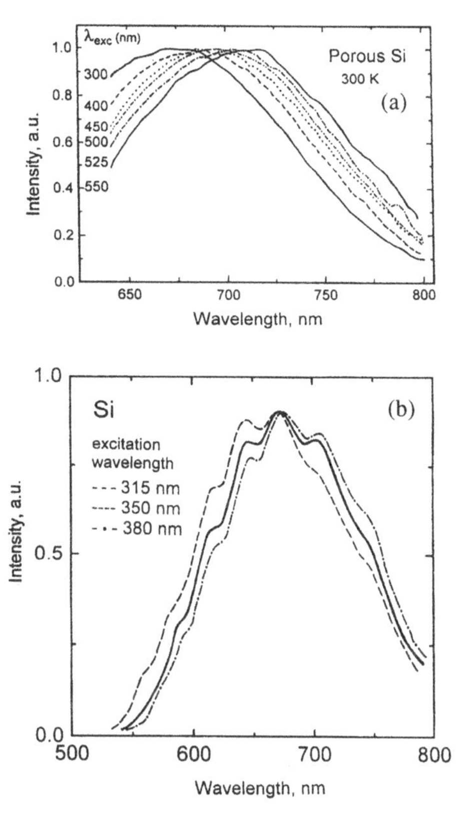

Fig. 5.39. Cavity effect on emission spectrum of a porous silicon film. Panel (a) shows typical emission spectra recorded at several excitation wavelengths (Gaponenko, Petrov et al. 1996). Cavity effect is negligible and emission spectrum exhibits monotonic shift with excitation wavelength tuning. Panel (b) presents emission spectra of a well-defined 5-μm thick porous silicon film with pronounced peaks due to the Fabry-Perot modes in the spectrum (Bondarenko et al. 1994). In this case the maximum of emission spectrum shows no shift in spite of the tunable excitation wavelength.

silicon technology. Porous silicon films on silicon substrates containing silicon nanocrystals connected via silicon oxide interfaces are developed by an electrochemical process. Luminescent properties of porous silicon are determined by the quantum confinement effects on silicon nanocrystals (Brus et al. 1995). Well-defined rigid films of a few micrometers thickness show pronounced Fabry-Perot modes in the emission spectrum because of reflection at air/film and film/silicon boundaries (Bondarenko et al. 1994). The emission spectrum is structureless and inhomogeneously broadened exhibiting a pronounced shift with excitation wavelength [Fig. 5.39(a)]. However, in the case of thin film mode structure develops in the spectrum, the maximum of emission being nearly independent of the excitation wavelength [Fig. 5.39(b)].

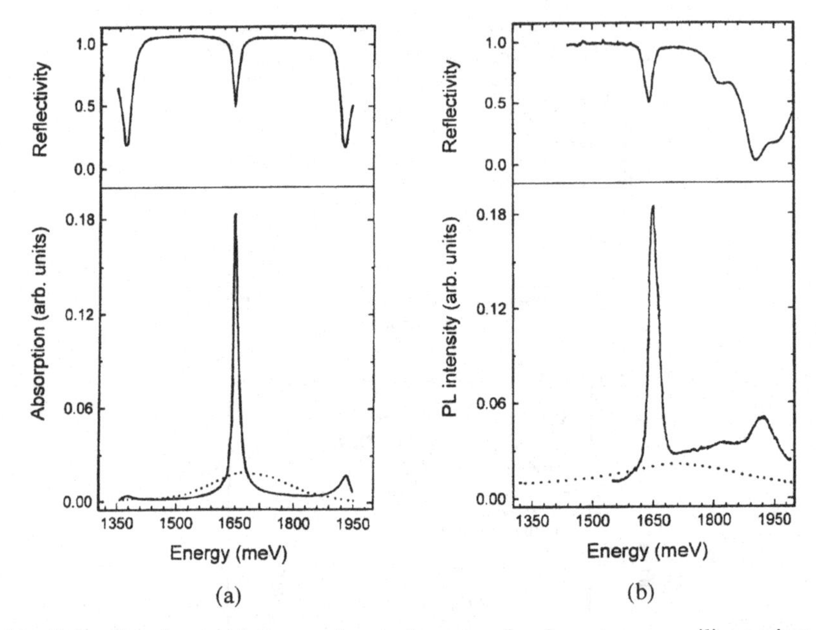

Fig. 5.40. Calculated and measured optical spectra of a planar λ porous silicon microcavity (Pellegrini et al. 1995). (a) Calculated reflectivity spectrum (upper panel) and absorption spectrum (lower panel) for a λ porous silicon microcavity. Also shown are the calculated absorption spectrum (dotted curve) for a λ porous silicon layer without reflectors. (b) Room-temperature reflectivity spectrum of a λ porous silicon microcavity (upper panel) and photoluminescence spectra of a porous silicon layer (lower panel) without any reflectors (dotted curve), and of a λ porous silicon microcavity (solid curve).

Thus, the spontaneous emission spectrum is controlled by the resonator to a large extent. The effect can be enhanced by a better quality of resonator. The refractive index of a porous film depends on the porosity, which in the case of porous silicon is controlled by the etching conditions. Alternating the current/time combination provides films of controllable thickness and porosity. Therefore, a microstructure can be developed on a silicon substrate in which alternating porous layers are used as active media containing quantum dots and Bragg reflectors (Pellegrini et al. 1995). This structure shows a very narrow emission spectrum as compared with that of a free porous film, the emission efficiency being enhanced by a factor of 16 [Fig. 5.40(a)]. These features are the consequence of the enhancement of spontaneous emission resonant with the optical modes of the Fabry-Perot resonator. Under these conditions no significant dependence of the FWHM on temperature was observed, confirming that the linewidth is determined only by the resonator finesse. The experimental findings are in good agreement with the calculated reflection and absorption spectra [Fig. 5.40(b)].

No doubt, the pioneering experiments by Pellegrini et al. will be followed by extensive studies of quantum dots in microcavities. Basically, a combination of size-dependent electron properties of nanocrystals with geometry-dependent photon density of states in microcavities and related microstructures is a very promising start towards the development of practical light-emitting devices with the complete control of emission spectrum, kinetics, and directionality. These devices are expected to possess an extremely high efficiency that cannot be achieved otherwise.

5.7 Recombination mechanisms

In the bulk semiconductor, electron-hole recombination processes described in terms of the recombination rate, R, occur by means of various mechanisms. These mechanisms include radiative recombination in electron-hole gas or plasma ($R \propto n^2$), radiative annihilation of excitons ($R \propto n$), Auger recombination ($R \propto n^3$), recombination via trap states (both radiative and nonradiative), recombination in donor-acceptor pairs, and others (see, e.g., Landsberg 1991 and references therein). The finite number of electron-hole pairs within a given nanocrystal as well as the finite number of defects and impurities result in certain peculiarities inherent in nanocrystals which are not relevant to the bulk solids. In Section 5.4 the principal features of radiative electron-hole transitions were discussed. In this section the competitive recombination paths will be considered.

In the most perfect crystallites of II-VI and III-V compounds possessing high quantum yield of intrinsic luminescence, the typical lifetimes are on the order of 10^{-6} s. On the other hand, most of the experiments on various crystallites in organic and inorganic matrices revealed population kinetics with characteristic times ranging from 10^{-11} to 10^{-7} s. The observed enhancement of the electron-hole population decay is due to the competitive radiative and nonradiative paths provided by the surface or volume defects and impurities as well as by the intrinsic surface states. This conclusion is proved by the well-known results that chemical and/or thermal treatments often lead to a drastic effect on the emission spectrum and kinetics even though the absorption spectrum remains unchanged.

Typically, most of the glasses containing II-VI nanocrystals exhibit a rather broad emission band related to defects and impurities dominating in the luminescence spectrum. With growing pump intensity the narrow-band edge emission increases and becomes dominating at higher excitation densities. A typical example relevant to the intermediate excitation level is presented in Fig. 5.41. At higher densities and/or radiation doses, however, a number of reversible and irreversible photoinduced phenomena occur resulting in laser annealing, photoionization, photodestruction, and other effects. These aspects

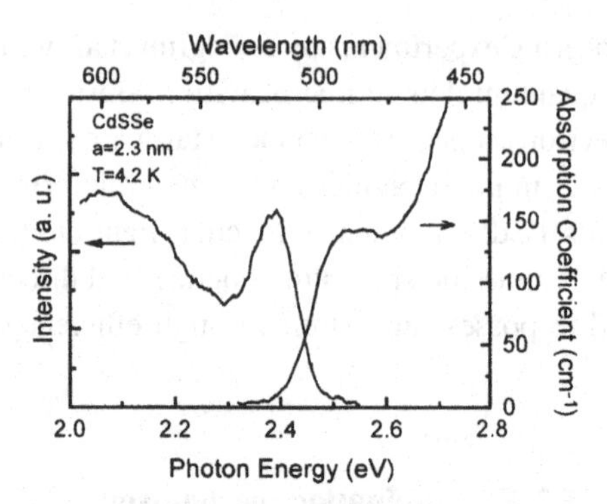

Fig. 5.41. Absorption and emission spectra of CdS_xSe_{1-x} nanocrystals in a glass matrix (Uhrig et al. 1991). The mean radius of crystallites is 2.3 nm, temperature is 4.2 K, the excitation wavelength is 337 nm (3.67 eV). The narrower emission band peaking near 2.4 eV is intrinsic luminescence, whereas the dominating wide band at 2.1 eV is defect-related emission. The large spectral shift between the first absorption maximum and the edge emission band is due to inhomogeneous broadening and nonresonant excitation.

will be the subject of Chapter 7. In what follows we consider some kinetic features related to the finiteness of the impurity states and to Auger processes.

In the case when only one electron-hole pair per crystallite is created, the radiative annihilation rate in an ensemble of the identical crystallites reads

$$R = \frac{n}{\tau_0} \tag{5.34}$$

as in the case of molecular or atomic ensembles. In Eq. (5.34) n is the number of the excited crystallites, and τ_0 is the intrinsic radiative lifetime. In the presence of trap states one has

$$R = n\left(\frac{1}{\tau_0} + k_{tr}n_{tr}\right), \tag{5.35}$$

where k_{tr} is the capture probability and n_{tr} is the number of traps per electron-hole pair. However, every given electron-hole pair experiences a specific capture rate because of the finite number of traps within a given crystallite. In the case of random trap distribution the latter is described by the Poisson function

$$P(n_{tr}, \langle n\rangle) = \exp(-\langle n\rangle)\frac{\langle n\rangle^2}{n!} \tag{5.36}$$

with $\langle n\rangle$ being the average number of traps per electron-hole pair over an ensemble under consideration. The population decay, $n(t)$ can be monitored by means of luminescence kinetics $I(t)$, which takes the form

$$I(t) = I_0 \exp\left\{-\frac{t}{\tau_0} - \langle n\rangle[1 - \exp(-k_{tr}t)]\right\}. \tag{5.37}$$

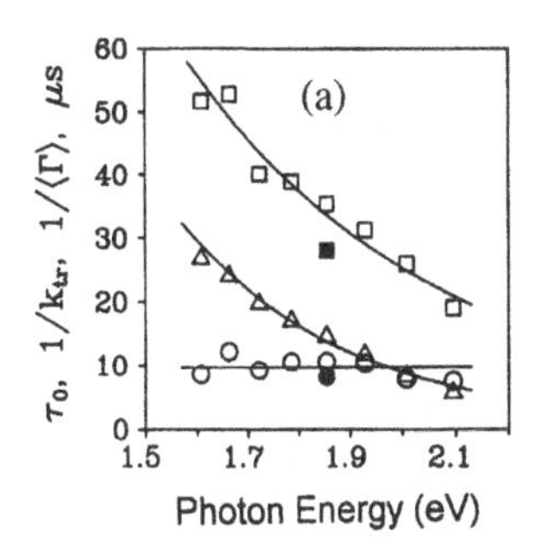

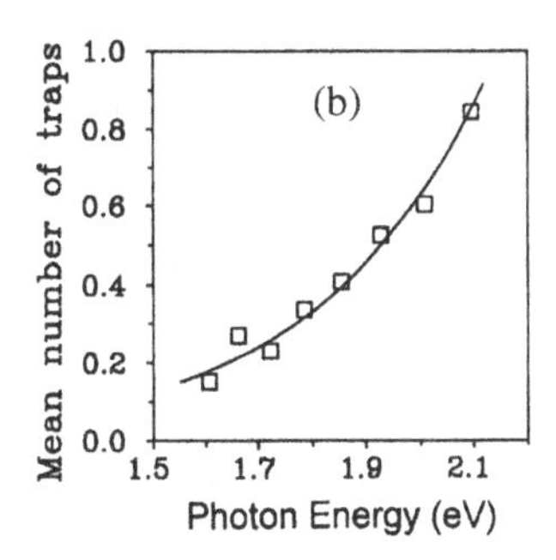

Fig. 5.42. Parameters of the model of micellar kinetics evaluated for silicon nanocrystals in oxidized porous silicon film versus energy of emitted photons (Gaponenko, Germanenko, Petrov et al. 1994). The intrinsic lifetime is τ_0, k_{tr} is the capture probability, and $\langle\Gamma\rangle$ is the mean decay rate, $\langle\Gamma\rangle = 1/\tau_0 + \langle n\rangle k_{tr}$, where $\langle n\rangle$ is the mean number of traps per electron-hole pair. Squares correspond to τ_0, triangles show mean decay time $\langle\Gamma\rangle^{-1}$, and circles are relevant to k_{tr}^{-1}.

The relation (5.37) is known as *micellar kinetics* because it was first proposed for an ensemble of organic micelles with a discrete number of quenchers (Dederen et al. 1981; Gehlen and de Schryver 1993). Micellar kinetics were found to describe the luminescence decay of Si nanocrystals (Gaponenko, Germanenko, Petrov et al. 1994; Kuznetsov et al. 1995) and AgBr nanocrystals (Masumoto et al. 1992). The parameters of the model evaluated for the case of silicon nanoparticles in strongly oxidized porous silicon films are presented in Fig. 5.42. Because of the inhomogeneous broadening of the emission band, every parameter of the model is wavelength-dependent. However, the trap probability, k_{tr}, which is the parameter of the defect rather than the crystallite, was found to be the same for various wavelengths and for the different samples. This is an additional confirmation that the model is relevant to the case under consideration. The value of τ_0, which can be considered as the intrinsic radiative lifetime, monotonically falls from 60 μs to 20 μs with growing photon energy (i.e., with decreasing size in terms of the quantum confinement effect). At the same time, the mean number of traps $\langle n\rangle$ grows from 0.15 to 0.85, yet being less than unity. The small $\langle n\rangle$ number is the reason for the high quantum yield, as was already discussed in Section 5.4.2. The model of micellar kinetics with respect to silicon nanocrystals has been verified not only by means of traditional nonlinear fitting but by decay time distribution analysis as well (Gaponenko, Germanenko, Petrov et al. 1994).

Another competitive process of electron-hole recombination with respect to radiative annihilation is the three-particle Auger process. In this case the energy of an electron-hole pair is carried away by another electron or a hole. Manifestations of Auger recombination in semiconductor nanocrystals embedded in a glass matrix have been reported by a number of groups (Roussignol, Kull et al.

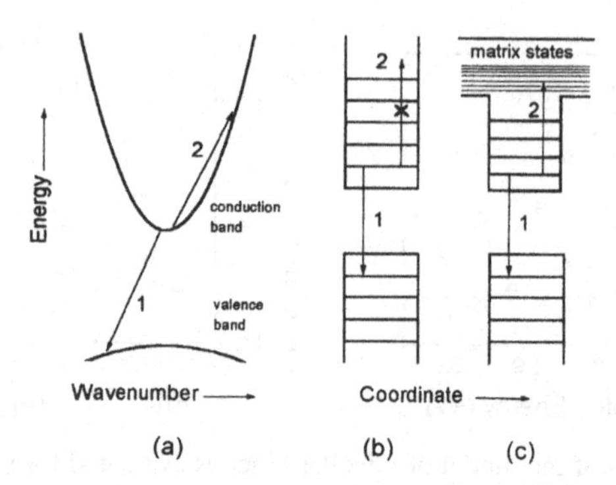

Fig. 5.43. Schematic presentation of Auger processes including two electrons and one hole in the case of (a) a bulk semiconductor and (b), (c) a nanocrystal. In the bulk solids electron-hole annihilation (1) with a simultaneous transition of another electron to the higher state (2) requires an exact matching of energy and momentum to satisfy conservation laws. In nanocrystals the momentum conservation law relaxes and only energy conservation remains. However, discreteness of the energy states restricts the probability of the Auger process because of the lack of the proper final states (b). In the case of the finite barrier (c) the probability of Auger processes is enhanced because a continuumlike multitude of the matrix states is available.

1987; Kull et al. 1989; Dneprovskii et al. 1990). In the case of bulk solids such a three-particle event needs an exact conservation of energy and momentum [Fig. 5.43(a)]. Additionally, the probability of a transition is proportional to the electron-hole function overlap. Therefore, the Auger recombination is essential only at rather high excitation intensities. Unlike bulk materials, in nanocrystals the momentum conservation law vanishes and only the energy conservation determines the probability of the Auger recombination. Electron-hole overlap is provided by spatial confinement. In the case of a high potential barrier at the dot-matrix interface the discreteness of energy states restricts Auger recombination rate because of the lack of the matching final states [Fig. 5.43(b)]. Numerical consideration of the Auger relaxation process in this case has been performed for CdS nanocrystals (Kharchenko and Rosen 1996). The rate of Auger recombination was found to increase from 10^7 to 10^9 s^{-1} when crystallite radius fell from 4 to 1.5 nm, with sharp enhancement corresponding to the energy levels matching. Numerical evaluation is in good agreement with experimental data (Chepik et al. 1990). Therefore, the Auger nonradiative process can be the dominating energy relaxation channel in a number of cases. The Auger recombination rate can be further enhanced when a continuum-like multitude of matrix states becomes available, which is the case for the low dot-matrix potential barrier [Fig. 5.43(c)]. Moreover, in this case the Auger process results in photoionization of a crystallite, which in turn leads to a change in electronic and optical properties. This interface-related effect will be discussed in more detail in Chapter 7.

Conversely, for some specific quantum-dot structures Auger processes may be inhibited as compared with the bulk structures. This is the case for narrow-gap semiconductor nanocrystals surrounded by a medium providing high potential barrier at the boundaries (e.g., InSb or PbSe inside CdS or ZnO). In this situation the energy of "interband" transitions is of the same order as the inter-band level spacing. Therefore, energy conservation is difficult and the Auger rate reduces. Furthermore, though momentum conservation relaxes, in the case of a spherical dot the total angular momentum must be conserved. By means of numerical calculations, the Auger lifetime for a 15-nm radius InSb quantum dot surrounded by CdTe barriers with a room temperature band-gap equal to 0.26 eV (4.8 μm) was found to be 135 ns, which is two orders of magnitude larger than in bulk semiconductors with similar bandgaps (Pan 1992b).

5.8 Electric field effect on exciton absorption

Electric field-induced effects on optical absorption are very important in the study of the optics of condensed matter. First of all, modulation of the absorption coefficient and refractive index by an external electric field (electroabsorption and electroreflection) is used as a kind of modulation spectoscopy to elucidate the fine features of the optical spectra (see, e.g., Cardona 1972). Second, field-induced absorption and reflection can be used in various electrooptical modulators, switchers, shutters, and other devices.

To calculate optical absorption of a semiconductor nanocrystal in the presence of an external uniform electric field one has to deal with the Schroedinger equation, which includes electron-hole Coulomb interaction, spatial confinement, dielectric confinement, and the electric field term $-e\boldsymbol{\varepsilon}\cdot\boldsymbol{r}$, where ε is the external electric field. This problem cannot be solved in a general form, and a number of approximations are applied to allow analysis. In what follows we consider first two limiting examples to give an idea of the expected phenomena, then we provide examples related to concrete experiments and calculations relevant to semiconductor nanocrystals.

In bulk crystals, electric field effects on exciton absorption have been examined both theoretically and experimentally (Blossey 1970; 1971; Perov et al. 1969; Lange 1971; Gribkovskii 1975; Tyagay and Snitko 1980). The important parameter is the ionization field ε_I, which provides a potential drop of $1Ry^*$ across the exciton Bohr radius a_B, that is, it is defined as

$$\varepsilon_I = Ry^*/ea_B = (\mu/m_0)^2\varepsilon^{-3} \times 2.59 \times 10^9 \text{ V/cm}. \qquad (5.38)$$

Ionization field values for a number of semiconductor crystals are listed in Table 5.2. In the case $\varepsilon/\varepsilon_I \approx 1$ exciton resonance is completely smeared. In the case $\varepsilon/\varepsilon_I \ll 1$ the problem can be analyzed by means of perturbative approach.

Table 5.2. *Ionization fields for several semiconductor crystals*

Crystal	ε_I, (10^3 V/cm)
GaAs	5.7
CdTe	31
CdSe	60
CdS	140
ZnSe	200
PbI_2	460

Source: After Blossey 1970.

In this case exciton peak experiences a quadratic Stark shift up to a field of about $0.25\varepsilon_I$. A minimum exciton energy is obtained at $\varepsilon \approx 0.5\varepsilon_I$, the shift being on the order of $0.1Ry^*$. For ε higher than $0.5\varepsilon_I$ the exciton peak moves to higher energies. For $\varepsilon/\varepsilon_I > 1$ the peak exists only as the first electric field-induced oscillation in the optical density of states. Along with the shift, the spectrum shows a considerable broadening. A typical electroabsorption spectrum at $\varepsilon < \varepsilon_I$ is presented in Fig. 5.44. One can expect similar behavior in the case of nanocrystals corresponding to the weak confinement regime when $a \gg a_B$ holds.

Another limit can be discussed it terms of electric field effects on molecular absorption. If the field is weak enough, the perturbational consideration gives the differential absorption spectrum (field-on, field-off) as the second derivative of the initial absorption in the case of a single isolated absorption band (Sacra et al. 1995). This result is valid for a series of absorption bands if the bandwidths Γ_i are less than separations ΔE_{ij}, i.e., $\Gamma_i \ll |\Delta E_{ij}|$. In the opposite limit, when $\Gamma_i \gg |\Delta E_{ij}|$, the differential absorption due to electric field effect is described by the zeroth and the first derivative of the original spectrum. Figure 5.45 shows an example of a Guassian absorption band along with its first and second derivatives. Basically, in the case of the first-derivative

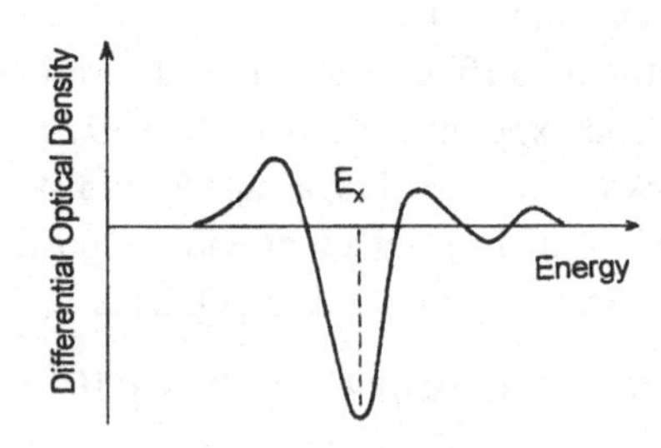

Fig. 5.44. A typical electroabsorption spectrum for an exciton absorption band when the external electric field is small as compared with the ionization field.

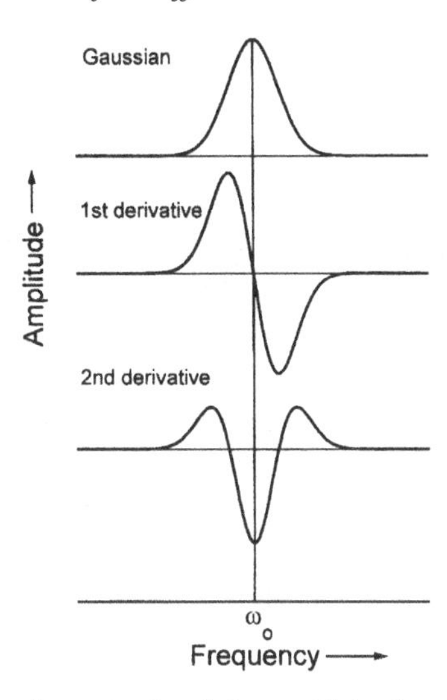

Fig. 5.45. A Gaussian absorption band along with its first and second derivatives.

modulation one can refer to the field-induced shift, whereas in the case of the second-derivative modulation the term field-induced broadening is appropriate. In the intermediate case both broadening and shift occur.

Theoretical and experimental studies of electric field effects on exciton absorption have been carried out for CdS_xSe_{1-x} nanocrystals corresponding to the intermediate and strong confinement. Numerical calculations performed by various authors have shown that in the case of a weak field when perturbation theory is valid the field-induced absorption modulation can be treated as a combination of spectral shift and broadening (Miller, Chemla, and Schmitt-Rink 1988; Hache, Ricard and Flytzanis 1989; Nomura and Kobayashi 1990b). Variational calculations and experiments have shown that for electric fields on the order of 10^4 V/cm the energy shift of the first absorption band is less than 0.1 meV (Nomura and Kobayashi 1990b). A typical example of the electric field effect on optical absorption of nanocrystals is presented in Fig. 5.46. Interestingly, electromodulation reveals resonances that are smeared by the temperature broadening and inhomogeneous broadening. Eventually, differential absorption spectra look similar to the second-derivative spectrum. Similar results have been reported for II-VI nanocrystals of $a < a_B$ in a glass matrix by other groups (Esch et al. 1990; Rossmann et al. 1990; Ekimov et al. 1990; Cotter, Burt and Girdlestone 1990; Sacra et al. 1995). According to Hache et al., the phenomenological broadening gains a microscopic explanation in terms of the field-induced state renormalization (Hache et al. 1989). Within the framework of perturbation theory (i.e., for $\varepsilon \ll \varepsilon_I$), the electric field mixes an nl state

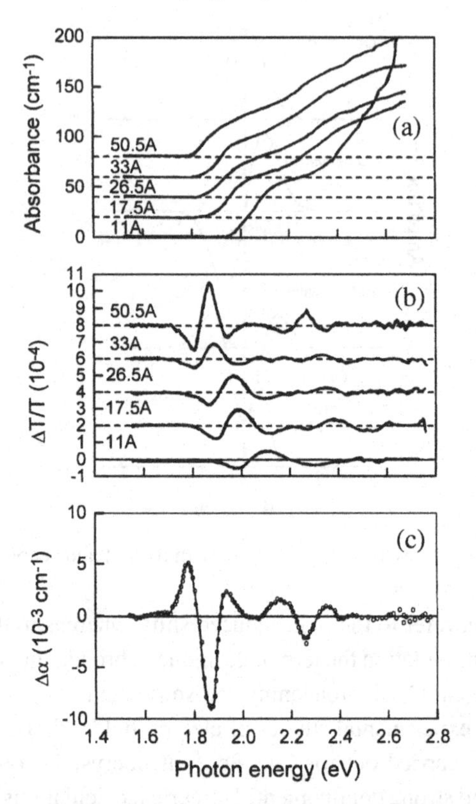

Fig. 5.46. Electroabsorption data for CdS_xSe_{1-x} in a glass matrix (Nomura and Kobayashi 1990a). Panel (a) shows the absorption spectra of a number of samples differing in the mean radius of nanocrystals. Panel (b) presents the normalized transmittance change spectra. Panel (c) shows the absorbance change of 50.5 Å microcrystallites (open circles) along with the fit (solid line).

with all $n'l \pm 1$ states. The $2l + 1$ degeneracy due to the quantum number m is partially broken and gives rise to $l + 1$ nondegenerate levels. As a result of the mixing of unperturbed states new transitions appear, for example, between the perturbed $1S$ and $1P$ states, which gives rise to additional absorption peaks. On the other hand, as the overlap of the electron and hole wave function is less perfect, the oscillator strength of the former resonance decreases. Thus, redistribution of the oscillator strength in the case of a negligible shift manifests itself as broadening. It should be noted that Cotter and co-workers examined electroabsorption of CdS_xSe_{1-x} nanocrystals with an external field up to 10^6 V/cm, that is, higher than the ionization field of the parent bulk crystal (see Table 5.2). For $a \leq a_B$ differential optical absorption spectra have a shape similar to that presented in Fig. 5.46(b). The quadratic dependence of the induced absorption change on the electric field applied becomes a linear dependence at the highest field. As the spectral shape of field-induced modulation was found to be the same throughout the whole range of ε applied, the results have been interpreted in terms of the quantum-confined Stark effect. Plausible arguments based on a dipole model have been put forward to show that this linear dependence at high field strengths may be related to saturable distortions in the wave functions because of confinement (Cotter, Girdlestone, and Moulding 1991).

For a larger than exciton Bohr radius at higher values of applied field the differential absorption spectrum exhibits oscillating shape with field-dependent extrema that cannot be assigned to the intrinsic resonances. This mode is similar to the Franz-Keldysh effect in bulk crystals, which occurs when $\varepsilon \gg \varepsilon_I$ (Gribkovskii 1975). This effect corresponds to unbound electron-hole pairs for which interband tunneling becomes possible in the presence of a high electric field.

To summarize, semiconductor nanocrystals exhibit noticeable absorption modulation in the presence of an external electric field. Though in certain cases electromodulation spectroscopy provides evaluation of some absorption features that are masked by thermal and inhomogeneous broadening, the use of this technique for identification of individual absorption bands is complicated because of the overlap of neighboring bands. Furthermore, unlike resonant absorption saturation or fluorescence line-narrowing at the red-edge excitation, electromodulation is not selective with respect to inhomogeneous broadening factors. Nevertheless, the application potential of the effect is rather promising and will be the subject of further studies. In situations of photoionization of crystallites, photo-induced surface charging or charge redistribution in the local environment or at the guest-host interface, a number of spectrally selective features may be distinguished. A manifestation of the local electric field effects has been mentioned already in connection with single dot spectroscopy (Section 5.5). Further selective effects of local electric fields will be the subject of Chapter 7.

5.9 Electroluminescence

Electroluminescence of nanocrystals is of great practical importance because it may be used for various device applications. In spite of the fact that luminescence of nanocrystals under optical excitation has been extensively studied, emission under excitation due to electric current has become the subject of investigation very recently. The main problem in obtaining efficient electroluminescence is to get a dense ensemble of nanocrystals in a proper conducting and transparent environment providing charge carrier injection and migration. To date CdSe and CdS nanocrystals embedded in a conducting polymer and porous silicon films consisting of silicon nanocrystals connected via silicon

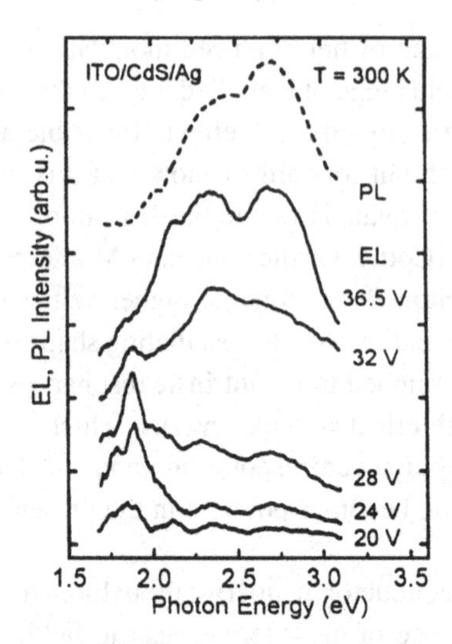

Fig. 5.47. Room temperature photoluminescence (dashed curve) and electroluminescence (solid curves) spectra of a thin film of close-packed CdS nanocrystals sandwiched between ITO and Ag contacts (Artemyev, Sperling and Woggon 1997). Numbers at the curves indicate the applied voltage.

oxide interface have been reported to show luminescence under excitation by electric current.

Colvin et al. (1994) and Dabbousi et al. (1995) have reported on electroluminescence from CdSe nanocrystals. A nearly monodisperse assembly of nanocrystals was embedded in a matrix of polyvinylcarbazole and an oxadiazole derivative and sandwiched between ITO and Al electrodes (Dabbousi et al. 1995). The active layer containing nanocrystals was 100 nm thick. Electroluminescence was observed both at room and at cryogenic temperatures featuring an intrinsic emission spectrum similar to that observed with the optical excitation. Typical voltage applied to the structure was about 20 V.

Artemyev et al. obtained electroluminescence from a thin film structure containing close-packed CdS particles of size 1–2 nm (Artemyev, Sperling and Woggon 1997). However, both photo- and electroluminescence spectra are related to the defects rather than to the intrinsic states. While the first absorption maximum is at $\hbar\omega = 3.75$ eV, the broad-band emission spectrum ranges from 2 to 3 eV (Fig. 5.47).

Electroluminescence from silicon-based nanostructures is of importance because it may potentially be used to fabricate light-emitting devices within the framework of silicon microelectronics. Electroluminescence of porous silicon embedded in an electrolyte has been observed and investigated by a number of groups (Kooij, Despo, and Kelly 1995 and references therein).

Electroluminescence from solid-state porous silicon structures has been obtained using two approaches. In the first case, porous silicon was encapsulated in aluminium and in aluminium oxide to get a Schottky diode, that is, a metal-semiconductor junction (Lazarouk et al. 1996). White light, visible in normal daylight, is emitted when a reverse bias is applied to the device promoting the junction breakdown. Excellent stability is reported for the device when tested for a month. In the second case, strongly oxidized porous silicon film was used and the possibility of integration of a silicon based light-emitting diode into a microelectronic circuit was demonstrated (Hirschman et al. 1996).

Generally, electroluminescence is a complex phenomenon that brings up a number of complicated aspects related to the charge transfer through a single nanocrystal and to carrier migration in an assembly of nanocrystals. In charge transfer processes the concept of capacitance becomes important (Macucci, Hess and Iafrate 1993). Tunneling of a single electron into a quantum dot increases the electrostatic energy of the system dot/contacts by an amount of $e^2/2C$, where C is the capacitance of the system. At low temperatures and small voltage when kT, $eU \ll e^2/2C$ holds, the resistance of the nanocrystal appears to be infinitely high. This phenomenon is known as the *Coulomb blockade* of single-electron tunneling. It is actively investigated for both metal and semiconductor nanostructures (Harmans 1992; Likharev and Claeson 1992; Geerligs, Harmans, and Kouwenhoven 1993 and references therein). A uniform increase in applied voltage results in an increase of the threshold current, which was the reason to treat such a system as a "single-electron transistor." Steady growth of the applied voltage gives rise to oscillations in the conductivity because of the sequential transport of electrons through the dot (Beenacker 1991; Wang, Zhang and Bishop 1994; Sakamoto et al. 1994; Tsu, Li and Nicollian 1994). The discreteness of the energies resulting from the confinement modifies the process significantly whenever the energy separation ΔE is $\Delta E \geq e^2/2C$.

Size and shape distribution inherent in most of the quantum-dot structures was found to result in a spectacular electroluminescence feature. Bsiesy and Vial (1996) observed a selective voltage-induced photoluminescence quenching and voltage-tunable electroluminescence of porous silicon. These effects are the manifestation of selective electro-excitation of the inhomogeneously broadened optical transition, resulting from the voltage distribution between the substrate and the silicon nanocrystals.

5.10 Doped nanocrystals

Most of studies performed on semiconductor nanocrystals are aimed at developing perfect crystallites and establishing their intrinsic features. However, purposeful utilization of resonant impurity transitions in nanocrystals is of great practical interest. Crystals and glasses doped with transition metals and rare-earth elements (Mn, Ti, Nd, Er) are widely used as phosphors and laser media. Therefore, it is reasonable to examine the radiative properties of these impurities when the latter are embedded in nanocrystals.

The first experiments on manganese-doped ZnS nanocrystals revealed a drastic change in spontaneous decay rate of Mn^{2+} ions embedded in nanocrystals as compared with ZnS monocrystals (Bhargava et al. 1994). In a bulk crystal after interband excitation an electron is captured by an ion in a period of microseconds with subsequent radiative decay, the lifetime being about 2 msec. In a nanocrystal electron capture occurs on a subnanosecond scale, and the radiative lifetime of Mn^{2+} ions was found to be no larger than a few nanoseconds. The enhancement of the radiative decay rate of excited Mn^{2+} ions in nanocrystals is attributed to the enhanced overlap of the s-p host electron wave function with the d-f electron wave function of the impurity. This sp-fd mixing promotes both enhanced energy transfer to the impurity and fast radiative decay. Hybridization of electron states is a direct consequence of spatial confinement. It occurs when nanocrystal size is less than the exciton Bohr radius, which is $a_B = 2.5$ nm for zinc sulfide. Further experiments on ZnS nanocrystals doped with Tb^{3+} ions have shown qualitatively similar results (Bhargava 1996). It is important that a rather fast capture process prevents both radiative and nonradiative energy relaxation via unwanted competitive channels such as surface states and other impurities. Therefore, the luminescence efficiency is rather high. In ZnS:Mn nanocrystals the quantum yield was found to be about 20 percent, showing a monotonic increase with decreasing size. Thus, doped nanocrystals are believed to offer a set of commercial phosphors and laser media in the near future.

6

Resonant optical nonlinearities and related many-body effects

6.1 Specific features of many-body effects in nanocrystals

In bulk semiconductors population-induced optical nonlinearities are described in terms of exciton gas, electron-hole plasma, and electron-hole liquid (Haug and Koch 1990; Klingshirn 1995). It is possible to distinguish three different excitation regimes corresponding to different concentrations of electron-hole pairs.[1]

For weak excitation, excitons move freely along the crystal as Bloch waves. Their concentration is so low that during their lifetimes they neither encounter nor interact with each other. In this case a consideration in terms of the gas of bosons is relevant. At finite temperature excitons are in equilibrium with free electrons and holes.

At the intermediate excitation level, the excitons still behave like quasi-particles; however, they begin to interact with each other. The interaction implies elastic scattering of excitons by excitons or by free charge carriers. Scattering processes result in intensity-dependent broadening of the exciton band, depending on the concentration of quasi-particles, as a result of decreasing the phase relaxation time [Fig. 6.1(a)].

Another mechanism leading to various nonlinear phenomena at the intermediate excitation level is related to the formation of *excitonic molecules* (*biexcitons*). These are bound states of two electron-hole pairs similar to the hydrogen molecule. An additional excitation-induced absorption band peaking at the energy

$$\hbar\omega = E_{\text{biex}} - E_{ex} \equiv E_g^0 - Ry^* - \delta E_{\text{biex}} \tag{6.1}$$

[1] Hereafter we ignore polaritonic aspects in nonlinear-optical processes in the bulk semiconductors, because they are not essential in the context of nanocrystals. A comprehensive review of polariton-related nonlinearities can be found in the recent book by Haug et al. (Haug, Ivanov, and Keldysh 1997).

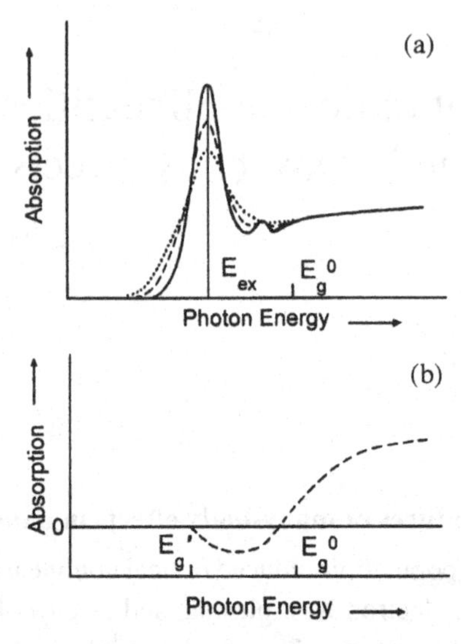

Fig. 6.1. Modification of the absorption spectrum of a bulk semiconductor in the case of (a) intermediate and (b) strong excitation. During intermediate excitation, exciton-exciton interactions result in the intensity-dependent broadening of the exciton band, the maximum of the band being independent of the intensity. At a high excitation level electron-hole interaction in plasma results in a significant band gap shrinkage and in optical gain.

appears. Here E_{biex} is the energy necessary for creation of one biexciton, $E_{ex} = E_g^0 - Ry^*$ is the energy necessary for creation of an exciton, and δE_{biex} is the biexciton binding energy. Biexcitons are created by means of the two-photon absorption

$$\hbar\omega_1 + \hbar\omega_2 = E_{\text{biex}}. \tag{6.2}$$

In the strong excitation regime, the concentration of nonequilibrium electron-hole pairs is so high that the average distance between them is comparable to the exciton Bohr radius. In this case the concept of an exciton as an isolated quasi-particle is not appropriate, and a new collective phase arises called *electron-hole plasma*. The formation of the plasma qualitatively changes the absorption spectrum. Interactions of electrons and holes in the plasma result in a uniform decrease in the band gap E_g' with increasing electron-hole pair concentration n_p. However, the energy of exciton resonance remains the same because band gap shrinkage is compensated by the decrease in the exciton binding energy resulting from screening of the Coulomb interaction. At some pair concentration n_p^M the difference between the "free" electron-hole pair and the electron-hole pair bound in an exciton vanishes, that is, exciton binding energy equals zero. In other words, a nonequilibrium dielectric-metal *phase transition* occurs (*Mott transition*). In the interval

$$E_g' < \hbar\omega < \Delta E_F$$

where ΔE_F is the difference of electron and hole quasi-Fermi levels, the absorption coefficient is negative, that is, optical gain occurs.

Unlike bulk crystals, in semiconductor nanocrystals only a finite number of electron-hole pairs can be created within a given crystallite. Therefore, the concepts of exciton gas and electron-hole plasma relevant to optical nonlinearities in the bulk crystals should be replaced by the correct consideration of interactions within the finite number of charged quasi-particles. In crystallites the concentration of excitons is attainable for which the average distance is close to the Bohr radius. Thus, nanocrystals can be considered as a microlaboratory for studies of many-body effects in a system of fermions. For larger crystallites relevant to the weak confinement range, exciton-exciton interactions in a system of hydrogenlike mechanical excitons (contrary to polaritons) can be examined. Evidently, when the average distance between excitons is comparable to the Bohr radius, excitons can no longer be treated as ideal noninteracting bosons. In nanocrystals exciton-exciton interactions are influenced neither by electron-hole plasma effects nor by polaritonic peculiarities. Exciton-exciton interactions should manifest themselves in nontrivial nonlinear optical effects, which have no analog either in atomic and molecular systems or in the bulk crystals.

In the cases of intermediate and strong confinement, the difference between free and bound electron-hole pairs vanishes and nonlinear optical properties are expected to resemble those of atomic and molecular systems to a large extent. The exception is the two-electron-hole-pair state, biexciton, which determines the nanocrystal excited-state absorption.

The nonlinear optical properties of semiconductor nanocrystals have been known for more than 30 years. Since the first publication on absorption saturation in commercial semiconductor-doped glasses (Bret and Gires 1964) this effect has been extensively used in ruby Q-switched lasers. Nowadays, the concept of quantum confinement, the intriguing issues related to many-body effects, and the clear quantum-mechanical formulation of the relevant problems in terms of particles-in-a-box models have led to extensive experimental and theoretical studies of nonlinear optical effects in nanocrystals (Woggon 1996).

6.2 Exciton-exciton interactions in large quantum dots

In the 1980s several authors predicted interesting nonlinear optical phenomena due to exciton-exciton interactions in the case of the weak confinement (Banyai and Koch 1986; Takagahara 1987; Hanamura 1988b). A number of experimental works were reported later using CuCl and CuBr nanocrystals possessing strong excitonic absorption and small exciton Bohr radius (see Table 1.2), providing the weak confinement limit to be examined. In the first experiments, the single-beam experimental performance provided a detection of nonlinear absorption, while the spectral shape of an excited sample absorption has not been revealed (Woggon and Henneberger 1988; Masumoto, Wamura and Iwaki 1989). Several groups have reported on nonlinear pump-and-probe experiments using independent excitation and probe light sources (Gilliot et al. 1989; Zimin et al. 1990; Wamura et al. 1991). In these experiments a clear photo-induced blue shift of the exciton band has been observed. The correct quantitative description of the effect has been proposed by Bellegie and Banyai (Bellegie and Banyai 1991; 1993). In further studies the size-dependent biexciton binding has been evaluated experimentally (Levy et al. 1991; Woggon, Wind et al. 1994; Masumoto, Okamoto, and Katayanagi 1994) and theoretically (Nair and Takagahara 1996). In what follows we consider these theoretical and experimental findings in more detail.

To describe the optical properties of an excited nanocrystal the concept of the many-particle state is of principal importance. Strictly speaking, to evaluate the absorption spectrum of a crystallite containing $(n-1)$ electron-hole pairs, it is necessary to solve the Schroedinger equation for $2n$ particles with the Hamiltonian

$$H(n) = -\sum_{i=1}^{n}\left(\frac{\hbar^2}{2m_e}\nabla_{e_i}^2 + \frac{\hbar^2}{2m_h}\nabla_{h_i}^2\right) + \frac{e^2}{2\varepsilon}\sum_{i,j=1;i\neq j}^{n}\left(\frac{1}{|\mathbf{r}_{e_i}-\mathbf{r}_{e_j}|}+\frac{1}{|\mathbf{r}_{h_i}-\mathbf{r}_{h_j}|}\right) - \frac{e^2}{\varepsilon}\sum_{i,j=1}^{n}\frac{1}{|\mathbf{r}_{e_i}-\mathbf{r}_{h_j}|} \tag{6.3}$$

where the first term describes the kinetic energy of electrons and holes, and the second and third terms describe, respectively, the Coulomb repulsive and attraction energy (see Section 2.1 for notations). The intensity-dependent absorption results from the absorption spectrum dependence on the number of excitons. The phenomenon has been treated with additional nontrivial aspects taken into account (Bellegie and Banyai 1991). These are exchange annihilation and exchange polarization. Exchange annihilation implies recombination of an electron and a hole belonging to the different excitons [Fig. 6.2(a)]. Exchange polarization accounts for processes like

$$\hbar\omega + E_{ex} = E'_{ex} + E''_{ex}, \tag{6.4}$$

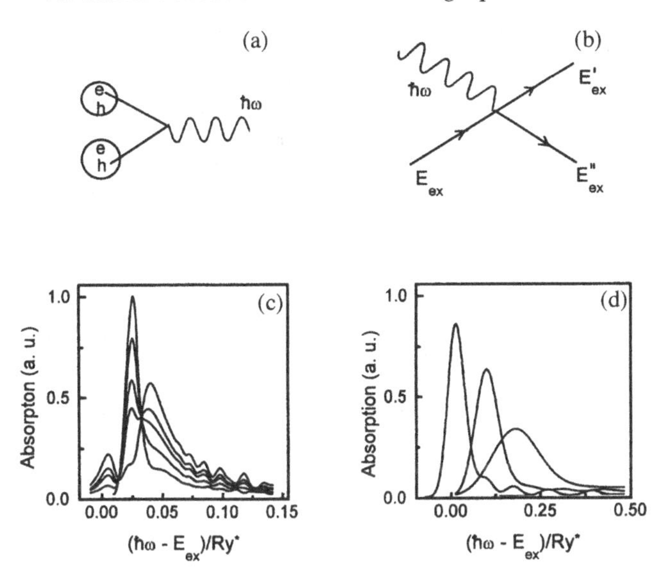

Fig. 6.2. Exciton-exciton interactions in larger quantum dots (Bellegie and Banyai 1991). (a) Exchange annihilation. (b) Exchange polarization. (c) Calculated nonlinear absorption with increasing occupation of one-exciton state in an ensemble of identical quantum dots with $a = 10\,a_B$. The fractions of the occupied one-exciton states are 0, 1/3, 1/2, 3/4, 1. Temperature is 10 K and homogeneous width is $\Gamma = 0.005Ry^*$. (d) Calculated absorption (arbitrary units) in a macrocanonically distributed quantum dots ($a = 10a_B$) with average number of excitons $\langle n \rangle = 0, 5$, and 10 at temperatures of $T = 0$, 100, and 200 K and widths $\Gamma = 0.03Ry^*$, $0.045Ry^*$, and $0.09Ry^*$, respectively.

that is, absorption of a photon with energy $\hbar\omega$ by a nanocrystal containing an exciton with energy E_{ex} results not only in the creation of the second exciton with energy E''_{ex} but may change the energy of the first exciton as well [Fig. 6.2(b)]. The results of computational analysis for the two model cases are presented in Fig. 6.2(c,d). Figure 6.2(c) corresponds to a situation in which every nanocrystallite in an ensemble under consideration contains either one exciton or no exciton at all. This is the case of intermediate excitation intensities. As the mean occupation number $\langle n \rangle$ varies from 0 to 1, the absorption band relevant to the unexcited nanocrystals ($n = 0$) bleaches, and simultaneously the new band develops corresponding to $n = 1$. The photo-induced band shows a pronounced shift towards higher energies by ΔE_{int}, the energy of interaction between two excitons confined within the same quantum box. Figure 6.2(d) shows absorption spectra of the ensembles of identical nanocrystals with the same number of excitons in every nanocrystal. Again, the increasing number of excitons results in a photo-induced blue shift of the absorption band. In these calculations the broadening of the exciton band has been accounted for

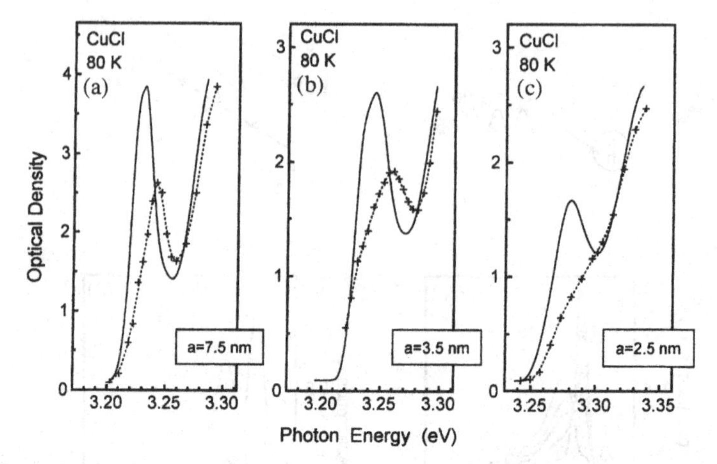

Fig. 6.3. Photo-induced blue shift of the Z_3 exciton band in CuCl-doped glass due to exciton-exciton interactions (Zimin, Gaponenko et al. 1990). Solid lines show absorption spectra of unexcited samples; crosses show absorption spectra of the samples excited by a dye laser radiation resonant with the maximum of the band for every sample; dashed lines are the guides for the eye. Temperature is 80 K, excitation intensity is about 10^7 W/cm^2, pulse duration is 15 ns, and the mean radius of crystallites is (a) 7.5 nm, (b) 3.5 nm, and (c) 2.5 nm.

phenomenologically, without microscopic consideration of intensity-dependent exciton decay and dephasing rates. With the growing number of excitons the temperature was supposed to increase resulting in the additional broadening of the absorption band.

Experimental data on photo-induced blue shift of the exciton band in the weak confinement limit are in a good agreement with the theory (Fig. 6.3). Under resonant excitation by a strong nanosecond laser pulse and simultaneous probing by another weak laser radiation, the Z_3 exciton band in CuCl nanocrystals exhibits a pronounced blue shift and broadening. As the absorption spectrum of the real sample possesses inhomogeneous broadening the observed absorption spectra do not strictly reproduce the computed ones, but the general behavior for both cases is the same. For larger nanocrystals [Fig. 6.3(a)] inhomogeneous broadening is less pronounced, and the exciton band seams to shift under excitation as a whole. For the intermediate size range [Fig. 6.3(b)] the inhomogeneous structure of the band is well pronounced. For the smaller size [Fig. 6.3(c)] the exciton band completely smears under conditions of strong excitation, because different subbands of the inhomogeneously broadened band experience different blue shifts. Intensity dependence of the optical density in the exciton band maximum examined for various mean sizes of crystallites shows valuable enhancement of the nonlinear response with decreasing size,

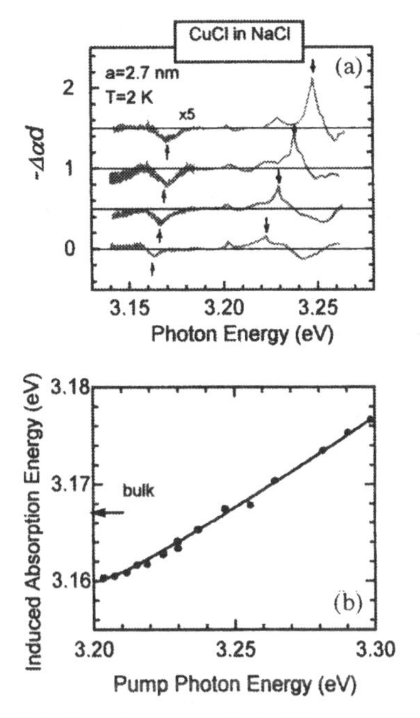

Fig. 6.4. Modification of the biexciton state in CuCl nanocrystals in NaCl (Masumoto et al. 1994). Panel (a) shows a set of the nonlinear differential absorption spectra recorded for several excitation wavelengths. The left-hand part of every spectrum contains induced absorption due to a transition from the exciton to the biexciton state. Panel (b) illustrates the size dependence of the biexciton binding energy.

which is attributed to the size-dependent lifetime of excitons and to the size-dependent exciton-exciton interaction.

Under conditions of resonant optical excitation not only does the exciton band exhibit intensity-dependent shift and bleaching but also an additional weak photo-induced absorption band develops at the long-wave side with respect to the excitation resonance [Fig. 6.4(a)]. This absorption is due to formation of the biexciton state, that is, the bound state of the two electron-hole pairs that resembles the ground state of the hydrogen molecule. The position of the biexciton band with respect to the relevant exciton resonance is indicative of the size-dependent biexciton binding energy δE_{biex}. The value of δE_{biex} was found to increase from 42 meV (0.22 Ry^*) at $a = 7.5$ nm to 64 meV (0.33 Ry*) at $a = 3.0$ nm [Fig. 6.4(b)]. Size-dependent biexciton binding has been considered theoretically for the case $a \gg a_B$, and the observed $\delta E_{\text{biex}}(a)$ dependence has been successfully explained (Nair and Takagahara 1996). The probability of a transition from the lowest exciton to the ground biexciton state, which is relevant to the oscillator strength of the induced-absorption band, has been calculated as well. The transition dipole moment was found to increase nearly linearly with size. Nair and Takagahara pointed to the significance of the weakly correlated exciton pair state in a quantum dot corresponding to the weak confinement regime. This four-particle state resembles the antibonding product state of the hydrogen molecule. As the size of the quantum dot is reduced, the two excitons overlap with each other and the antibonding product state acquires a repulsive energy. Its wave function can be expressed as

$$\Psi_{xx}^{--} = (1/\sqrt{2})\left[\varphi_x(\mathbf{r}_{e_1}, \mathbf{r}_{h_1})\varphi_x(\mathbf{r}_{e_2}, \mathbf{r}_{h_2}) - \varphi_x(\mathbf{r}_{e_1}, \mathbf{r}_{h_2})\varphi_x(\mathbf{r}_{e_2}, \mathbf{r}_{h_1})\right], \quad (6.5)$$

were φ_x is the envelope function of the exciton ground state. This state is responsible for the exciton band blue shift at low excitation intensities when a quantum dot ensemble can be described in terms of the average population number of a single exciton state [Fig. 6.2(c)]. The exciton pair bonding product state

$$\Psi_{xx}^{++} = (1/\sqrt{2})\left[\varphi_x(\mathbf{r}_{e_1}, \mathbf{r}_{h_1})\varphi_x(\mathbf{r}_{e_2}, \mathbf{r}_{h_2}) + \varphi_x(\mathbf{r}_{e_1}, \mathbf{r}_{h_2})\varphi_x(\mathbf{r}_{e_2}, \mathbf{r}_{h_1})\right] \quad (6.6)$$

with decreasing size gets more and more mixed with and repelled by the biexciton state.

When the medium reacts to the change in the incident electromagnetic field $\mathbf{E}_i$ *instantaneously*, the polarizability of the medium $\chi = \varepsilon - 1$ can be expanded in a power series in the field

$$\chi\mathbf{E} = -\frac{1}{\varepsilon_0}\mathbf{P}_i = \chi^{(1)}E_i + \chi^{(2)}E_iE_j + \chi^{(3)}E_iE_jE_k + \cdots. \quad (6.7)$$

The first term in this expansion describes the linear response; the second term corresponds to frequency doubling, summation, and subtraction; and the third term describes four-wave mixing and other processes. All these processes are coherent; that is, there is a definite phase relationship between **P** and **E**. If the light field is monochromatic, a similar expansion is also possible for the index of refraction

$$\tilde{n} = \tilde{n}_0 + \tilde{n}_2 I + \cdots. \quad (6.8)$$

One can see that the nonlinear susceptibility $\chi^{(3)}$ is the important parameter of a nonlinear medium. E. Hanamura has predicted the mesoscopic enhancement of $\chi^{(3)}$ in quantum dots in the case of weak confinement due to size-dependent exciton oscillator strength (Hanamura 1988b). The degenerate four-wave mixing experiments confirmed valuable third-order nonlinearity in CuCl and CuBr nanocrystals with the pronounced resonant enhancement for Z_3 and Z_{12} exciton bands (Nakamura et al. 1994). However, the interpretation of these studies still remains somewhat questionable. First, the results of the Hanamura study have become the subject of critical discussion by other authors (Ishihara 1990; Ishihara and Cho 1990; Bellegie and Banyai 1993). Nonlocal light-matter interaction and population-dependent phase relaxation of excitons[2] have been pointed out as the important aspects to be introduced into consideration. The exciton-exciton interactions have been outlined as the only origin of nonlinear susceptibility observed in the experiment. In this connection, it is reasonable to note that in the case of the real rather than virtual excitations in the medium interacting with the electromagnetic field, the power expansion (6.7) is not justified. Indeed, in this case the response is essentially noninstantaneous and the absorption and refraction is controlled by the population dynamics rather than by the light intensity. In the case of weak excitation, which is usually used to evaluate nonlinear susceptibility, the interplay between biexciton and exciton states determines the nonlinear response of an ensemble of nanocrystals pumped by the resonant laser radiation.

The exciton-biexciton dynamics of larger CuCl and CuBr nanocrystals has been the subject of extensive research (Levy et al. 1991; Woggon et al. 1994; Edamatsu et al. 1995; Ikezawa and Masumoto 1996; Yano et al. 1996; 1997). The increase of exciton luminescence was found to correlate with the decay of biexciton luminescence, the characteristic lifetimes being about 1 ns for excitons and about 100 ps for biexcitons. The three-level scheme including the nanocrystal ground state, the exciton state, and the biexciton state, and the stochastic treatment of exciton-biexciton dynamics in terms of the Poisson distribution of excitons, provided a reasonable explanation of the nonlinear absorption and emission. At a lower temperature ($T \approx 2$ K) a dipole transition between a bound biexciton state and a bound exciton state was found to occur. Note that in the case of CuCl nanocrystals in NaCl the saturation intensity was found to be as low as 2 kW/cm^2.

6.3 Genuine absorption saturation in small quantum dots

In this section we consider a bright manifestation of absorption saturation in semiconductor-doped glasses containing II-VI nanocrystals of size a comparable with or somewhat less than the exciton Bohr radius a_B. Absorption saturation in semiconductor crystallites with $\bar{a} \approx a_B$ was studied intensively in the 1960s in connection with the use of commercial glasses colored with CdS_xSe_{1-x} crystallites as passive shutters in ruby lasers during the initial stages of quantum electronics (Bret and Gires 1964; Schmackpfeffer and Weber 1967; Bespalov and Kubarev 1967; Bonch-Bruevich, Razumova, and Rubanova 1967; Lisitsa

[2]Population-dependent phase relaxation time was evaluated in the experiments on nondegenerate four-wave mixing (Woggon and Portune 1995).

et al. 1967). It is obvious that at that time the sizes of crystallites were not measured, and the fact that commercial glasses correspond to the case $\bar{a} \approx a_B$ became known from late publications. Since during this period there were no tunable dye lasers, absorption saturation could be studied only with a fixed energy of the exciting photon, which corresponds to the principle harmonic of the ruby laser (694 nm) or to the second harmonic of the neodymium laser (532 nm). A survey of the studies relevant to this period was made by Pilipovich and Kovalev (1975).

In the early 1980s interest in nonlinear absorption in selenocadmium glasses reappeared, since it was shown that they are a convenient model for investigating different manifestations of the quantum-size effects in quasi-zero-dimensional structures. The first experimental studies of absorption saturation in selenocadmium glasses using tunable dye lasers (Gaponenko et al. 1982; Gaponenko, Zimin and Nikeenko 1984) showed that, in contrast to single crystals, glasses have an unusually wide spectral interval (0.2–0.3 eV) in which bleaching is observed in single-beam experiments and in pump-and-probe measurements. These studies were followed by extensive experiments (see, e.g., the review by Brus 1991) that revealed the unique features of semiconductor-doped glasses as saturable absorbers.

Absorption saturation in semiconductor-doped glasses is due to population of the one-electron-hole-pair state in an ensemble of nanocrystals. Therefore, the physical mechanism of the phenomenon is clear and in the context of this section the specific features of the effect rather than the saturation itself should be the subject of consideration. Particularly, it is reasonable to discuss the nonlinear properties of semiconductor-doped glasses in terms of the following parameters:

(i) the saturation intensity, I_{sat} if bleaching is described in terms of the simple models discussed in Chapter 4 or an analogous value I^*, which corresponds to a noticeable change in absorption coefficient;
(ii) the recovery time of bleaching;
(iii) the ratio of the saturable absorption coefficient to the nonsaturable background;
(iv) photostability;
(v) spatial diffusion rate of excited species, which is important when a saturable absorber is used in optical processing; and
(vi) spectral range, where nonlinear response is available.

Depending on the specific functions that are desirable in the device containing saturable absorbers, priority can be given to the specific parameters of a nonlinear medium. For example, if a synchronization of a number of optical pulses with different wavelengths is desirable, one needs a saturable medium with broad-band nonlinear response. However, if a saturable absorber is used to provide discrimination of several beams, narrow-band spectral features are necessary. Nevertheless, fast recovery time, high saturable-to-nonsaturable absorption contrast, and good photostability are necessary in all applications.

Glasses doped with semiconductor nanocrystals possess a number of advantageous features as saturable absorbers compared with other media, such as organic dyes, bulk semiconductors, and glasses doped with ions. Organic dyes possess high nonsaturable absorption that is, first of all, due to excited singlet state absorption, which spectrally overlaps the main absorption band and possesses a significant oscillator strength. Furthermore, these media exhibit low photostability because of photodestruction (photobleaching). Finally, high diffusion rate of molecules in a liquid solution is crucial for a large number of applications. Although the latter two drawbacks can be avoided by embedding organic molecules in inorganic glasses, the high nonsaturable losses remain (Gaponenko, Gribkovskii et al. 1993). Bulk semiconductor crystals show high nonsaturable absorption as well due to population induced band-gap shrinkage (see, e.g., Klingshirn 1990; 1995). Moreover, excitonic nonlinearities in bulk crystals manifest themselves at lower temperatures and very often diminish at room temperature. Nonlinear response occurs within a narrow spectral range in the vicinity of the fundamental absorption edge.

Considering glasses doped with ions versus semiconductor-doped glasses, one can see that the latter make it possible to tune the absorption resonance towards a desirable wavelength by means of the quantum-size effect, which is not possible for ions.

Semiconductor-doped glasses exhibit genuine absorption saturation at room temperature in a wide spectral range with low nonsaturable background and fast recovery time. The spectral range in which absorption saturation occurs may be as wide as 100 nm, an increase in transmission being as large as 10^4–10^6 times, while the recovery time is in the subnanosecond range (Zimin, Gaponenko, and Lebed 1988). The typical examples are given in Figs. 6.5 and 6.6, which represent intensity-dependent optical density and time-dependent bleaching spectra of the commercial glass containing CdS_xSe_{1-x} crystallites.

As most of the commercial glasses at room temperature possess rather large homogeneous bandwidth (see Section 5.1), the effect of inhomogeneous broadening can be neglected in the context of this section. The intensity-dependent absorption coefficient may be written as

$$\alpha(I) = \frac{\alpha_{\text{sat}}^0}{1 + I/I_{\text{sat}}} + \alpha_{\text{non}}, \tag{6.9}$$

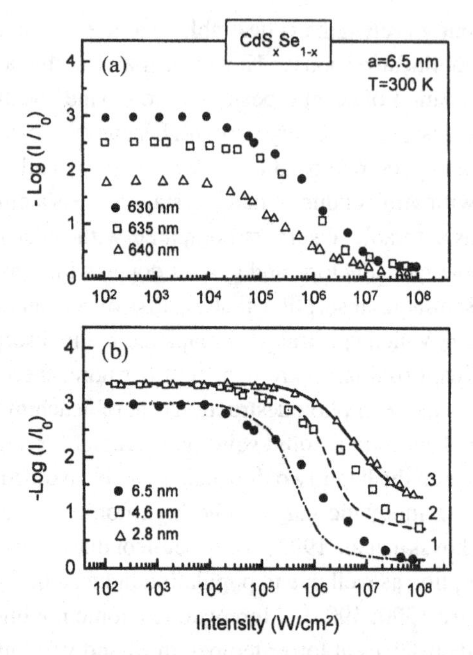

Fig. 6.5. Absorption saturation in commercial glasses doped with CdS_xSe_{1-x} nanocrystals at room temperature measured with nanosecond laser pulses (Gaponenko 1996). Panel (a) presents three saturation curves for the sample with $\bar{a} = 6.5$ nm at different wavelengths within the first absorption feature. Panel (b) shows saturation curves for three samples with the different mean size of crystallites, $a = 6.5$ nm (circles), 4.6 nm (squares), and 2.8 nm (triangles). The laser wavelength corresponds to the first absorption maximum of every sample. Dashed curves are the result of fitting for the absorption versus intensity dependence in the form of Eq. (6.9) with the finite sample thickness taken into account. The saturation intensity is $I_s = 0.16$ (1), 0.30 (2), and 1.05 (3) MW/cm^2. The ratio of saturable to nonsaturable absorption coefficients is 0.06 (1), 0.22 (2), and 0.40 (3). The thickness of the samples is 3 mm.

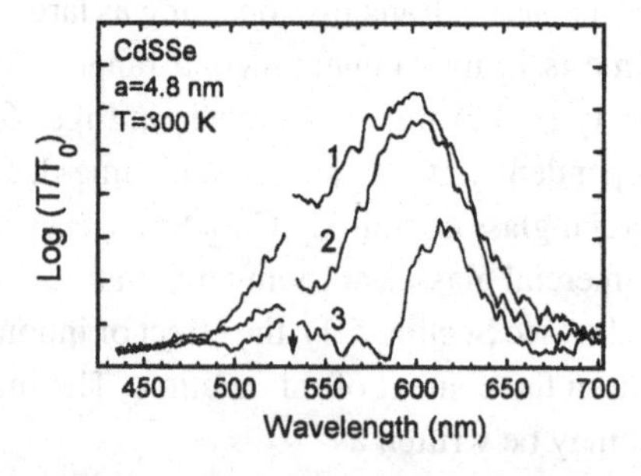

Fig. 6.6. Nonlinear bleaching spectra of commercial glass doped with CdS_xSe_{1-x} nanocrystals recorded at room temperature with excitation by a picosecond pulsed laser ($\lambda = 532$ nm, $\tau = 8$ ps) with different delay times (Gribkovskii et al. 1988). Delay time is (1) 2, (2) 15, and (3) 530 ps. The mean size of crystallites is $\bar{a} = 4.8$ nm.

where the first term describes the absorption saturation in the case of the ideal two-level system and α_{non} accounts for some residual background absorption connected with excited state absorption or with competing absorption channels. In the case of nanocrystals, excited state absorption is due to a transition from the one electron-hole pair state (exciton) to the two electron-hole pair state (biexciton). The saturation intensity I_{sat} reads

$$I_{sat} = \hbar\omega N_{sat}\left(\alpha_{sat}^{0} T_1\right)^{-1}, \tag{6.10}$$

where N_{sat} is the saturation density of electron-hole pairs, α_{sat}^{0} is the absorption coefficient in the saturable absorption channel at $I \to 0$, and T_1 is the lifetime of electron-hole pairs. If excited state absorption is the only reason of nonsaturable losses, the intensity dependence of the absorption coefficient reads

$$\alpha(I) = \frac{\alpha_{sat}^{0}}{1 + I/I_{sat}} + \text{const}\, \frac{\alpha_{sat}^{0}}{1 + I/I_{sat}} I, \tag{6.11}$$

where the second term relevant to the excited state absorption implies that the latter is proportional to the number of crystallites containing one electron-hole pair. One can see that for $I \gg I_{sat}$ the second term tends to a constant value that manifests itself as nonsaturable background absorption.

Systematic studies of absorption saturation in commercial glasses doped with CdS_xSe_{1-x} crystallites of size ranging from 2.5 to 7.0 nm (note that the exciton Bohr radius for CdSe is 5.0 nm) have shown that the lowest saturation intensity and the lowest nonsaturable losses are inherent in the larger nanocrystals within the size range examined. The saturation intensity is about 10^5 W/cm^2, and the $\alpha_{sat}^{0}/\alpha_{non} \geq 10$ (Fig. 6.5). The negligible nonsaturable losses are possibly due to the small oscillator strength of the exciton-biexciton transition and the spectral shift of the absorption band relevant to this transition with respect to the exciton resonance. The bleaching spectrum shows subnanosecond decay time (Fig. 6.6), the lifetime being longer for longer wavelengths, because of the fast intraband relaxation of electron-hole pairs with the subsequent recombination. Pump-and-probe studies of glasses with $\bar{a} \approx a_B$ using tunable picosecond excitation laser and time-resolved probing provided additional evidence that for this size range at room temperature phonon bottleneck is absent, and the relaxation time is small as compared with the recombination time. Under prolonged laser irradiation the recombination time decreases by one or two orders of magnitude to as low as 10^{-11}. This effect is known as laser annealing and will be the subject of Chapter 7. Ultrafast dynamics of absorption saturation has been the subject of many publications to which the reader is referred for further details (Hsu and Kwok 1987; Mitsunaga, Shinojma, and Kubodera 1988; Williams et al. 1988; Peyghambarian et al. 1989; Nakano, Ishida, and Yanagawa 1991;

Dneprovskii et al. 1992; Hunsche et al. 1996). Size dependence of nonlinear response of CdS_xSe_{1-x}-doped glasses has been examined also by means of refractive index studies of the excited samples (Shinojima, Yumoto, and Uesugi 1992; Schanne-Klein et al. 1995), and the decrease in nonlinear response with decreasing size has been outlined.

The intensity dependence of the absorption coefficient cannot be determined unambiguously from nonlinear optical transmission measurements. In the case of the finite sample thickness, attenuation of light intensity propagation along the x-axis obeys the differential Buger's law

$$dI(x) = -\alpha[I(x)]I(x)dx, \tag{6.12}$$

which leads to a relation

$$\int_{I_0}^{I_T} \frac{dI}{\alpha(I)I} = -\int_0^d dx, \tag{6.13}$$

which couples the incident intensity $I_0 \equiv I(0)$, the transmitted intensity $I_T \equiv I(d)$, and the sample thickness d. In the linear case ($\alpha(I) \equiv \alpha_0$) one has

$$I_T = I_0 \exp(-\alpha d)(1 - R)^2 \tag{6.14}$$

where R is the reflection coefficient. In the nonlinear case the $I_T(I_0)$ dependence can be evaluated only numerically. In this connection it is noteworthy that in a number of publications on nonlinear absorption of semiconductor-doped glasses the logarithm of *nonlinear* transmission is identified with the αd value. This is justified only in the limit $\Delta\alpha d \ll 1$, which implies either a very thin sample or very small nonlinearity. Otherwise, to elucidate the intrinsic $\alpha(I)$ dependence by means of $I_T(I_0)$ measurements, one has to deal with the proper inverse problem (6.13). Therefore, the term "optical density" in the case of significant nonlinearity in absorption should be treated explicitly as $\log(I_0/I_T)$ or $\ln(I_0/I_T)$ and by no means as the αd value. Likewise, differential optical density by no means can be treated as $\Delta\alpha d$.

Since the inverse problem cannot be solved explicitly, the measured $I_T(I_0)$ curve is considered usually versus functions derived for some model $\alpha(I)$ dependencies.

Note that the simple formula (6.9) does not describe the absorption saturation of larger crystallites accurately (Fig. 6.5). The $(1 + I/I_{sat})^{-1}$ term implies a monomolecular recombination. The observed deviation of the experimental saturation curve from that calculated according to Eq. (6.9) means that the recombination process is more complicated in case of larger crystallites. In other words, for smaller crystallites the population number of excitons takes either the value 0 or 1, whereas for larger $\bar{a}$ higher occupation numbers should be considered. A deviation from monomolecular recombination

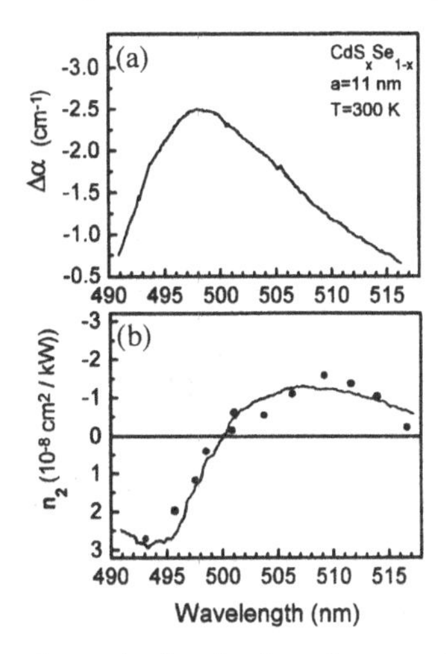

Fig. 6.7. Nonlinear absorption and refraction in a glass sample containing $CdS_{0.9}Se_{0.1}$ crystallites with size equal to 11 nm (adapted from Olbright and Peyghambarian 1986). Data were taken at room temperature. Upper panel shows the measured change of absorption coefficient, $\Delta\alpha$, for two pump intensities: 3 MW/cm^2 and 200 kW/cm^2. Lower panel shows the nonlinear refraction index, $n_2(\lambda)$, computer generated by Kramers-Kronig transformation (curve) and measured by the interferometric technique (dots).

also may be due to randomly distributed trap centers and Auger processes (see Section 5.7).

Absorption saturation in semiconductor-doped glasses is accompanied by noticeable nonlinear refraction in the region of the absorption onset. The change in the refraction index can be evaluated from the absorption saturation data by means of the relation

$$\Delta n(\omega') = \frac{c}{\pi} \int_0^\infty \frac{\alpha(\omega, I_1) - \alpha(\omega, I_0)}{\omega^2 - \omega'^2} d\omega \tag{6.15}$$

based on the Kramers-Kronig relation (Olbright et al. 1987). In Eq. (6.15) $\Delta\alpha = \alpha(\omega, I_1) - \alpha(\omega, I_0)$ is the frequency-dependent change in the absorption coefficient for constant intensities I_1 and I_0. The integration interval in Eq. (6.15) can be restricted to the finite range where $\Delta\alpha \neq 0$. The values of $\Delta\alpha$ outside this range do not contribute to the integral.

The value of nonlinear refraction can be evaluated experimentally using a number of nonlinear interferometric techniques. Figure 6.7 presents the data on nonlinear absorption and refraction of the same glass sample containing large $CdS_{0.9}Se_{0.1}$ crystallites (Olbright and Peyghambarian 1986). Nonlinear refraction was evaluated by means of the Kramers-Kronig transformation and measured using a nonlinear Twyman-Green interferometer. The results show

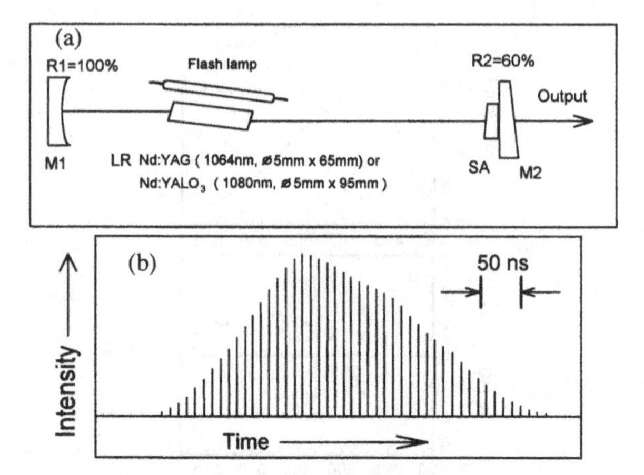

Fig. 6.8. Passive mode-locking in Nd laser using CuInSSe-doped glass (Yumashev et al. 1994). (a) Scheme of the passively mode-locked laser cavity with a saturable absorber, SA. M1 and M2 are resonator dielectric mirrors with reflectivities R1 and R2, respectively. LR is laser rod. (b) Typical pulse train from a passively mode-locked Nd:YAG laser with the CuInSSe-doped glass saturable absorber.

reasonable agreement of the measured and deduced values of nonlinear refractive index n_2 indicating the defocusing nonlinearity at the absorption long-wave tail and the self-focusing nonlinearity at shorter wavelengths.

A number of authors have reported on the pronounced absorption saturation in other nanocrystals. A partial list of other substances includes Si nanocrystals (Klimov, Dneprovskii et al. 1994; Henari et al. 1995; Dneprovskii et al. 1995), Cu_xS nanocrystals (Klimov et al. 1995), and CuInSSe-doped glasses (Yumashev et al. 1994). The latter was shown to be a suitable saturable absorber for passive mode-locking in a neodymium laser operating at 1.06 μm. A sequence of ultrashort pulses with pulse width of about 20 ps was reported. (Fig. 6.8).

6.4 Biexcitons in small quantum dots

In smaller nanocrystals relevant to strong confinement ($a < a_B$), every discrete electron and hole energy state can be populated either by one or by two particles with different spin orientations. If the Coulomb interaction is neglected, a creation of the first electron-hole pair results in absorption saturation, whereas a creation of two pairs with respectively oriented spin corresponds to population inversion, that is, to optical gain. However, Coulomb interaction modifies this simple scenario significantly, giving rise to the concept of the many-particle state. As we agreed to refer to the one-electron-hole-pair state as exciton, it is reasonable to classify the two-electron-hole-pair state as biexciton. Unlike the cases of bulk crystals and large nanocrystals, where the lowest biexciton state corresponds to the hydrogenlike molecule (see Section 6.2), this type of four-particle state cannot exist in small nanocrystals, where even the hydrogenlike atom state cannot occur because $a < a_B$ holds.

To calculate the energy spectrum of biexciton states in smaller nanocrystals one has to deal with the four-particle Schroedinger equation with Hamiltonian similar to that of Eq. (6.3). Moreover, in smaller crystallites, the finite potential barrier should be considered, because in the case of $a < a_B$ the zero-energy of a particle can be comparable with the potential barrier attainable in real structures. Likewise, dielectric confinement should be involved as well. Therefore, the resulting Hamiltonian takes the form

$$H = H_e + H_h + V_{ee} + V_{hh} + V_{eh} + \delta V(\varepsilon_1, \varepsilon_2, \mathbf{r}_e, \mathbf{r}_h) + V^{\mathrm{conf}}, \qquad (6.16)$$

where H_e and H_h describe the electron and hole kinetic energy, respectively; $V_{ee}(V_{hh})$ and V_{eh} are the electron-electron (hole-hole) repulsive and electron-hole attractive potentials, respectively; δV describes the dielectric confinement effects due to different dielectric constants $(\varepsilon_1, \varepsilon_2)$ of semiconductor and the host medium (see Section 2.1); and V^{conf} is the barrier potential. The important concept in this case is the value of binding energy

$$\delta E_{\mathrm{biex}} = E_{\mathrm{biex}} - 2E_{ex}, \qquad (6.17)$$

which measures the difference of the biexciton ground-state energy E_{biex} and the double exciton ground-state energy E_{ex}. In other words, the sign of δE_{biex} indicates whether the dominating interaction between the two electron-hole pairs confined in a quantum box of size less than the exciton Bohr radius is repulsive ($\delta E_{\mathrm{biex}} > 0$) or attractive ($\delta E_{\mathrm{biex}} < 0$). The problem has been extensively examined by various authors using various theoretical approaches and calculation techniques. Applying the perturbation theory to this problem has led to somewhat ambiguous results. In the earlier work $\delta E_{\mathrm{biex}} > 0$ was reported, which was attributed to the repulsive interaction between two holes (Bânyai et al. 1988), whereas other work reported $\delta E_{\mathrm{biex}} < 0$, the absolute δE_{biex} value being larger for smaller dots (Bryant 1990). The calculations performed by means of variational technique yielded δE_{biex} with variable value and sign depending on nanocrystal size and the ratio of the effective masses m_e/m_h (Takagahara 1989). The numerical matrix diagonalization technique developed by Koch and co-workers provided an unambiguous result on the sign of the binding energy as well as on the size dependence $\delta E_{\mathrm{biex}}(a)$ (Hu, Lindberg, and Koch 1990; Hu et al. 1990). The biexciton binding energy δE_{biex} was found to be a negative,

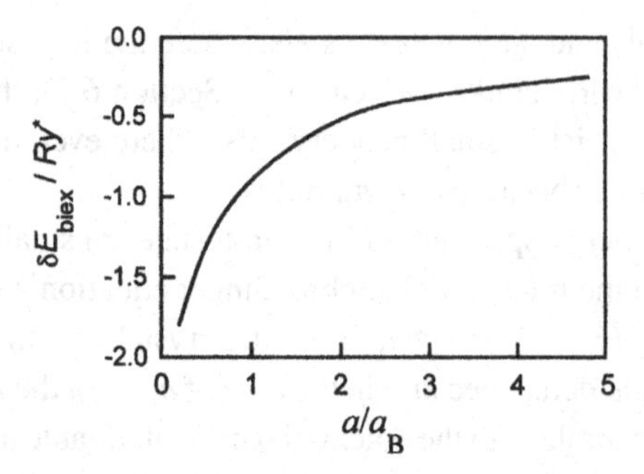

Fig. 6.9. Calculated biexciton binding energy versus nanocrystal size (adapted from Hu, Koch et al. 1990).

unvarying function of the quantum dot radius a. The absolute value of δE_{biex} increases with decreasing a up to values several times larger than the biexciton binding in the bulk crystal (Fig. 6.9).

Similar to biexciton states discussed in Section 6.2 in the context of the hydrogenlike molecule state inherent in the larger crystallites, biexcitons in small quantum dots can be established experimentally by means of pump-and-probe experiments. While at $\hbar\omega_{\text{pump}} = E_{ex}$ the absorption saturates, at $\hbar\omega_{\text{probe}} = \hbar\omega_{\text{pump}} + \delta E_{\text{biex}}$ an induced absorption must be observed due to the creation of the biexciton state when the pump-and-probe quanta are absorbed by the same dot. Complications can be expected, however, if along with the many-band bleaching of the exciton states, a whole spectrum of induced absorption bands appears due to the biexciton ground and excited states as well. Furthermore, to resolve a contribution from the biexciton ground state in the pump-and-probe experiment, the width of the spectral hole from exciton bleaching has to be less than the biexciton binding energy.

To reveal the biexciton state in II-VI nanocrystals with size smaller than the exciton Bohr radius, a number of pump-and-probe studies has been performed (Peyghambarian et al. 1989; Spiegelberg et al. 1991; Uhrig et al. 1991; Klingshirn and Gaponenko 1992). In these measurements an induced absorption at the high-energy rather than at the low-energy side with respect to the pump energy was observed systematically. The separation between the excitation photon energy and the induced absorption peak position is close to 50 meV. This induced absorption band can be attributed to the first excited state of the biexciton in which the two electrons are in the $1s_e$-state, one of the holes is in the $1S_{3/2}$-state, and the other hole is in a higher excited state (most probably, in $2S_{3/2}$-state). The ground biexciton state is masked due to the contribution of the larger nanocrystals to the bleaching spectrum at the long-wave tail of the inhomogeneously broadened absorption spectrum of the ensemble.

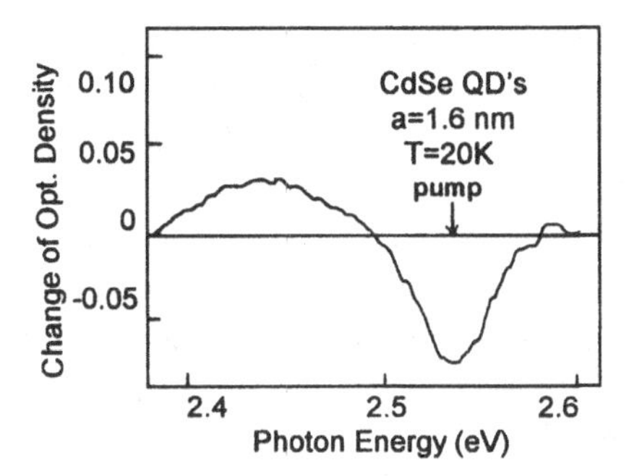

Fig. 6.10. Differential optical density spectrum of CdSe nanocrystals in a glass matrix (Woggon, Gaponenko et al. 1994). Negative signal at the excitation energy is due to the absorption saturation related to the population of the exciton state. Red-shifted induced absorption is due to the transition from the ground exciton to the ground biexciton state.

Two experimental approaches have been used to study the biexciton ground state in II-VI nanocrystals of $a < a_B$. The first is based on the three-beam configuration in the pump-and-probe scheme. To reveal the induced absorption signal from the biexcition ground state on the background of the bleaching from the larger dots, an additional pump beam was applied to saturate the exciton absorption of larger dots at the long-wave tail of absorption onset. A pronounced manifestation of the biexciton ground state was observed at low energy with respect to the pump photon energy, which is indicative of the negative δE_{biex} value in agreement with the theory (Kang et al. 1993). Another possibility to reveal the biexciton ground state is to apply the conventional two-beam pump-and-probe spectroscopy scheme to the perfect nanocrystals with narrow homogeneous linewidth and small size providing the spectral hole width to be less and the distance between the two lowest exciton bands to be larger than δE_{biex} value. Under these conditions the biexciton ground state does not smear in the nonlinear differential absorption spectrum and can be identified unambiguously (Fig. 6.10) (Woggon, Gaponenko et al. 1994). Further manifestations of thc bicxciton ground state have been observed by means of the transient femtosecond pump-and-probe technique (Klimov, Hunsche and Kurz et al. 1994).

In several experiments (Woggon, Gaponenko et al. 1993; 1994; Gaponenko, Woggon et al. 1994) a qualitative difference in the exciton and biexciton lineshapes was observed. While the bleaching at the excitation wavelength is satisfactorily fitted by a Lorentzian, the shape of induced absorption due to biexciton state was found to obey a Gaussian (Fig. 5.18). At least two mechanisms of this difference should be discussed. The first is connected with a possibility that a set of quantum dot parameters that determines the biexciton energy may not coincide with that determining the exciton energy. For example, biexciton binding energy may be sensitive not only to the size but to the shape or to the surface structure of the dot as well. The non-Lorentzian shape of the biexciton absorption band is then determined by the *static* distribution function of the parameters that control δE_{biex} value but do not affect the E_{ex} value. The second mechanism is related to the dynamics of biexciton dephasing process. Non-Lorentzian shape in this case shows that phase relaxation cannot be described in terms of the constant dephasing rate or time $T_2 \propto \Gamma_0^{-1}$. A *dynamic* function should be used instead that accounts for the non-Markovian statistics of the energy fluctuations responsible for biexciton dephasing. Particularly, the Kubo-Anderson jump process leads to the Gausssian lineshape. A consistent numerical calculation of the differential absorption spectrum in pump-and-probe experiments was performed with the different dephasing mechanisms of exciton and biexciton taken into account and a good agreement with the experiment was obtained. A reason of the different dephasing statistics for one- and two-pair states is not clear. One possible mechanism may be suggested in terms of the different interaction with phonons. As soon as a nonnegligible dipole moment is expected for the exciton state due to surface polarization effects (Banyai and Koch 1993), for the biexciton the detectable quadrupole moment may occur, leading to a more complicated interaction with the phonon system.

6.5 Optical gain and lasing

Semiconductor lasers play an important role in modern life and technology. Being the most widespread laser devices, semiconductor injection lasers are the principal components of optical communication circuitry, laser printers, and laser CD players. Quantum confinement effects are purposefully used in quasi-two-dimensional quantum well lasers, providing a number of advantageous features as compared with conventional semiconductor lasers (Gribkovskii 1995; Murray and Koch 1995). The unique properties of semiconductor nanocrystals, such as discrete density of states and controlled energies of the optical transitions due to their quasi-zero dimensionality, offer new possibilities for semiconductor lasers. Along with the desirable emission spectrum and the extremely small size, there is one more essential feature that is peculiar to quantum dot injection lasers as compared with other quantum-size lasers. This is the temperature-independent threshold current.

The threshold current is the main technical parameter of an injection laser. The rapid growth of the threshold current with temperature is inherent in all types of semiconductor lasers, preventing a development of highly effective semiconductor lasers and arrays operating at room temperature. The value of the threshold current can be expressed as

$$J_{th} = edR_{sp}/\eta, \tag{6.18}$$

where d is the active layer thickness, η is the quantum yield, and R_{sp} is the total spontaneous emission rate calculated by the energy integral of the spontaneous emission rate $r_{sp}(E)$. In the majority of cases the threshold current can be calculated only numerically. Its temperature dependence obeys a relation

$$J_{th}(T) = J_{th}^{0} \exp\left(\frac{T}{T_0}\right), \tag{6.19}$$

where the J_{th}^{0} value corresponds to $T = 0°\mathrm{C}$ and T_0 is the characteristic temperature. Arakawa and Sakaki examined the temperature dependence of the J_{th} value for the models of low-dimensional lasers including quantum well (2D), quantum wire (1D), and quantum dot (0D) structures (Arakawa and Sakaki 1982). They found that lower dimensionality results in weakening of the $J_{th}(T)$ dependence; that is, T_0 is larger for lower dimensionalities. Remarkable is the fact that a zero-dimensional laser was found to possess the temperature-independent threshold current; that is, $T_0 = \infty$ for a quantum dot laser. The temperature dependence of J_{th} arises from the thermally induced spread of electrons and holes over a wider energy interval. Therefore, in quantum dot lasers this dependence vanishes because of the delta-functionlike density of states allowing the thermally induced population of the higher states to be inhibited. The pioneering result reported by Arakawa and Sakaki was one of the cornerstones that promoted systematic studies of optics of semiconductor nanocrystals afterwards.

Optical gain and lasing in semiconductor nanocrystals were observed for the first time a decade later by Vandyshev et al. (1991) using CdSe nanocrystals in a glass matrix of size $a \approx a_B$. Lasing was observed at $T = 80$ K at a wavelength of 640 nm under pumping with second-harmonic radiation of a picosecond YAG:Nd laser (532 nm). This report was followed by further experimental studies relevant both to II-VI and I-VII nanocrystals (Faller et al. 1993; Masumoto, Kawamura, and Era 1993; Hu, Koch, and Peyghambarian 1996; Woggon et al. 1996). In Sections 6.2 and 6.4 we saw that the nonlinear response of semiconductor nanocrystals is largely determined by the interplay of the population and depopulation of the exciton and biexciton states. This remains true for the origin of the optical gain. The latter was found to result from the stimulated decay of a biexciton into a photon and exciton.

The variety of the dipole allowed optical transitions from biexciton to exciton states for smaller quantum dots is presented in Fig. 6.11. The qualitative features

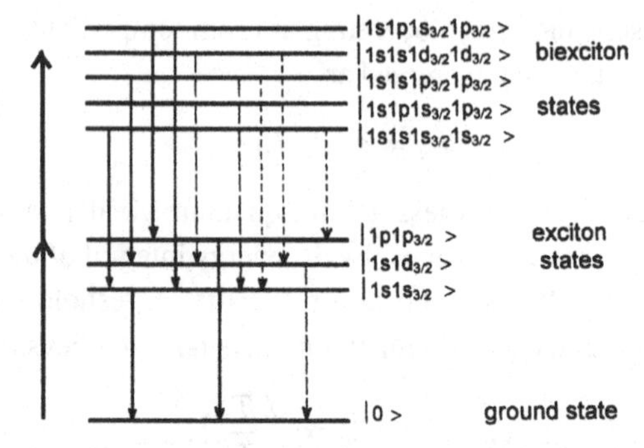

Fig. 6.11. A variety of the optical transitions involving biexciton and exciton states in a nanocrystal of size less than the exciton Bohr radius (Hu et al. 1996). The upward arrows indicate the optical pump process and solid downward arrows indicate the stimulated emission, which is dipole allowed without the Coulomb interaction or valence band mixing effect. The dashed arrows show the stimulated recombination induced by the valence band mixing. The dotted arrows indicate the transitions induced by the Coulomb interaction.

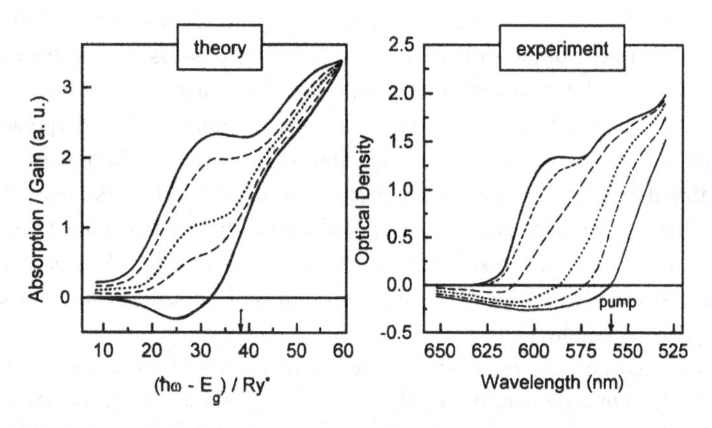

Fig. 6.12. Computed (left) and observed (right) optical absorption spectra of CdSe nanocrystals at various population of exciton and biexciton states (adapted from Hu et al. 1996). Upper curves in both panels are the linear absorption spectra. The other curves correspond to successively growing population and excitation intensity, from top to bottom. The mean nanocrystal radius is $\bar{a} = 2.5$ nm, the Gaussian size distribution in the calculation was given to be $\Delta a = 0.1\,\bar{a}$.

of the expected gain spectrum are as follows. The gain can occur in a rather broad spectral range including photon energies far below the absorption onset. The lower edge of the red-shifted emission is determined by a transition from the lower biexciton state $|1s1s1S_{3/2}1S_{3/2}\rangle$ to the higher exciton state $|1p1P_{3/2}\rangle$, which is possible because of the Coulomb interaction. Figure 6.12 presents the computed and the measured absorption spectra of CdSe nanocrystals with the

Gaussian size distribution taken into account (Hu et al. 1996). The build-up of the gain spectrum under successively growing pump intensity observed in the experiments is qualitatively in good agreement with the computational result. Thus, the phenomenon of optical gain in nanocrystals can actually be attributed to the biexciton-to-exciton recombination.

The most important prospects of quantum-dot lasers are expected for electrically pumped devices. Currently, breakthroughs towards quantum dot injection lasers are occurring. III-V nanocrystals developed on a crystal substrate by means of the self-organized growth in strained heterostructures are considered as the most promising active medium for this application. The first successful experiments on electrically pumped quantum dot lasers have been reported recently (Fafard et al. 1996; Alferov in press). In the experiments by Fafard et al. self-assembled nanocrystals of highly strained InAlAs have been grown by molecular beam epitaxy on a GaAs substrate. Carriers injected electrically from the doped regions of a separate confinement heterostructure thermalized efficiently into the zero-dimensional quantum dot states. Lasing has been observed at $\sim$707 nm at 77 K. These findings may result in the practical realization of a new class of effective commercial injection quantum dot lasers in a few years.

6.6 Two-photon absorption

The selection rules for one-photon and two-photon transitions are essentially different. The two-photon process generates a state with total angular momentum zero or two, whereas one-photon transitions correspond to an angular momentum transfer of one (Banyai and Koch 1993). Therefore, one-photon and two-photon spectroscopies access completely separate manifolds of transitions, and two-photon experiments can be considered as a useful complementary technique to probe exciton states that cannot be examined otherwise. Within the framework of the simple particle-in-a-box problem in a case of a spherical box smaller than the exciton Bohr radius, one-photon absorption creates $1s_e1s_h$-,$1p_e1p_h$-, ... exciton states, whereas the two-photon process results in $1s_e1p_h$-, $1p_e1s_h$-states. Taking into account the valence-band mixing, the one-photon process corresponds to $1s_e1S_{3/2}$, $1s_e2S_{3/2}$, $1s_e1S_{1/2}$, and $1s_e2S_{1/2}$ excitons (see Section 5.2), whereas the two-photon absorption is relevant to $1s_e1P_{3/2}$, $1s_e1P_{5/2}$, $1s_e2P_{5/2}$, and $1p_e1S_{3/2}$ states, and so forth. These states have been elucidated for II-VI nanocrystals by means of the two-photon absorption spectroscopy (CdS nanocrystals in glass) (Kang et al. 1992) and two-photon fluorescence excitation spectroscopy (CdSe nanocrystals in glass and in organics and CdS nanocrystals in organics) (Tommasi, Lepore, and Catalano 1992;

1993; Blanton et al. 1996; Schmidt et al. 1996). These studies confirmed the predictions in terms of the different selection rules for the two-photon absorption and provided evidence of excited exciton states that are not pronounced in one-photon spectroscopy. Additionally, two-photon scanning microscopy setup promoted evaluation of the single-dot contribution to the total fluorescence spectrum (Blanton, Hines, Schmidt and Guyot-Sionnest 1996). This became possible because of the I^2-dependence of the cross section of two-photon absorption. This nonlinearity provides automatic volume selection when the laser beam is focused into a sample.

The nonlinearity arising due to the two-photon absorption can be described in terms of nonlinear susceptibility $\chi^{(3)}$, which can be monitored by means of nonlinear refraction. Experiments performed on GaAs and CdS_xSe_{1-x} nanocrystals at the Nd:YAG-laser wavelength (1.06 μm) revealed strong defocusing nonlinearity that is considerably stronger than that of the relevant bulk crystals (Justus et al. 1992; Cotter, Burt, and Manning 1992). Moreover, the ratio $|Re\chi^{(3)}/Im\chi^{(3)}|$, which is a vital parameter for nonlinear optical devices, was found to be enhanced in nanocrystals as well. These findings may have far-reaching consequences for practical devices such as ultrafast optical switches.

6.7 Optical bistability and pulsations

Optical bistability implies a hysteresislike dependence of output versus input light intensity. A many-valued output-input response function arises due to the complicated interplay of optical nonlinearity and feedback. Most of the schemes possessing optical bistability are based on nonlinear refractive or saturable media and an external Fabry-Perot or ring resonator providing optical feedback [Fig. 6.13(a)]. This type of optical bistability was predicted in 1969 (Szoke et al. 1969) and has become a subject of extensive research (Gibbs 1985).

Another type of optically bistable element does not include any external feedback source but needs specifically strong induced absorption nonlinearity. For this type of optical bistability to occur, the absorption coefficient of the medium should be an increasing function of the number of absorbed photons rather than light intensity. This means that nonlinearity with the instantaneous response in the form of Eq. (6.7) will never result in bistable response independently of the absolute values of nonlinear susceptibilities. Bistability becomes possible only if the real excitations with finite lifetime are created due to light-matter interaction. This type of bistability possesses a clockwise hysteresis response function [Fig. 6.13(b)]. Intrinsic resonatorless bistability was predicted in 1982 (Kochelap et al. 1982) on the basis of the population-induced band gap shrinkage in bulk semiconductors. This type of bistability is

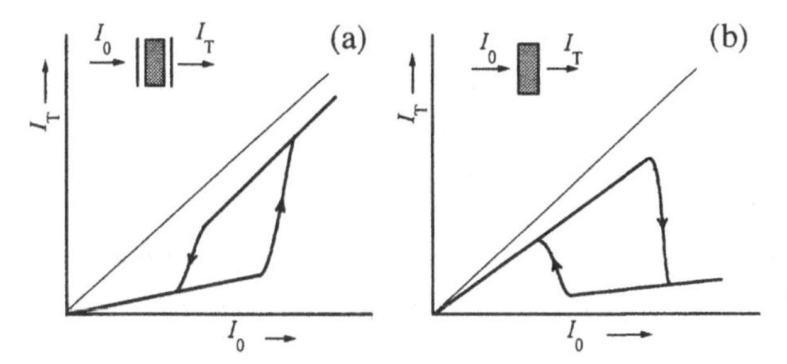

Fig. 6.13. Transmitted intensity I_T versus incident intensity I_0 in the case of optical bistability. Case (a) corresponds to a resonator filled with a saturable absorber or with a nonlinear refractive medium. Case (b) corresponds to a resonatorless bistability due to increasing absorption nonlinearity. Fine lines correspond to $I_T = I_0$.

a specific kind of nonequilibrium phase transition and instability arising due to generation-recombination dynamics in semiconductors (see, e.g., Schoell 1987; Haug 1988). A peculiar feature of intrinsic optical bistability is formation of the spatial longitudinal domain of a high concentration of nonequilibrium quasi-particles separated from the rest of a sample by a transition region commonly referred to as *kink*. The thickness of the kink is on the order of the diffusion length.

Since the first observation of optical bistability in semiconductors (Karpushko and Sinitsyn 1978) different types of optical bistability in the bulk semiconductors have been examined extensively (Henneberger 1986; Klingshirn 1990; Gaponenko, Malinovsky et al. 1992 and references therein). However, the studies of optical bistability and related instabilities in systems consisting of semiconductor nanocrystals are still at the preliminary stages.

Resonator schemes of optical bistability have been performed using CdS_xSe_{1-x}-doped glasses (Yumoto, Fukushima, and Kubodera 1987) and CuBr-doped glasses (Woggon and Gaponenko 1995) as nonlinear absorptive/refractive media. In the first case the mechanism of nonlinearity is refractive, resulting in more than a single hysteresis loop due to sequential tuning of a nonlinear Fabry-Perot resonator. The switching rise and fall times were measured to be as small as 25 ps, the switching intensity being about 300 kW/cm^2. Assuming the active area of a single optical trigger to be on the order 10×10 μm^2, the switching energy was estimated to be 10 pJ. The plane Fabry-Perot cavity was 600 μm long with two 90 percent-reflectivity mirrors. The geometric cavity length was changed by using a piezo pusher to adjust the initial detuning. Bistability was observed at room temperature at laser wavelengths about 530 nm.

In the second case bistability was realized using similar 85 percent-reflection plane mirrors with a slab of glass doped with 16 nm-size CuBr nanocrystals. Bistability has been observed at room temperature, the switching intensity being about 500 kW/cm^2. The nonlinearity is combined absorption and refraction change due to population induced shift of the exciton band. The operation wavelengths are in the range 400–410 nm.

Resonatorless bistability in CdS_xSe_{1-x}-doped glasses was observed due to thermal nonlinearity, that is, absorption red-shift induced by continuous-wave laser radiation of a wavelength corresponding to the absorption onset (Gibbs et al. 1985). Because of the slow switching thermal bistability, it is of minor practical importance. However, the negligibly small diffusion rate in the case of nanocrystals in a dielectric host provided the basic features of the resonatorless absorptive bistability to be examined. Using a commercial semiconductor-doped glass sample Gibbs et al. observed saw-tooth shaping of a smooth incident laser pulse due to formation and step-like motion of a kink throughout the sample. The experimental findings confirmed the theoretical results obtained in the limit of zero diffusion length of nonequilibrium excitations (Lindberg, Koch, and Haug 1986).

Coexistence of electronic and thermal nonlinearities under excitation by laser radiation corresponding to the absorption onset in semiconductor-doped glasses was used to analyze nonlinear dynamics in two competitive nonlinearities with drastically different relaxation times (Zheludev, Ruddock, and Illinworth 1987). In this case bleaching due to population-induced absorption saturation relaxes on a scale of electron-hole lifetime, whereas darkening due to red-shift promoted by heating relaxes with time determined by thermoconductivity. Coexistence of these competitive processes results in instabilities and pulsations including generation of subharmonics.

Finally, one more mechanism of resonatorless optical bistability relevant to nanocrystals in a dielectric matrix should be outlined. It is based on the interplay of electronic nonlinearity and dielectric confinement. Population of electron states results in significant change in absorption and refraction, that is, the complex dielectric function of a given nanocrystal. This modifies the dielectric confinement because the guest/host dielectric function changes. Modification of dielectric confinement results in turn in absorption change because of the energy shift of electron states. Thus, an internal feedback develops resulting in certain situations in bistable response. This type of intrinsic optical bistability is peculiar to nanocrystals and has no analog in the bulk materials. It was proposed a decade ago (Miller and Chemla 1986) but has not been performed experimentally to date.

7
Interface effects

In quasi-zero-dimensional structures under optical excitation there are, along with reversible processes that decay over the recombination time of electron-hole pairs, processes that result in a persistent change in the optical properties. These processes are controlled by the integral dose of the absorbed radiation rather than radiation intensity. Numerous examples of similar behavior can be found in photophysics and photochemistry of molecular structures. Similar to molecular structures, semiconductor nanocrystals embedded in a matrix or precipitated in a solution exhibit a variety of guest-host effects. Some of the phenomena related to the photo-induced modifications in absorption and/or emission features will be the subject of the present chapter. The main attention will be given to laser annealing, photodarkening, persistent spectral hole-burning, and photochemical reactions resulting in permanent spectral hole-burning. Finally, we consider the intercrystallite migration of carriers and its effect on luminescence kinetics.

7.1 Laser annealing, photodarkening, and photodegradation

Semiconductor-doped glasses exposed to prolonged illumination by laser light of a wavelength corresponding to resonant absorption by nanocrystals were found to exhibit systematically a number of photo-induced modifications. These include, first of all, a sharp decrease in the intrinsic edge luminescence versus impurity and defect related emission. Second, the lifetime of electron-hole pairs decreases by several orders of magnitude and reaches 10^{-11} s. Finally, additional structureless absorption with a coefficient on the order of 1 cm^{-1} appears in a wide spectral interval. The initial properties of the samples can sometimes be restored by heating to temperatures of 400–500°C. The above photo-induced phenomena are pronounced at room temperature, being negligible at $T < 80$ K. Roussignol et al. were the first to outline the importance of the persistent photo-induced phenomena in glasses containing nanocrystals of II-VI compounds (Roussignol, Ricard, Lukasik, and Flytzanis 1987). Later extensive research has shown that the three abovementioned effects on luminescence, lifetime, and absorption show systematic correlation (Mitsunaga et al. 1988; Kull et al. 1989; Tomita and Matsuoka 1990; Chepic et al. 1990; Horan and Blau 1990; Malhotra, Hagan, and Potter 1991; Nakano et al. 1991; Yanagawa et al. 1992).

Two models have been proposed to explain the entire set of phenomena occurring under prolonged irradiation of CdS_xSe_{1-x}-doped glasses. According to one model, new recombination centers appear in the crystallites (Miyoshi and Miki 1992; Jin et al. 1992). This becomes possible, for example, as a result of photostimulated migration of defects out of the volume into the surface of the crystallites. In the other model the escape of an electron from the crystallite into the matrix and subsequent trapping of the electron on localized states in the band gap of the glass is considered (Grabovskis et al. 1989; Chepic et al. 1990; Malhotra et al. 1991). Photoionization of crystallites is greatly facilitated by the Auger process, the probability of which increases as the size decreases due to the breakdown of the translational symmetry and increase in overlapping of the electron and hole wave functions (see Section 5.7). The filling of localized levels in the glass matrix produces induced absorption in the impurity-band channels, which results in the appearance of structureless additional absorption in the entire visible range of the spectrum. At the same time the cross sections for the capture of quasi-particles by defects in a crystallite change as a result of the presence of a charged defect near the surface of a crystallite in the matrix or as a result of the breakdown of electrical neutrality. In this context it is worthwhile to refer to the studies in which glass matrices containing Cd, S, and Se in the form of ions and not in the form of CdS_xSe_{1-x} crystallities were examined to obtain additional information about the mechanism of photodarkening in glasses containing nanocrystals (Gaponenko, Germanenko et al. 1992). It was found that photodarkening appears in a glass with no crystallites. The only peculiarity is that for photodarkening to appear the glass with no crystallites must be irradiated with light of a wavelength at which the matrix itself has some absorption. Specifically, effective darkening was observed with a single pulse of the second harmonic of a ruby laser (347 nm) with a power density on the order of 10^8 W/cm^2. These results point to the second model, which is based on photoionization of the crystallites. A summary of the photoinduced processes is given in Fig. 7.1. Manifold optical excitation promotes ionization of a crystallite resulting in a photodarkening and in enhanced defect-related recombination. Heating of a sample facilitates the reverse, deionization process resulting in the recovery of the original sample properties. However, migration

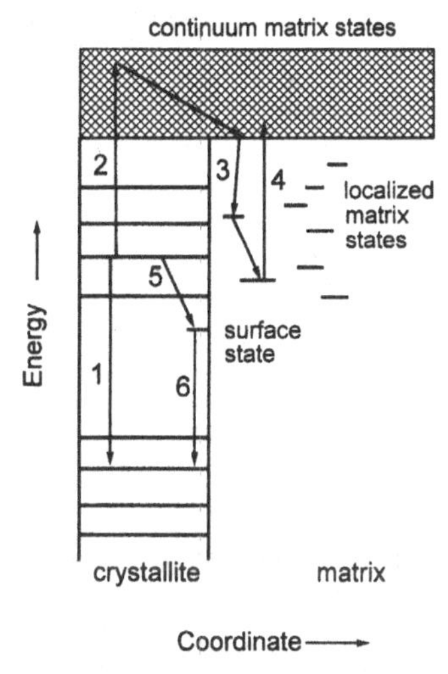

Fig. 7.1. Photoinduced processes in semiconductor-doped glasses. 1 – Annihilation of an electron-hole pair; 2 – a transition of another electron to the upper states promoted by the Auger process; 3 – relaxation within matrix states; 4 – absorption due to transition between matrix states (photodarkening); 5, 6 – recombination via defect or surface state.

of electrons over the localized states in the matrix may result in nonrecovery of the initial state under certain conditions, because the probability for a given particle to visit a given site in the random-walk process strongly depends on the proper scaling factors.

Photoinduced absorption is not inherent in semiconductor-doped glasses only but occurs in colloidal semiconductors as well. Likewise, photoionization promoted by the Auger process is being considered in similar ways (see, e.g., Henglein 1989 and references therein). The same mechanism has been proposed also for the degradation effect on the visible photoluminescence of porous silicon consisting of silicon nanocrystals connected via a SiO_2 skeleton (Gaponenko, Petrov et al. 1996). It is interesting that photoinduced modification of the emission spectrum under prolonged irradiation by a nitrogen laser radiation (337 nm) is spectrally selective (Fig. 7.2). This effect may be considered as spectral hole-burning in the emission spectrum due to selective photodegradation of the crystallites resonant to the illumination wavelength. Persistent spectral hole-burning in the luminescence spectrum has been observed also for CuI nanocrystals embedded in a glass matrix (Masumoto 1996).

A local electric field present because of a noncompensated charge in the close vicinity of the photo-ionized crystallite should result in a modification of the absorption spectrum of the crystallite. Moreover, charging itself may lead to a change in absorption spectrum. Manifestations of local electric field effects

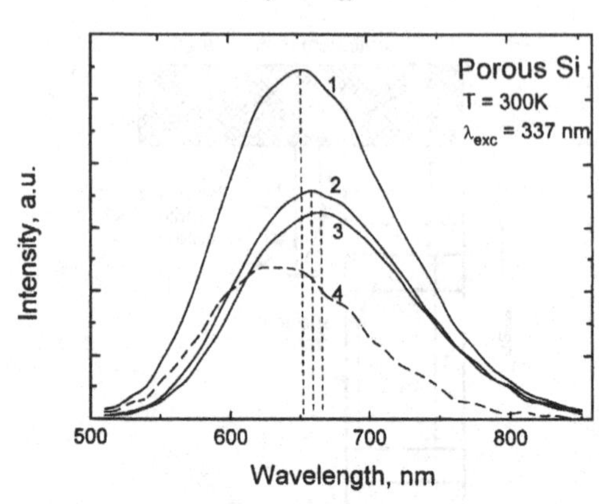

Fig. 7.2. Degradation of the photoluminescence of porous silicon during exposure to pulsed nitrogen laser radiation ($\lambda = 337$ nm, $\tau = 10$ ns, E = 10 μJ). (Gaponenko, Petrov et al. 1996). Curve 1 is initial spectrum of the fresh sample; curves 2 and 3 show the emission spectra after illumination by 3.6×10^5 and 7.2×10^5 pulses, respectively. Curve 4 is the difference of curves 1 and 3.

on the intrinsic absorption of CdS_xSe_{1-x} nanocrystals in glass matrices have actually been observed, and a correlation of the persistent differential absorption spectrum with that inherent to the electroabsorption effect have been outlined (Esch et al. 1990; Norris et al. 1994; Kang et al. 1994; Flytzanis et al. 1996).

It is important to note that the laser annealing effect on an electron-hole recombination lifetime may result in a number of artifacts when nonlinear optical response is measured and the absolute values of the saturation intensity and nonlinear susceptibility are evaluated. In these experiments the interplay of the intensity and dose effects should be accounted for carefully. The laser annealing effect should be minimized either by low intensities or by a smaller number of laser pulses applied for data acquisition. Laser annealing also can be excluded when lower temperatures are used. Note that experimental data given in the figures in Chapters 6 and 7 related to population-induced nonlinearities are taken with proper consideration of the annealing effects.

On the other hand, the laser annealing effect on electron-hole recombination time can be purposefully used to get fast nonlinear response. For example, Mitsunaga et al. demonstrated in laser-induced grating experiments a decrease in the recovery time from 100 to 7 ps (Mitsunaga et al. 1988).

It is evident that laser annealing and photodarkening strongly depend on the matrix in which crystallites are embedded. Kang et al. (1994) compared these effects using similar CdS nanocrystals in various glasses. Studies performed for conventional silicate glasses, sodium borosilicate glasses, and organically modified silicates (ormosils) have revealed the inhibited degradation processes in sol-gel glasses as compared with conventional melt-quenched glasses.

7.2 Interface effects on the properties of copper halide nanocrystals

Since CuCl, CuBr, and CuI nanocrystals can be developed in different types of matrices, these structures can be used to study a variety of guest-host interface related processes. In what follows we discuss the distinct differences in nonlinear optical properties, in photoluminescence, and in photodegradation of CuCl crystallites that systematically correlate with the lower potential barrier in the case of glass matrices as compared with the NaCl host.

CuCl crystallites in glass and in NaCl show strong population-induced nonlinearity at resonant excitation in the exciton band (Section 6.2). However, optical excitation above the exciton resonances leads to different results for different matrices. In the case of a glass matrix, excitation at $\hbar\omega = 3.57$ eV (second harmonic of ruby laser) does not lead to any detectable change in the exciton spectrum, even if the incident intensity is close to the damage threshold of the glass (Zimin et al. 1990). The excitation spectrum of the exciton luminescence likewise exhibits a pronounced drop at photon energies higher than that of exciton resonance, in spite of the fact that the absorption spectrum shows a steady rise in this energy range (Fig. 7.3) (Gaponenko, Germanenko et al. 1993). These features are absent in the spectra of CuCl nanocrystals developed in NaCl. These samples show excitonic nonlinearity both under resonant and interband excitation (Wamura et al. 1991), and the excitation spectrum of luminescence reproduces the absorption without any drop for $\hbar\omega \rightarrow E_g$ (see Fig. 5.9).

Further differences in the properties of CuCl crystallites in different matrices were revealed in the luminescence degradation phenomenon. Itoh et al. examined systematically photodegradation of exciton luminescence under exposure to continuous-wave 325-nm laser radiation (Itoh et al. 1994). Crystallites in glass exhibit systematically lower luminescence quantum yield and considerably higher degradation rate compared with NaCl matrix. Finally CuCl-doped glasses were found to show wide-band structureless photodarkening similar to CdS_xSe_{1-x}-doped glasses (Gaponenko, Germanenko et al. 1992).

An explanation of these peculiarities can be found in terms of guest-host interface related processes. The embedding of the crystallites in different matrices is connected with large differences in the height of the confining potential barrier for electrons and holes as well as with modifications of the crystallite-matrix boundary. In the case of the glass matrix, the lower potential barrier and the presence of a large number of localized states within the forbidden gap

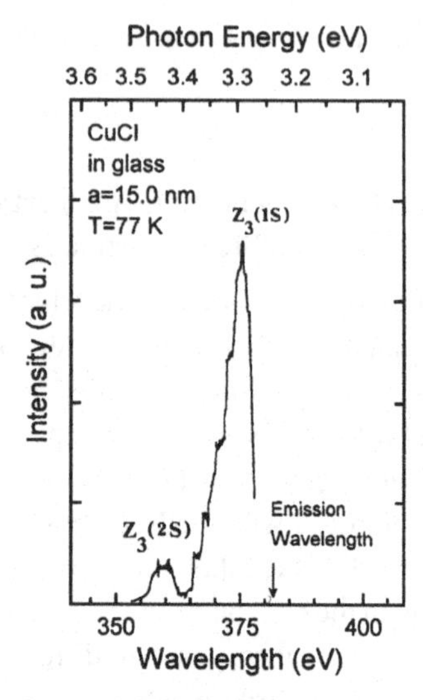

Fig. 7.3. Luminescence excitation spectrum of CuCl nanocrystals in glass (adapted from Gaponenko, Germanenko et al. 1993). Mean size of nanocrystals is 15 nm, temperature is 77 K. The registration wavelength is 382 nm. The main excitation peak is relevant to the resonance absorption corresponding to the $1S$ state of a Z_3 exciton. The second peak at 360 nm corresponds to the $2S$ state of a Z_3 exciton.

promote the capture and transfer processes of photoexcited electrons or holes into the matrix. These competitive processes are responsible for the lack of any noticeable exciton population when samples are excited with photon energies above the CuCl band gap. Therefore, neither emission nor population-induced nonlinearities were observed under these conditions. The enhanced degradation rate inherent in CuCl-doped glasses as compared with CuCl in NaCl can be explained by the lower potential barrier. Similar to the discussion in Section 7.1, the low potential barrier enhances photoionization probability because of the intensive Auger processes.

Glasses containing nanocrystals of copper halides are known to exhibit a pronounced photochromic effect due to photostimulated formation of $(Cu)_n$ clusters (Tsekhomskii 1978). This property makes them a convenient photochromic material along with glasses and films containing silver halide nanocrystals. The mechanism of photochromic effect is as follows. First, illumination of a sample with near UV radiation corresponding to the absorption onset of the crystallites results in exciton formation. Because of the low confinement potential the probability of finding an exciton at the interface region becomes rather high, and the exciton can be trapped by some defects at the interface region. Further dissociation of exciton results in charging of a nanocrystal. Charged nanocrystals containing Cu-ions at the surface then serve as nucleation centers promoting growth of $(Cu)_n$ clusters at the surface of a nanocrystal. Note that further illumination by the visible light corresponding to the photoinduced absorption band results in the further growth of $(Cu)_n$ colloids. This phenomenon of photostimulated growth of metal colloidal particles also occurs for metal colloids in solutions (Henglein 1989).

Another interesting example of interface related phenomena is the fine thermal annealing effect reported for CuCl-doped glasses (Vasiliev et al. 1995). It is known that nanocrystals possess size-dependent melting temperature T_{melt} that can be considerably lower than that of the parent bulk crystal (Ekimov 1996; Tolbert and Alivisatos 1995). Size-dependent phase transition is not a peculiarity of semiconductor nanoparticles but a general property of all small particles. Another specific feature is a hysteresis with freezing temperature T_{freez} being noticeably lower than melting temperature. In the case of smaller CuCl crystallites T_{melt} falls down to values about 300°C, and the difference of $T_{\text{melt}} - T_{\text{freez}} \approx 100$°C. A continuous thermal annealing at the temperature

$$T_{\text{freez}} < T_{\text{anneal}} < T_{\text{melt}} \tag{7.1}$$

was found to result in a noticeable sharpening of the exciton absorption band [Fig. 7.4(a)]. There is a rather narrow range of T_{anneal} for every size. If the annealing temperature is too low, the effect is absent. If the annealing temperature is too high, the crystallites experience reversible melting-freezing during annealing and subsequent cooling to room temperature.

The mechanism of the fine annealing effect is not clear at present. Taking into account that at temperatures close to T_{melt} the average amplitude of ion oscillations is close to the interatomic distance, it is reasonable to suggest that during fine annealing migration of the surface atoms and surface reconstruction promote elimination of defects responsible for extra line broadening.

A correlation of the optimal annealing temperature with the size makes it possible to perform size-selective annealing by means of the proper choice of T_{anneal} for the given mean size of the crystallites exhibiting inhomogeneous broadening. This is illustrated in Fig. 7.4(b) in which the optical spectrum of a glass sample of the mean radius $a = 3.5$ nm is shown versus the spectrum of the same sample after fine annealing at a temperature relevant to $a = 15$ nm. A manifestation of the substructure of the inhomogeneously broadening band revealed by means of the selective thermal treatment is evident. Therefore, the fine thermal annealing applied to crystallite in a glass matrix is useful not only in enhancement of crystallite quality but also can serve as a selective technique providing elucidation of the inhomogeneous broadening of optical spectra.

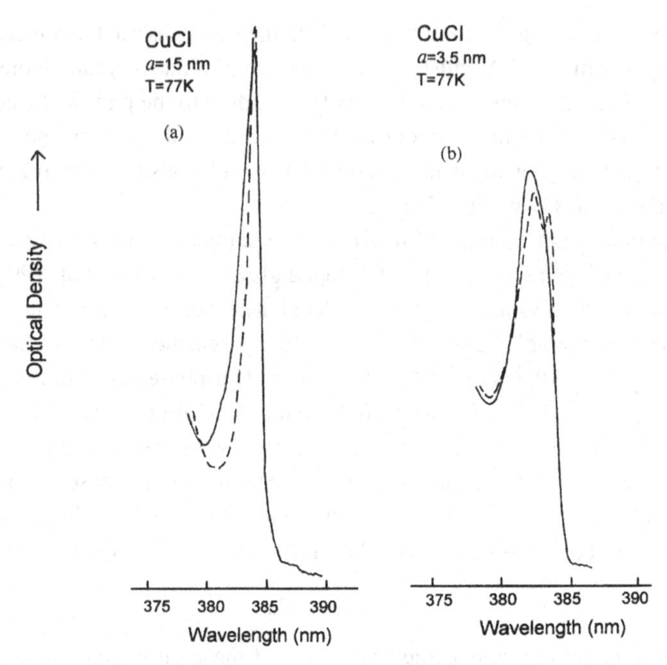

Fig. 7.4. Absorption spectra of the two glass samples containing CuCl nanocrystals before (solid lines) and after (dashed lines) annealing at temperature 300°C during 70 hours. The mean radius of crystallites is (a) 15 nm and (b) 3.5 nm (Vasiliev et al. 1995).

7.3 Persistent spectral hole-burning

In Chapter 5 (Section 5.8) we discussed the influence of an external electric field on exciton absorption. An electric field applied to a sample containing semiconductor nanocrystals causes spectral shift and broadening of the absorption bands. Every ensemble of semiconductor nanocrystals possesses significant inhomogeneous broadening due to randomized size, shape, and local environment of nanocrystals. In Section 7.1 of this chapter the evidence for photoinduced ionization of crystallites exposed to laser light was provided. Therefore, it is reasonable to pose a question: is it possible to get spectrally selective response due to site-selective photoionization? To do this, one needs to perform continuous exposure of a nanocrystal ensemble with noticeable inhomogeneous broadening to monochromatic light. A variety of spectrally selective effects relevant to multiple site-selective ionization-deionization acts in the case under consideration is nothing but persistent spectral hole-burning which has been known for many years to occur for molecules and ions embedded in solid matrices.

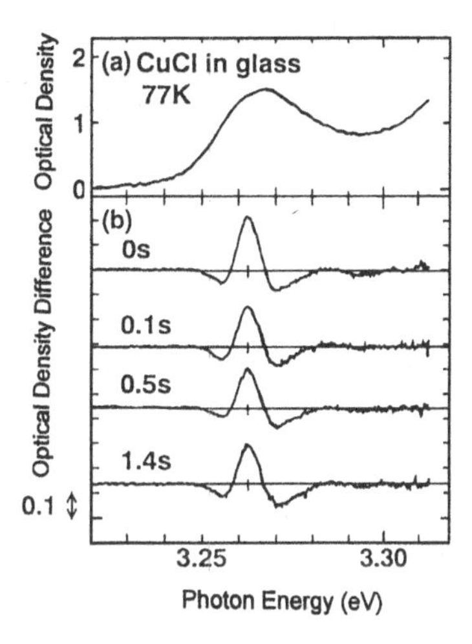

Fig. 7.5. Persistent spectral hole-burning in CuCl-doped glass (Naoe et al. 1994). (a) Linear absorption spectrum at 77 K; (b) time-dependent differential absorption spectra after illumination at 3.263 eV.

In the case of semiconductor nanocrystals persistent spectral-hole burning was established and has been thoroughly examined in recent years (Naoe, Zimin, and Masumoto 1994; Masumoto et al. 1995; Masumoto 1996; Valenta et al. 1997). The phenomenon was observed for I-VII nanocrystals in various matrices at low temperatures. Though persistent changes in the absorption spectrum are inherent in most nanocrystals, the spectrally narrow holes are most pronounced in copper halide crystallites relevant to the weak confinement range because they possess rather narrow homogeneous width.

Pronounced spectral hole-burning occurs at low temperatures after prolonged illumination by laser light corresponding to the long-wave shape of the exciton absorption band (Fig. 7.5). Typically, the shape of the differential absorption spectrum is characterized by a narrow bleach band ("hole") and two side-bands of induced absorption ("antiholes"). It is remarkable that the integral square of two antiholes is equal to that of the hole. This corresponds to the broadening and shift of individual exciton subbands within an inhomogeneously broadened spectrum of the nanocrystal ensemble. Naturally, not only the exciton ground state but also the excited states contribute to the differential absorption spectra. As the illumination wavelength is tuned far from the absorption edge, the differential absorption spectrum becomes more complicated. It is more complicated also in the case of intermediate and strong confinement modes. Persistent change in absorption remains for several hours. It can be erased either

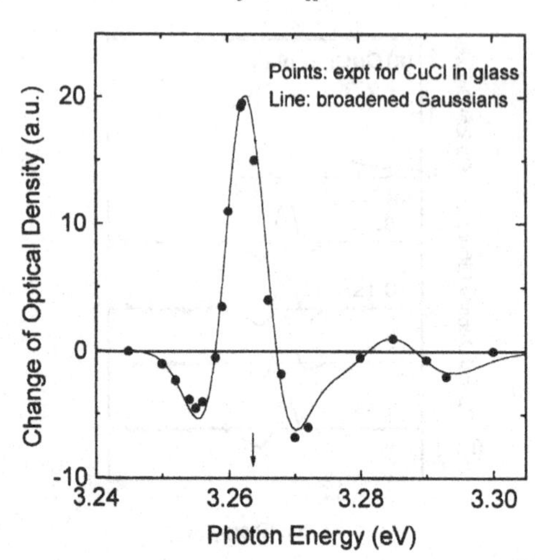

Fig. 7.6. Differential absorption spectrum of CuCl-doped glass recorded during persistent spectral hole-burning (dots), along with fit by the two broadened Gaussians (line) (Gaponenko 1995b).

by heating or by illumination with light nonresonant to the burned hole. With successive illumination by several wavelengths, several holes can be burned.

The whole variety of these phenomena can be consistently explained in terms of the local electric field effect due to selective photoionization of resonantly excited crystallites (Woggon and Gaponenko 1995; Masumoto et al. 1995). Therefore, the spectral shape of differential absorption is typical of electric field effects [compare Fig. 7.6 with Fig. 5.46(c)], whereas other features, such as the relation between the hole square and the radiation dose or erasing by heating, are similar to photodarkening (Fig. 7.1). In this connection it is not surprising that persistent hole-burning was observed also in the luminescence spectrum (Masumoto 1996) similar to selective photodegradation reported by other authors (Itoh et al. 1994; Gaponenko, Petrov et al. 1996).

Along with other spectrally selective phenomena, persistent reversible hole-burning provides a possibility of revealing the substructure of an absorption spectrum that is masked by inhomogeneous broadening. However, persistent hole-burning does not provide any evidence on the origin of inhomogeneous broadening. Evidently, the differential absorption spectrum due to photoionization of crystallites will reproduce its basic features if no inhomogeneous broadening occurs at all. In this case it simply reduces to the response inherent in the case of the external electric field or in the case of the charging of nanocrystals.

7.4 Photochemical hole-burning

Optical excitation of semiconductor nanocrystals may result, in some situations, in a sequence of chemical reactions. A photogenerated electron trapped at the surface of a nanocrystal or photoemitted in the solution promotes a reduction process of organic molecules presenting in the same solution. Likewise, a photogenerated hole at the surface or emitted in the solution gives rise to an oxidation reaction involving other components of the solution. Afterwards, a number of secondary chemical reactions can be initiated. These photocatalytic processes are extensively studied using liquid solutions containing semiconductor nanoparticles (see reviews by Henglein 1989 and Kamat 1993). The quantum size effect plays a considerable role in the efficiency of photochemical reactions, because it provides high surface to volume ratio as well as the necessary initial potentials of electrons and holes that are required for an effective electron transfer. Photocatalytic reactions involving semiconductor nanoparticles are of significant practical importance. They can be used in photocatalytic degradation of undesirable chemical contaminants. The principle of such a treatment lies in initiating an oxidation or reduction process at the semiconductor surface. Addition of other reactants then drives the secondary reactions to give a desirable product. For example, H_2S, which is a byproduct in the coal and petroleum industry, can be oxidized to S at the CdS semiconductor surface. Photoassisted catalytic degradation of a number of organic materials in TiO_2 suspension have been reported. For example, complete mineralization of chlorphenols into CO_2 and HCl can be achieved by means of band gap excitation of TiO_2 particles with sunlight for a few hours. Another well-known photochemical process initiated by photoexcitation of nanometer-size semiconductor particles is the whole set of photographic reactions, which has been considered recently in the context of solid state physics by Sviridov et al. (1987). Generation of electron-hole pairs in submicron-size silver bromide colloidal particles in a photographic emulsion results in ripening of silver colloids at the final stage of the process. The secondary processes involving organic and inorganic compounds promoted by photoexcitation of semiconductor nanocrystals are beyond the scope of this book. In what follows we restrict ourselves to consideration of the irreversible permanent photostimulated effects relevant directly to semiconductor nanoparticles. These include photodissolution of nanocrystals of binary compounds, which manifests itself in a pronounced permanent spectral hole-burning.

Photodissolution of semiconductor nanoparticles in aerated solutions has been reported by Henglein and co-workers (Henglein 1982; Gallardo et al. 1989). These authors observed photolysis of CdS and PbS colloidal particles after prolonged illumination by ultraviolet light. The photochemical process can be monitored by optical transmission measurements, which revealed a uniform decrease in the concentration of colloids on exposure. The absorption spectrum diminishes with time as a whole without any selective feature in the course of degradation.

The above process can be site-selective in some situations. First of all, the illumination wavelength should be relevant to the long-wave tail of the first absorption band. Second, the illumination photon energy should be sufficiently high to promote photoionization, which is the first step of the photochemical process. Third, the crystallites should be small enough to enhance the Auger process and provide a high surface-to-volume ratio, which is necessary for high chemical activity. Finally, the solvent should be of the proper chemical composition to promote the necessary chemical reactions.

These conditions were satisfied in the experiments by Artemyev et al. (Artemyev et al. 1995; Gaponenko, Germanenko et al. 1996). In the experiments smaller CdS nanoparticles ($\bar{a} \approx 1$ nm) in polyvinyl pyrrolidone exhibiting absorption onset in the near-ultraviolet range were exposed to prolonged illumination by radiation of tunable nanosecond pulsed lasers.

The effect is attributed to the two-stage process stimulated by resonant optical excitation. At the first stage selective photoionization of the crystallites gives rise to the broadening of the relevant components in the integral absorption spectrum due to the local electric field effect. At the second stage a continuous photolytic process occurs, providing an irreversible destruction of the crystallites and a characteristic dip in the absorption spectrum.

The initial optical absorption spectrum of the samples (Fig. 7.7) is shifted towards higher energies compared with the bulk CdS by about 1 eV, which is typical of semiconductor quantum dots in the case of the strong confinement regime ($a \ll a_B$). Under prolonged laser illumination a pronounced dip in the absorption spectrum appears, indicating permanent photobleaching. The position of the dip follows the laser wavelength λ_{las}, the maximum nearly coinciding with λ_{las}. Generally, for smaller laser wavelengths the dips appear at smaller intensities and/or number of pulses, which simply correlates with the growth of the absorption coefficient with decreasing wavelength. Fig. 7.8 (a,b) shows changes in the absorption spectrum with increasing dose of the absorbed radiation for two selected laser wavelengths. Two peculiarities are worthy of discussion. First, with a small radiation dose not only is bleaching at the excitation wavelength observed, but also an induced absorption at both sides is clearly seen with a tendency to decrease with increasing total radiation energy absorbed. Second, for longer laser wavelengths there is a pronounced tendency

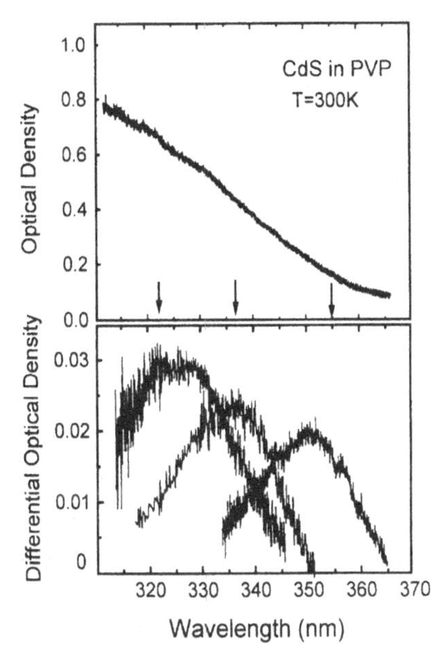

Fig. 7.7. Permanent photochemical spectral hole-burning in an ensemble of CdS crystallites in a polyvinyl pyrrolidone film (Gaponenko, Germanenko et al. 1996). Mean radius of crystallites is 1.1 nm, the temperature is 300 K. Upper panel shows the initial absorption spectrum. Lower panel shows the change in optical density recorded in the vicinity of the laser wavelength after illumination. Positive optical density change corresponds to photobleaching. Vertical arrows indicate position of the laser wavelengths. Note a change in the scale for the differential spectrum as compared with the initial one.

of the induced bleaching to shift towards shorter wavelengths with growing absorbed energy.

The observed permanent hole-burning can be explained in terms of selective photoionization of the resonantly excited quantum dots followed by their photodestruction. The primary physical process resulting in a persistent change of optical properties is photoionization of crystallites. Photoionization occurs due to the Auger process, in which one electron-hole pair recombines providing the extra electron enough energy to leave the crystallite. This process is enhanced with decreasing size for the following reasons: (i) stronger overlapping of electron and hole wave functions and (ii) the lack of translational symmetry. Additionally, the potential barrier at the guest-host interface becomes smaller with growing confinement energy, providing easier electron escape from smaller crystallites. This phenomenon is considered to be the main reason of the laser annealing and photodarkening effects occurring in semiconductor-doped

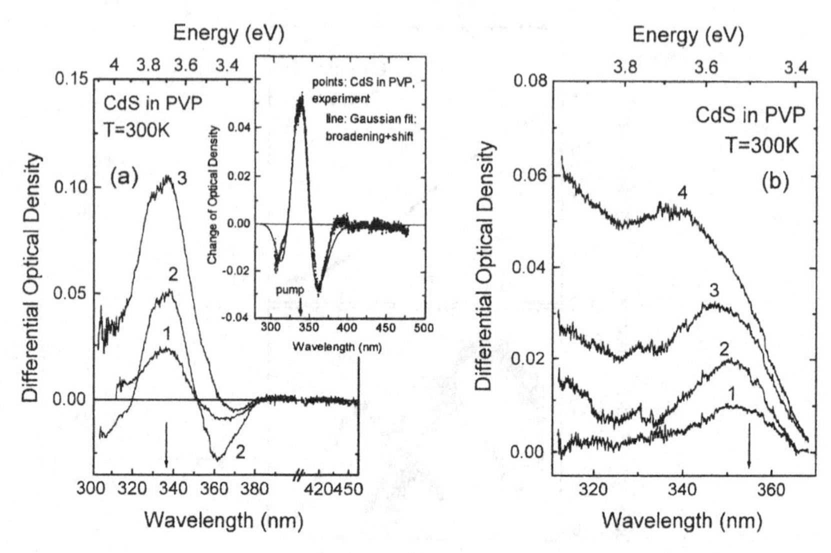

Fig. 7.8. Permanent changes in optical density of an ensemble of CdS crystallites in a polyvinyl pyrrolidone film under illumination by (a) 337 nm and (b) 355 nm laser radiation (Gaponenko, Germanenko et al. 1996). Mean radius of crystallites is 1.1 nm, the temperature is 300 K. Total number of incident laser pulses is $1.5 \cdot 10^4$ (1), $4.5 \cdot 10^4$ (2), and $1.8 \cdot 10^5$ (3) in panel (a) and $1 \cdot 10^3$ (1), $2 \cdot 10^3$ (2), $1 \cdot 10^4$ (3), and $2 \cdot 10^4$ (4) in panel (b). Positive change of optical density corresponds to bleaching. Pulse duration is 10 ns. Pulse intensity is about 10 μJ/mm^2 (a) and 1 mJ/mm^2 (b). The insert shows experimental data (dots) along with the fit by the broadened Gaussian.

glasses (see discussion in Section 7.1). There is the non-negligible probability that an electron may be captured by the surface state or may occupy some localized site in the close vicinity of the crystallite. In both cases, the excited dot experiences the influence of the local electric field with an inevitable shift and broadening of the exciton absorption band, which must be called bleaching in the maximum of the band and induced absorption at the band wings. In the second case (matrix state), the selective broadening of the spectrum can be permanent. Similar considerations and observations have been reported for CdS crystallites in glasses (Kang et al. 1994; Flytzanis et al. 1996) except that broadening was not sensitive to excitation wavelength.

Local electric field effect is just quantum confined Stark effect, which has been investigated for the strong confinement regime in CdS and CdSe quantum dots both theoretically and experimentally (see Section 5.8). Particularly, for the strong confinement regime electric field induced broadening was found to dominate the Stark shift of the absorption band. Likewise, the differential absorption spectrum at small doses can be fitted by broadened Gaussian (insert in Fig. 7.7). The only difference between this and the external electric field experiments is the local character of the former. Thus, under conditions of resonant photoionization of nanocrystals the term "site-selective quantum confined Stark effect" might be appropriate.

Photoionization then stimulates a sequential corrosion of semiconductor particles and photoinduced absorption is replaced by a steady photobleaching monotonically growing with absorbed energy. This is a manifestation of size-selective *photolysis*, which became possible due to small homogeneous width compared with inhomogeneous width and due to small changes in the optical spectra examined. Eventually, as optical density changes, the shape of photobleaching tends to the shape of the initial absorption spectra because of the saturating contribution of the crystallites, with absorption peaking at λ_{las} and the larger weight of crystallites possessing a long-wave edge in the same spectral region. This is the well-known effect of dose broadening, which is similar to the power broadening inherent in the case of selective absorption saturation.

The relevant sequence of chemical reactions is the following (Henglein 1989). First of all, photoionization by means of the Auger process yields an excess electron outside the crystallite and an excess hole inside or at the surface that is,

$$\begin{aligned}(\text{CdS})_{\text{n}} &\xrightarrow{h\nu} (\text{CdS})_{\text{n}}(e^{+} + h^{-}) \xrightarrow{h\nu} (\text{CdS})_{\text{n}}(e^{+} + h^{-})_{2} \\ &\to (\text{CdS})_{\text{n}}(h^{+}) + e^{-}. \qquad (7.2)\end{aligned}$$

After photoionization, the remaining hole may act as an oxidizing agent and initiate the secondary chemical reactions at the CdS surface. Sulfide anions can be oxidized to the radical anions S^{-} by h^{+} with further oxidation by air oxygen to SO_2^{-} or SO_4^{2-}; that is,

$$h^{+} + \text{S}^{2-} \to \text{S}^{-} \qquad (7.3)$$

$$\text{S}^{-} + \text{O}_2 \to \text{SO}_2^{-} \xrightarrow{\text{O}_2} \text{SO}_4^{2-}. \qquad (7.4)$$

Such anions have been detected after photolysis of CdS colloidal solution (Kamat 1993). Another less intensive reaction of direct photochemical reduction of Cd^{2+} ions by an excited electron can occur inside the CdS nanocrystals prior to photoionization,

$$\text{CdS} + e^{-} \to \text{Cd}^{0} + \text{S}^{2-}. \qquad (7.5)$$

The presence of Cd^{0} has been detected in thin CdS films after illumination on air (Kamat 1993). The S^{2-} ions released are oxidized then by h^{+} in the reaction (7.3). The fate of the emitted electron is as follows. A number of polymeric functional groups may effectively accept the photoelectrons. Also, the electron released may be scavenged by oxygen, especially when the matrix, like polyvinyl pyrrolidone, is permeable partly to air

$$e^- + O_2 \rightarrow O_2^-. \tag{7.6}$$

In addition to a number of other techniques, the observed phenomenon is an effective complementary tool for analysis of quantum dot ensembles with inhomogeneously broadened spectrum. Within the limit of a small integral dose of absorbed radiation, the permanent photobleaching has a shape of a dip in the absorption spectrum with the full half-width twice as large as the homogeneous half-width Γ_0 of the destroyed component of the ensemble (Moerner 1988). Thus, one can evaluate $2\Gamma_0 \cong 100$ meV for $\lambda = 355$ nm and $2\Gamma_0 \cong 120$ meV for $\lambda = 322$ nm. These resonant wavelengths correspond to $a = 1.4$ nm and 1.25 nm, respectively, if estimated in terms of the effective mass approximation. Both values of homogeneous width are in reasonable quantitative agreement with the theory of homogeneous line broadening due to exciton-phonon interaction (see Section 5.3) as well as with experimental studies of II-VI quantum dots in the strong confinement regime. Remarkably, $2\Gamma_0 \cong 100$ meV found for $\lambda = 355$ nm coincides with the data reported for a highly monodisperse ensemble of 55-atom CdS clusters developed in zeolites (Wang and Herron 1991).

Taking into account that the size of quantum dots examined is rather small and no more than 10^2 atoms correspond to $a \cong 1$ nm, one more question arises. Generally, at small a value a continuous quantum dot approach fails and should be replaced by the careful consideration of clusters with the finite number of atoms obeying a series of some "magic" numbers. For these clusters one should expect photoinduced bleaching at certain concrete wavelengths rather than a continuous shift of the bleaching peak following the laser wavelength. Probably, this is just the case for the smaller sizes examined. For $\lambda_{las} = 322$ nm the bleaching peak is shifted from the excitation wavelength by about 5 nm (see Fig. 7.7). For longer excitation wavelengths (and larger crystallites) there seems to be a continuous shift of bleaching with strict coincidence with λ_{las}. This observation means either that the continuous quantum dot approach is still meaningful for $a \cong 1$–1.5 nm, or that quantum-size effect is not the only reason of the inhomogeneous broadening, that is, that crystallites of the same size may exhibit variable exciton energies due to fluctuations in shape, surface structure, or environment.

Recently, Nirmal et al. (1996) observed the site-selective photochemical process by means of spatial selection of a CdSe nanocrystal in single-dot spectroscopy. Photostimulated selective oxidation manifests itself in a continuous blue shift of the single dot fluorescence spectrum with observation time.

Under certain conditions the anion and cation ions released in the course of photodissolution may subsequently reform molecules on larger crystallites.

This process can be referred to as photo-ripening of semiconductor colloids. It was observed for PbS in a colloidal oxygen-free solution (Gallardo et al. 1989). There is also some evidence that photo-induced growth of II-VI nanocrystals may occur in a glass matrix during prolonged illumination (Yanagawa, Nakano, and Sasaki 1991; Miyoshi et al. 1996).

7.5 Classification of the spectral hole-burning phenomena in quantum dot ensembles

In the previous chapters and in the present chapter the experimental data on spectral hole-burning in semiconductor nanocrystals have been considered, providing evidence of various types of hole-burning including transient, persistent, and even irreversible phenomena. It is reasonable to summarize the relevant experimental findings as well as the underlying microscopic mechanisms.

In terms of the relaxation time of the burned spectral hole the following classification can be proposed (Gaponenko 1995a and 1995b).

(i) Transient spectral hole-burning due to selective population of the higher electron-hole states within the same crystallite with consequent relaxation to the lower levels. The typical relaxation times are in the subnanosecond range. To observe this effect the following condition should be satisfied

$$\tau_{\text{pulse}} < \tau_{\text{rel}}, \tag{7.7}$$

where τ_{pulse} is duration of the pump-and-probe pulses and τ_{rel} is the energy relaxation time of excitons from the upper to the lower states. In the specific case of a large energy level separation when the phonon bottleneck effect is pronounced, a condition

$$\tau_{\text{rec}} < \tau_{\text{rel}}, \tag{7.8}$$

where τ_{rec} is recombination lifetime, makes it possible to observe hole-burning of this type in a quasi-steady-state regime, that is, independently of the pump pulse duration. The relevant effect in molecular spectroscopy is often considered in terms of "hot luminescence." In bulk semiconductors a similar phenomenon is possible on the femtosecond time scale when Fermi-Dirac distribution of electron-hole gas is not achieved. This effect is readily observable in nanocrystals of II-VI compounds in glass and in porous Si.

(ii) Quasi-steady-state hole-burning due to absorption saturation of resonantly excited crystallites in an inhomogeneously broadened ensemble. The recovery process is controlled by electron-hole recombination with characteristic times from subnanosecond to microsecond range. This effect has been reported by a number of groups for II-VI and I-VII quantum dots in dielectric matrices. If relaxation time is comparable to or longer than the recombination time, then a selective steady-state population of the higher states is possible, similar to case (i). Spectral hole-burning as a result of creation of the first electron-hole pair in a crystallite is accompanied by an induced absorption due to formation of the biexciton state when the pump and the probe quanta are absorbed by the same crystallite. In the case of larger dots exciton-exciton interactions result in more complicated deformation of the absorption spectrum.

(iii) Persistent reversible hole-burning due to photoionization of a crystallite and/or surface localization of the electron occurs under certain conditions. In this case optical absorption experiences modifications due to the local electric field effect (quantum confined Stark effect) with differential absorption spectrum typical of exciton absorption in the external electric field. For small quantum dots surface localization of one of the carriers leads to a problem of biexciton, and the differential absorption spectrum may be similar to case (ii). The effect has a recovery time from microseconds to hours depending on the temperature and/or dot-matrix interface.

(iv) Permanent spectral hole-burning is possible due to an irreversible sequence of photochemical reactions in resonantly excited crystallites stimulated by a manifold photoionization. The phenomenon was observed in semiconductor-doped polymer and may be interpreted in terms of size-selective photolysis of nanocrystallites. In the early stages photoionization results in the local electric field effect with differential absorption spectrum similar to case (iii).

Unlike case (i), cases (ii)–(iv) are possible only under conditions of inhomogeneous broadening. Therefore, they have no analog among the intrinsic bulk semiconductors. Analogs can be found, however, in a number of systems possessing inhomogeneous broadening, like doped semiconductors, atoms in gases, ions in dielectrics, and molecules in a gas phase, in solutions and in solid matrices.

In terms of the underlying microscopic mechanisms, spectral hole-burning in nanocrystals can be classified as follows:

(i) population-induced effects (selective absorption saturation);
(ii) local electric field-induced effects (selective photoionization); and
(iii) photochemical effects (selective photolysis).

One can see that different hole-burning phenomena may sometimes have the same underlying mechanism, such as surface localization or local electric field effect, and accordingly, the same differential absorption spectrum differing only in the recovery time. Vice versa, hole-burning due to different mechanisms can show similar recovery times. For example, population induced hole-burning may have recovery time on the same order as photoionization-induced hole-burning. This is the case when the former experiences slowing down due to a long-lived triplet or surface state and the latter has enhanced recovery due to a low potential barrier.

In all cases the shape of the burned hole is connected with the homogeneous linewidth of the resonantly excited dot, although in every case the proper theoretical analysis is necessary to evaluate intrinsic parameters from the observed photoinduced changes in absorption. However, observation of a certain type of hole-burning cannot lead to an unambiguous conclusion on the origin of inhomogeneous broadening. To distinguish between inhomogeneous broadening due to quantum size effect and due to secondary factors such as fluctuations of the nanoenvironment other experiments should be performed. Note that all of the above discussed processes (absorption saturation, photoionization, and photolysis) occur in the absence of inhomogeneous broadening as well. The specific feature is that in the case of inhomogeneous broadening these processes can lead to spectrally selective effects due to site-selective optical impact. In this connection the discussion by Masumoto et al. (1995; Masumoto 1996) on the possibility of evaluating the role of nanoenvironment fluctuations by means of persistent hole-burning experiments is not justified. To do this one needs to eliminate size- and shape-related effects as in the case of a molecular ensemble in a solid matrix where all molecules are *intrinsically identical* differing only in nanoenvironment. This is not the case when nanocrystals are considered because until now there is no technique to prepare an absolutely monodisperse ensemble of nanocrystals.

This line of reasoning becomes evident if one considers, for example, a mixture of intrinsically different molecules in a matrix or in a solution. In this case inhomogeneous broadening is a priori due to intrinsic and extrinsic factors. Evidently, the persistent hole-burning, if any, will remain in the case of the mixture. However, it by no means points to the minor contribution of the broadening due to chemical (i.e., *intrinsic*) differences as compared with the environmental (i.e., *extrinsic*) fluctuations. Likewise, an ensemble of nanocrystals when being compared with molecules and ions implies necessarily that there is a distribution of intrinsic properties (size, shape, defect concentration, lattice modification, surface morphology) involved.

The specific feature of population-induced effects as compared with photoionization and photolysis is the intensity dependence of the former and the dose dependence of the latter. This means that photoionization and photolysis may be accumulated during the experiment in proportion to the number of photons absorbed. For photoionization this is strictly valid only for the constant intensity. Larger intensities result in higher photoemission yield because the underlying Auger process involves two electron-hole pairs, and its probability is sensitive to the time interval between the successive absorption acts. At lower intensities the first electron-hole pair relaxes or recombines prior to when the second one is created and the photoionization process diminishes.

In the context of the proposed classification, it is evident that various types of spectral hole-burning may coexist. Observation of the specific hole-burning effect does not eliminate a manifestation of the other possible hole-burning phenomena. For example, for CdS nanocrystals three types of spectral hole-burning were demonstrated (Gaponenko et al. 1997). For CuBr nanocrystals, coexisting population-induced and local electric field-induced hole-burning was established (Valenta et al. 1997). It seems reasonable to describe various photoinduced phenomena in terms of the relevant quantum yield as is commonly accepted in photochemistry.

7.6 Tunneling and migration of carriers and their influence on luminescence decay

At a high concentration of crystallites, not only do electrons escape into the matrix but migration of electrons and excitons within the ensemble of crystallites also becomes possible. Migration occurs by means of site-to-site hopping, which brings about a number of characteristic features inherent in other disordered media like amorphous solids and molecular aggregates. The hopping process is described in terms of random walks in space of noninteger dimensionality, that is, *fractal* space (Feder 1988; Kopelman 1988 and references therein). A transition from integer (Euclidian space) to noninteger (fractal space) dimensionality results in an important physical issue. Unlike conventional space with integer dimensionality ($d = 1, 2,$ or 3), different physical properties and processes (e.g., mass, diffusion, mobility, conductivity) in fractal space are described by means of the different dimensionalities. It is of principal importance to evaluate dimensionalities relevant to every specific process and to find the relationship between them.

Porous silicon is a typical example of the fractal-like microstructure consisting of a fractal network of nanocrystals connected via silicon oxide interface (see Fig. 5.25). Due to the close arrangement of nanocrystals and the finite potential barriers, tunneling and migration of electrons and excitations become possible. Size distribution and the quantum size effect give rise to the distribution of potential barriers. Because of the random distribution of potential barriers and distances between crystallites, the transport of excitations is essentially dispersive. Dispersive transport results in peculiar kinetic features of luminescence.

Vial et al. (1992) were the first to highlight the importance of tunneling in luminescence of porous silicon. These authors suggested that the experimentally observed exponential decrease of the mean lifetime with the emitted photon energy is due to exponential increase in the transparency of intercrystallite potential barrier. Several authors have proposed the stretched-exponential function

$$I(t) = I_0 \exp\{-(t/\tau)^\beta\} \tag{7.9}$$

with the scaling exponent β to approximate kinetics of the red-orange emission band of porous silicon (Ookubo 1993; Pavesi and Ceschini 1993). This model is known to describe the kinetics of various processes in disordered media due to random distribution of lifetimes as well as due to random walks of reacting components in space of noninteger dimensionality (Kopelman 1988), namely

$$d = 2\beta. \tag{7.10}$$

The scaling factor $\beta = d/2$ appears because of the specific dependence of the number of sites visited in the course of random walks versus number of jumps made, as well as due to the dependence of the probability to visit the same site more than once during a finite time interval.

Later the modified stretched-exponential function was found to describe adequately the decay of porous silicon photoluminescence in a wide spectral range (Gaponenko, Petrov et al. 1996). This function has the form

$$I(t) = I_0 \exp\{-t/\tau_0 - (t/\tau)^\beta\}. \tag{7.11}$$

It implies a local decay process with time τ_0 and quenching associated with random walks in a fractal space (Lianos 1988).

Figure 7.9 presents the typical luminescence decay kinetics along with the weighted residuals, and chi-square (χ^2) and Durbin-Watson parameters (DW), which were examined to ensure the goodness of fit. Chi-square is close to unity in the case of the successful fit, and DW is close to 2.0 in the case of completely random residuals and tends to zero when the residuals are systematic.

The parameter τ_0 is assumed to reflect the radiative or nonradiative decay of excitations within the emitting centers, whereas the τ and β values describe the radiative or nonradiative decay due to random walks of electrons and/or holes in the space formed by silicon crystallites. Therefore, τ_0 can be referred to as *local*

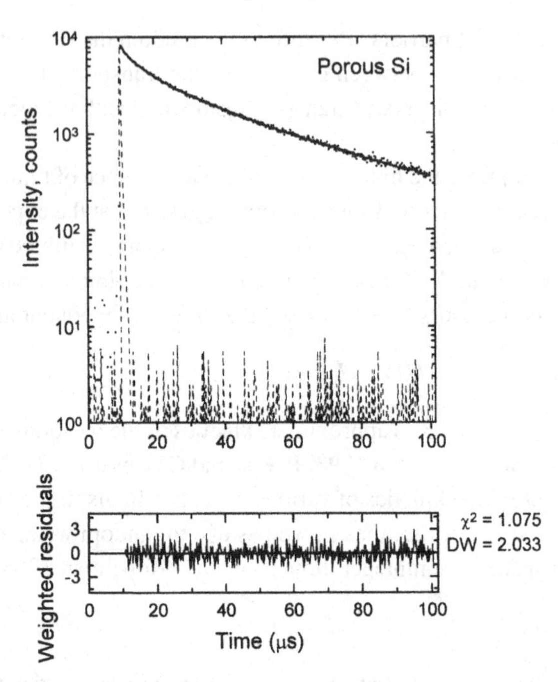

Fig. 7.9. Photoluminescence decay of porous silicon (Gaponenko, Petrov et al. 1996). Upper panel presents experimentally measured decay (dots), its approximation using the modified stretched exponential (solid line) and the apparatus response function (dashed line). Lower panel shows weighted residuals along with the χ^2 and Durbin-Watson parameters.

lifetime, and τ as *migration* lifetime. The value of τ_0 was found to be nearly independent of the emitted photon energy [Fig. 7.10(a)]. This result, along with the fact that the specific sample under consideration was fabricated from low-resistant p-type silicon, makes possible the conclusion that a photoexcited electron, independent of its energy, always has equal probability to recombine with a hole within the same crystallite, since the concentration of equilibrium holes is much higher than that of nonequilibrium electrons.

Remarkable are the values of the random walk parameters evaluated from experimental kinetics. First of all, the migration time falls exponentially with energy, indicating that migration takes place via tunneling escape of carriers through potential barriers. This is the confirmation of the model proposed by Vial et al. (1992). Additionally, the dimensionality of space in which random walks occur is as according to (7.10) $d = 0.86$, that is, close to but less than the dimensionality of a quantum wire. It is reasonable to suppose that the spatial arrangement of crystallites in porous silicon possesses some symmetry

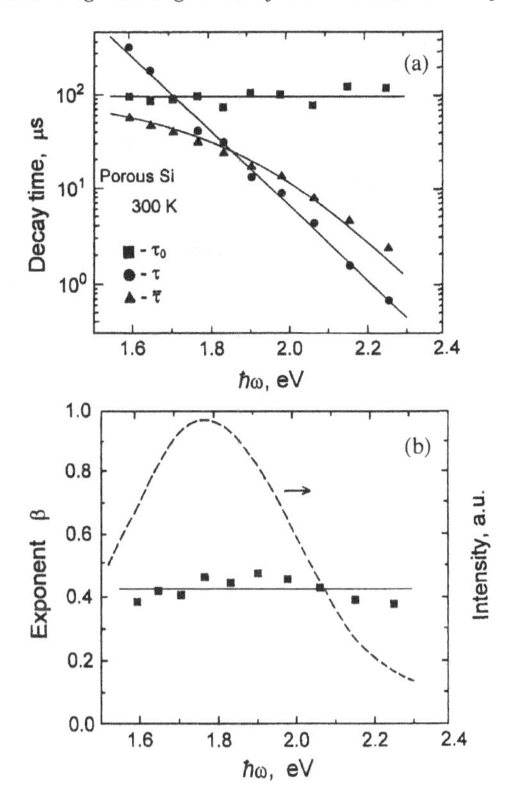

Fig. 7.10. Kinetic parameters of porous silicon (adapted from Gaponenko, Petrov et al. 1996). The emission spectrum in panel (a) corresponds to the recovered time-resolved spectrum collected within a few microseconds after excitation. Kinetic parameters in panel (a) are the local time τ_0 (squares), the migration lifetime τ (full circles), and the estimate of the mean time $\bar{\tau}$ (triangles). Note independence of τ_0 on $\hbar\omega$ and exponential dependence of τ on $\hbar\omega$. Panel (b) presents spectral dependence of the scaling exponent of the stretched-exponential term (squares), along with the time-resolved emission spectrum recorded within a few microseconds after excitation. All data correspond to the same sample.

due to the axial etching geometry. Thus, the absolute values of the parameters τ_0, τ, β and their spectral dependencies provide an additional justification of the model.[1]

Pavesi (1996) considered the similar model that includes on-site recombination and intersite hopping. He found that in the range of porosity $\rho = 67 \cdots 87$ percent the migration time versus photon energy obeys the same exponential

[1] In strongly oxidized aged samples of porous Si, migration is inhibited and the modified stretched-exponential decay is replaced by micellar kinetics with randomly distributed trap states (see Section 5.7 for details).

relation independent of ρ. It is interesting that for lower porosity, $\rho \approx 50 \cdots 60$ percent the scaling exponent β is nearly independent of $\hbar\omega$, the value of β being close to 2/3, which is the case of a *percolation cluster* on a fractal embedded in Euclidian space with $d = 2$ or $d = 3$ (Kopelman 1988 and references therein).

Brus (1996) considered migration processes for dry and wet porous silicon. When pores are impregnated with water, migration and mobility of electrons increase considerably due to strong electron-polar molecule dynamic coupling. This approach is similar to that previously used in molecular and protein electron dynamics.

8
Spatially organized ensembles of nanocrystals

The advances in physics and the technology of semiconductor nanocrystals that were summarized in Chapters 2–7 provide comprehensive knowledge on the optical and electronic properties of nanocrystals and make it possible to create novel mesoscopic materials with desirable parameters by means of stoichiometry and size control. In these chapters the intrinsic properties of nanocrystals were discussed, implying the absence of any cooperative effect on the properties of a given nanocrystal ensemble. In recent years significant progress has been made in moving from randomized nanocrystals towards spatially organized structures like nanocrystal superlattices, quantum dot solids, and photonic crystals. The principal results obtained in the field will be reviewed in this chapter.

8.1 Superlattices of nanocrystals: quantum dot solids

There are several ways to develop a nanocrystal superlattice, that is, a structure consisting of identical nanocrystals with regular spatial arrangement. The first is to use zeolites, which form a skeleton with regular displacement of extremely small cages, the size of a cage being typically about 1 nm. A number of clusters such as Cd_nS_m and Zn_nS_m can be embedded in these cages, the cluster size distribution and geometry being controlled by the topology of the three-dimensional host surface (Wang et al. 1989; Stucky and MacDougall 1990; Bogomolov and Pavlova 1995 and references therein). Using various zeolites as frameworks for semiconductor clusters makes possible the study of regular three-dimensional cluster lattices with variable intercluster spacing. A pronounced shift to the red of absorption spectrum with decreasing spacing has been observed (Stucky and MacDougall 1990).

Several authors reported on successful realization of periodic two-dimensional arrays of nanocrystals. Bestwick et al. (1995) developed a two-dimensional array of free-standing GaAs quantum dots from GaAs/AlGaAs single quantum well material by electron-beam lithography and low damage electron cyclotron resonance plasma etching. Each quantum dot had a pillar shape of approximately 50 nm diameter, the typical interdot distance being about 200 nm. Neither quantum confinement effects nor collective phenomena for this size/spacing scale are expected. Nanocrystals were found to exhibit excitonic luminescence similar to bulk GaAs crystals.

Tersoff et al. (1996) examined the self-organization effect in strained heterostructures, resulting in free-standing nanocrystals (see Section 3.3 for more detail). They investigated the growth of multilayer arrays of coherently strained islands and showed theoretically and experimentally a pronounced vertical correlation between islands in successive layers. The island size and spacing grow progressively more uniform. In effect, the structure organizes into a more regular three-dimensional arrangement, providing a possible route to obtain the size uniformity important for electronic applications. Using Si-$Si_{0.25}Ge_{0.75}$ superlattices, nanocrystals of size on the order of 10 nm were fabricated.

Motte et al. (1995) used reverse micelles in organic solution to fabricate a close-packed monolayer of Ag_2S particles. Particle size was controlled to range from 3 to 6 nm. A pronounced self-organization of particles in a hexagonal network was observed. The network formed on a long-range domain, which is usually on the order of 10^6 nm^2. Self-assembly was mainly due to Van der Waals and dispersion forces. When solid support was removed, a multilayered structure developed, featuring face-centered cubic symmetry.

Basically, self-assembling of nanosize particles promoted by Van der Waals interaction in a monodisperse concentrated solution seems to be the most challenging route to close-packed three-dimensional structures. These objects belong to a class of materials known as *colloidal crystals*. Since the first identification of natural colloidal crystals, namely a specific type of *virus* (Williams and Smith 1957), the features of colloidal crystals were found to be inherent in a number of natural and artificial objects (Pieranski 1983 and references therein). With respect to nanocrystals self-organized to form a macroscopic colloidal crystal, the term *quantum dot solid* was introduced (Murray, Kagan and Bawendi 1995). First, Spannel and Anderson (1991) found that in concentrated colloidal solution, growth of ZnO nanoparticles due to Ostwald ripening is not the only process. Cluster-cluster aggregation resulting in a three-dimensional network was found to occur as well. Later on, Weller and Bawendi with co-workers reported on well-defined close-packed superstructures consisting of II-VI nanocrystals. In the first case, CdS-based clusters were found to form a double-diamond superlattice (Vossmeyer et al. 1995). X-ray analysis revealed a superlattice framework built up of covalently linked clusters consisting of about 100 atoms each. The superlattice is best described as two enlarged and interlaced diamond or zinc blende lattices. Because both

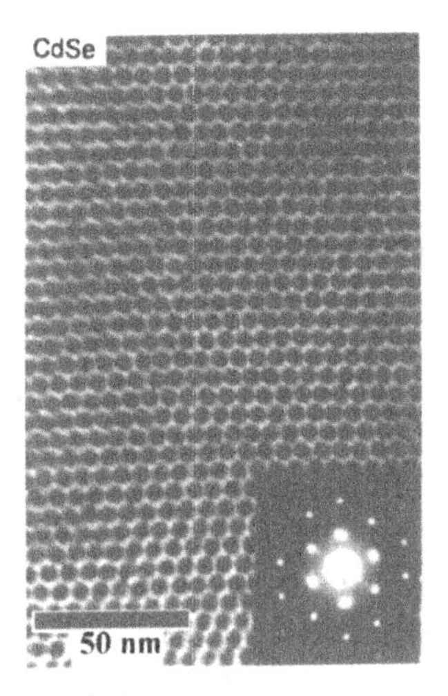

Fig. 8.1. High-resolution transmission electron microscopy image and small-angle electron diffraction pattern (insert) of a three-dimensional quantum dot superlattice (Murray, Kagan, and Bawendi 1995). The superlattice consists of CdSe nanocrystals of 4.8 nm diameter each. Nanocrystals are assembled in a face-centered cubic lattice. Image corresponds to (101) crystallographic plane of the array.

the superlattice and the clusters display the same structural features, the crystal structure possesses the self-similarities inherent in fractal geometry.

In the second case, high-quality monodisperse CdSe nanoparticles of variable size were organized in a face-centered cubic lattice (Murray et al. 1995). Perfect supercrystal structure is evident from TEM and X-ray studies (Fig. 8.1). Optical properties were found to be modified due to interdot interactions. Comparison of optical spectra for nanocrystals close-packed in the solid with those in a dilute matrix revealed that, although the absorption spectra are essentially identical, the emission line shape of crystallites in the solid is modified and red-shifted, indicating interdot coupling. Systematic studies of electronic energy transfer in CdSe quantum dot solids (Kagan et al. 1996) revealed long-range resonance transfer of electronic excitation from the more electronically confined states of the small crystallites to the higher excited states of the large dots. Foerster theory for long-range resonance transfer through dipole-dipole interdot interaction was used to explain electronic energy transfer in these close-packed nanocrystal structures.

No doubt, studies of periodic superstructures of quantum-confined crystallites and the concept of the quantum dot solid will result in a number of challenging findings in the near future. One can foresee such issues as formation of electron and hole minibands, *Anderson localization* of electrons on potential fluctuations, and "impurity states" in a supercrystal when the regular structure contains some imperfections.

8.2 Photonic crystals

The concept of the quantum dot solid is relevant to the case of a three-dimensional superstructure with a period compared with the de Broglie wavelengths of an electron and a hole. Even more interesting with respect to its optical properties seems to be a three-dimensional superstructure with a period on the order of the photon wavelength, that is, 10^2–10^3 nm when the optical range is considered. Structures with a periodic space modulation of dielectric function with a period of the photon wavelength are referred to as *photonic crystals*. Formation of allowed bands of particle states and appearance of the forbidden band gap in between, which is well-known for conventional solids (see the problem of an electron in a periodic potential considered in Section 1.1), is not due to the charge carried by an electron but is basically the common property of all waves in a periodic potential. In the case of optical waves, space modulation of the refractive index is an analog of electrostatic potential in the case of electrons. Therefore, the *photon density of states* is significantly modified when alternating layers, needles, or balls are arranged in a one-, two-, or three-dimensional lattice, respectively. One can see that a one-dimensional photonic crystal is just the well-known dielectric mirror with alternating $\lambda/2$ layers. Because of the Bragg condition, certain wave numbers (namely, $k_n = 2\pi n/\lambda$) are relevant to the standing waves and thus have to be reflected backward because they cannot propagate throughout a medium. This is shown schematically in Fig. 8.2.

Two- and three-dimensional photonic crystals are much more difficult to fabricate. However, three-dimensional structures promise an opportunity to create a photonic dielectric structure with a forbidden gap for photons in all directions. The concept of omnidirectional photonic band gap was advanced a decade ago (Yablonovich 1987; John 1987) and great progress in the theory has been achieved (see, e.g., Yablonovich 1994 and references therein). A number of basic issues like photon effective mass, Anderson localization of photons, inhibition and enhancement of spontaneous emission, coupled atom-field state, and so on are currently being discussed in relation to three-dimensional dielectric lattices. However, in spite of significant progress in theoretical analysis of these problems, experimental studies are still rather fragmentary. The main problem is fabrication of three-dimensional solid-state lattices with a period on the order of 10^2 nm. Colloidal crystals assembled from semiconductor nanocrystallites seem to be the most promising way toward real photonic dielectric. The first step has been made using artificial gem opals (Bogomolov et al. 1996; 1997). Gem opals are known to consist of spatially arranged silica microspheres organized either in cubic or hexagonal lattices. Iridescent properties of natural and artificial opals are due to interference in the spatially periodic structure.

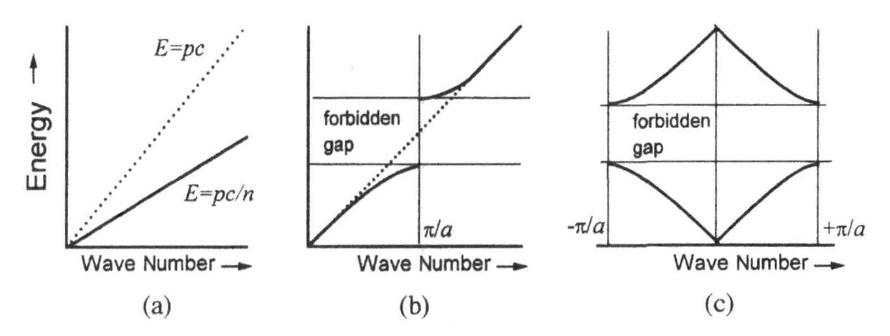

Fig. 8.2. Formation of photonic bands in a one-dimensional periodic structure. (a) Dispersion curve of photons in vacuum ($n = 1$, dotted line) and in a continuous medium ($n > 1$, solid line). (b) Dispersion curve of photons in a periodic medium with alternating layers of the same thickness, $a/2$ but different refraction indices, n_1 and n_2. A forbidden gap appears at wave number $k_m = m\pi/a$, where m is an integer. At these values, standing waves occur and no propagating waves are possible. The larger the n_1/n_2 ratio, the wider is the forbidden gap. (c) Reduced band structure. Because of the periodicity of the medium, wave number values differing by $2m\pi/a$ are equivalent and plot in panel (b) can be presented in reduced form. An interval of wave numbers

$$\left[-\frac{\pi}{a}; +\frac{\pi}{a}\right]$$

is the first Brilloin zone for photons. Compare with Fig. 1.5, which presents similar plots for electrons in a periodic one-dimensional medium.

A typical electron microscopy image of the artificial opal is presented in Fig. 8.3. Close-packing of monodisperse submicron globules in a face-centered cubic lattice is evident. Note that every globule has its internal structure. The globule consists of smaller SiO_2 nanoparticles. All the lattices show a dip in the optical transmission spectrum, with the spectral position depending on the lattice period and propagation direction (Fig. 8.4). The nature of a spectrally selective transmission of a disperse medium with vanishing dissipation is just multiple scattering and the interference of light waves. This can be intuitively understood in terms of the Bragg diffraction of optical waves. In other words, formation of a pronounced stop band is indicative of a reduced density of photon states within the sample, and thus, can be classified as a photonic pseudogap phenomenon. By analogy with solid crystals, the structure under consideration resembles features of a gapless semiconductor or a semimetal. The gap does exist but it is not omnidirectional.

The observed spectrally selective characteristics of closed-packed SiO_2 microspheres were succesfully reproduced by means of the multiple wave scattering theory using a quasi-crystalline model (Bogomolov et al. 1997). The model implies a finite number of equally spaced monolayers, each consisting of identical close-packed spherical particles (Fig. 8.5). Assuming the statistical

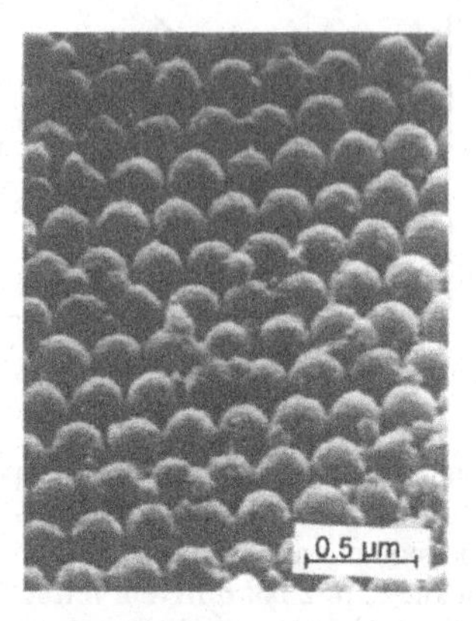

Fig. 8.3. Microphotograph of an artificial opal sample (Bogomolov et al. 1996). Opal is a face-centered cubic lattice built of spherical SiO_2 globules. Each globule has internal structure consisting of smaller silica nanoparticles.

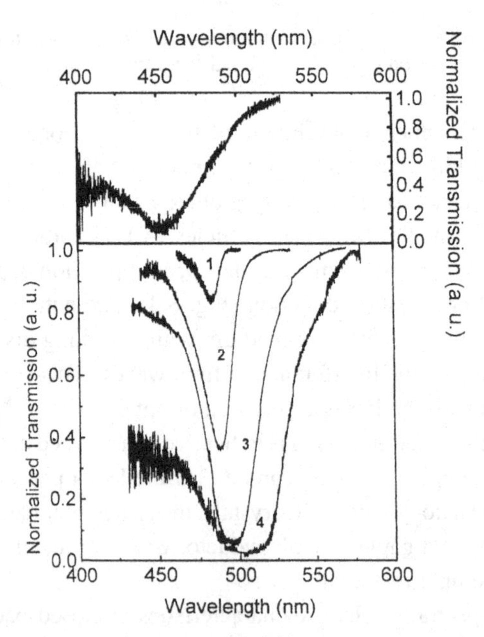

Fig. 8.4. Optical transmission spectra of an opal sample with empty voids (upper panel) and with voids impregnated with various liquids: methanol (1), $n = 1.3284$; ethanol (2), $n = 1.3614$; cyclohexane (3), $n = 1.4262$; and toluene (4), $n = 1.4969$ (Bogomolov et al. 1996).

Quasi-crystalline approximation

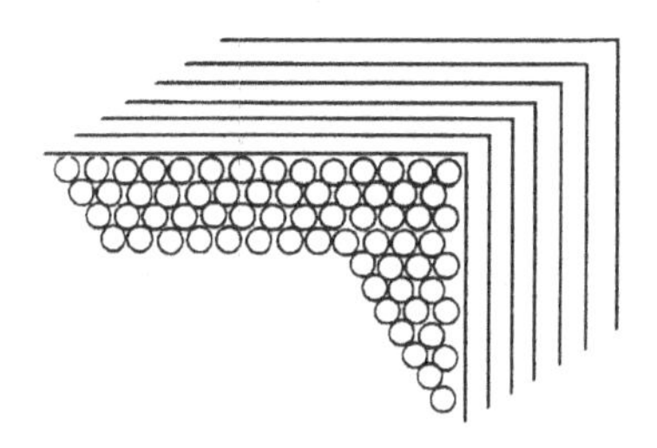

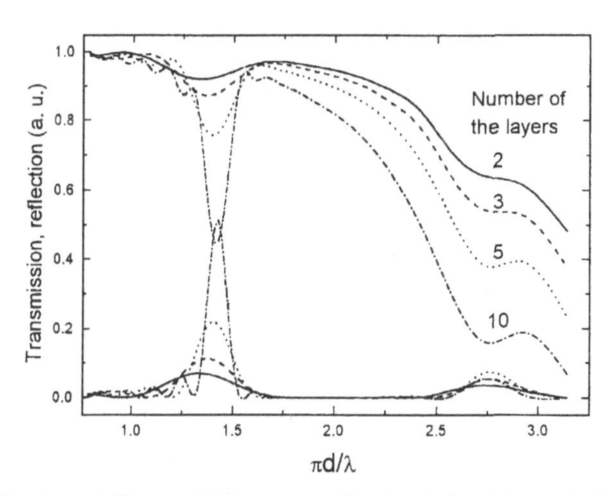

Fig. 8.5. Quasi-crystalline model (upper panel) and calculated transmission/reflection spectra (lower panel) for number of the layers ranging from 2 to 10 (Bogomolov et al. 1997). Transmission and reflection are plotted versus a dimensionless parameter $\pi d/\lambda$, where d is the sphere diameter that is equal to interlayer distance and λ is the photon wavelength. Numbers at the curves indicate the number of layers. Refractive index of spheres versus voids is 1.26, which is typical for an opal with empty voids.

independence of the individual monolayers, it is possible at first to find the scattering amplitude of a single monolayer, taking into account multiple rescattering of the particles within the layer, and then to account for the irradiation between the different monolayers within the samples under consideration. The results are presented in Fig. 8.5, showing a build-up of selective transmission/reflection with the increasing number of layers.

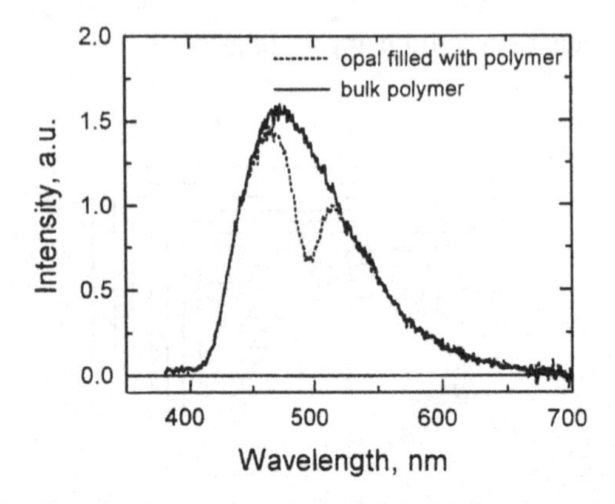

Fig. 8.6. Modification of spontaneous emission of dye molecules embedded in an artificial opal lattice (Petrov et al. in press). The solid curve corresponds to the luminescence spectrum of 1, 8-naphthoylene-1′, 2′-benzimidazole in polymethyl metacrylate. The dashed curve shows the luminescence spectrum of the same molecules in the same polymer embedded in an opal. A pronounced dip in the emission spectrum is evidence of inhibited spontaneous emission due to the lack of photon density of states in the spectral region relevant to the pseudogap.

The voids between close-packed microspheres form a sublattice that can be impregnated with a liquid or polymer. This provides the ability to enhance band gap due to enhancement of the refraction index modulation (see lower panel in Fig. 8.4). In a similar manner, resonant species like molecules and nanocrystals can be embedded as well. The first experiments on dye molecules in opals show a pronounced effect of the modified photon density of states on spontaneous emission (Fig. 8.6).

Recently the first experimental results have been obtained on synthesis and properties of a photonic crystal doped with quantum dots (Gaponenko et al. in press). Well-defined CdTe nanocrystals were developed in an aqueous solution and then embedded in a silica-based photonic crystal. Nanocrystals embedded in a photonic crystal exhibit a noticeable change of the luminescence spectrum when the latter overlaps with the photonic pseudogap (Fig. 8.7). The experiments clearly show a dip in the emission spectrum correlating with the spectral position of the gap. The appearence of the dip is indicative of the inhibited spontaneous emission.

The observed modification of the spectrum of spontaneous emission of quantum dots in silica-based photonic crystals is more significant than the modification observed for organic molecules (Fig. 8.7). This is because of the intrinsically narrower emission spectrum of quantum dots as compared to organic molecules that provides good overlapping of the emission spectrum and

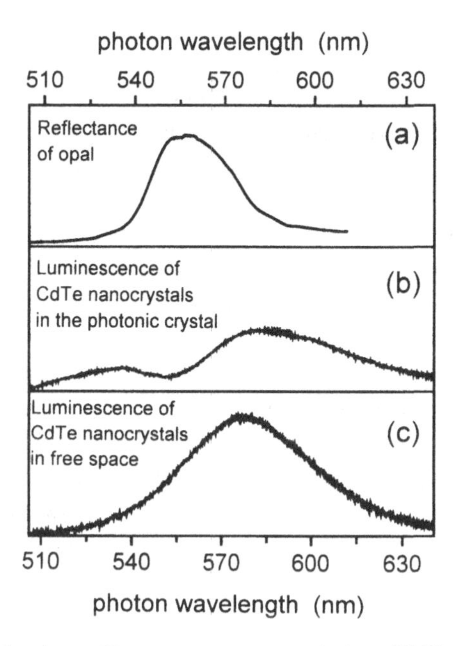

Fig. 8.7. Photonic band gap effect on spontaneous emission of CdTe nanocrystals. Panel (a) shows the reflection spectrum of the silica colloidal crystal doped with 2.4 nm size CdTe crystallites. The pronounced peak in reflection is indicative of the dip in the spectral distribution of the photon density of states. In the spectral range relevant to the photonic pseudogap the luminescence spectrum has a minimum (panel b) which is indicative of inhibited spontaneous emission. Panel (c) presents the reference emission spectra of the same nanocrystals in solution.

the gap. Every individual component within the inhomogeneously broadened spectrum reacts individually to the modified density of photon states. Therefore the total modification of spontaneous emission enhances in the case of inhomogeneously broadened spectrum as compared to a homogeneous band of the same total bandwidth.

These experiments demonstrate the novel mesostructure with separately controllable densities of photon and electron states and show that spontaneous emission of nanocrystals which is controlled at large by quantum confinement effect experience strong modification due to modified photon density of states in a photonic crystal with respect to the free space. Thus the interplay of electron and photon confinement within the same structure opens a way toward novel light sources with controllable spontaneous emission.

Further attempts towards omnidirectional photonic band gap in the optical range are necessary that can offer a modulation of the refractive index as large as 100 percent and even higher. This means that an opal lattice consisting of silica globules with the effective refraction index $n \approx 1.2 \ldots 1.3$ should be impregnated with a medium possessing $n \approx 2.5$ and higher. This can be realized, in principle, by means of sol-gel processes to get amorphous oxides. Recently a successful fabrication of a SiO_2/TiO_2 superlattice in this way has been reported (Kapitonov et al. 1997). Another possibility is to develop opal-like structures based on oxides other than silica.

Photonic crystals containing resonant inclusions are expected to show a number of interesting phenomena of great scientific and practical importance. Inhibited spontaneous emission makes it possible to reduce the threshold energy in lasers. In principle it may lead to a thresholdless laser in the sense that it is possible to fabricate a thresholdless device that spontaneously emits light of the desirable wavelength and directionality. Furthermore, these structures are expected to possess enhanced resonant nonlinearities due to the interplay of electronic and morphological resonances. Additionally, they offer efficient laser harmonics generation. This becomes possible due to separation of nonlinearity (provided by the resonant inclusions) and phase synchronism (provided by periodicity of the medium), which is not possible in conventional media.

References

Abraham, F. F. 1974. *Homogeneous Nucleation Theory*. New York: Academic Press.

Agekyan, V. F., and Serov, A. Yu. 1996. Optical properties of PbI_2 microcrystals in a glass matrix. *Solid State Phys.* (translated from Russian). 38: 122–6.

Alferov, Zh. I., Gordeev, N. Yu., Egorov, A. Yu., Zhukov, A. E., Zaitsev, S. V., Kovsh, A. R., Kop'ev, P. S., Kosogov, A. O., Ledentsov, N. N., Maksimov, M. V., Saharov, A. V., Ustinov, V. M., Schernyakov, Yu. M., and Bimberg, D. (In Press)

Alivisatos, A., Harris, A., Levinos, N., Steigerwald, M., and Brus, L. 1988. Electronic states of semiconductor clusters: Homogeneous and inhomogeneous broadening of the optical spectrum. *J. Chem. Phys.* 89: 4001–11.

Andreas, R. P., Averback, R. S., Brown, W. L., Brus, L. E., Goddard, W. A., Kaldor, A., Louie, S. G., Moskovits, M., Peersy, P. S., Riley, S. J., Siegel, R. W., Spaepen, F., and Wang, Y. 1989. Research opportunities on clusters and cluster-assembled materials. *J. Mater. Res.* 4: 704–36.

Andrianov, A. A., Kovalev, D. I., Shuman, U. B., and Yaroshetskii, I. D. 1992. Blue-green emission band in porous silicon. *JETP Lett.* 56: 943–6.

Andrianov, A., Kovalev, D. I., Zinoviev, N. N., and Yaroshetskii, I. D. 1993. Anomalous polazition of porous silicon photoluminescence. *JETP Lett.* 58: 417–20.

Apanasevich, P. A. 1977. *Principles of the Theory of the Interaction of Light with Matter.* (in Russian). Minsk: Nauka i Technika.

Arakawa, Y., and Sakaki, H. 1982. Multidimensional quantum well laser and temperature dependence of its threshold current. *Appl. Phys. Lett.* 40: 939–41.

Artemyev, M. V., Gaponenko, S. V., Germanenko, I. N., and Kapitonov, A. M. 1995. Irreversible photochemical spectral hole burning in quantum-sized CdS nanocrystals embedded in polymeric film. *Chem. Phys. Lett.* 243: 450–5.

Artemyev, M. V., Yablonskii, G. P., and Rakovich, and Yu. P. 1995. Luminescence spectra of quantum-sized CdS and PbI_2 particles in static electric field. *Acta Physica Polonica A* 87: 523–7.

Artemyev, M. V., Sperling, V., and Woggon, U. 1997. Electroluminescence in thin solid films of closely-packed CdS nanocrystals. *J. Appl. Phys.* 81: 6975–7.

Bahel, A., and Ramakrishna, M. V. 1995. Structure of the Si_{12} cluster. *Phys. Rev. B* 51: 13849–51.

Baldereshi, A., and Lipari, N. O. 1973. Hole Hamiltonian in spherical approximation. *Phys. Rev. B* 8: 2697–701.

Bandaranayake, R. J., Wen, G. W., Lin, J. Y., Jiang, H. X., and Sorensen, C. M. 1995.

Structural phase behavior in II-VI semiconductor nanoparticles. *Appl. Phys. Lett.* 67: 831–3.

Banyai, L., and Koch, S. W. 1986. Absorption blue shift in laser excited semiconductor microspheres. *Phys. Rev. Lett.* 57: 2722–4.

Banyai, L., and Koch, S. W. 1993 *Semiconductor Quantum Dots*. Singapore: World Scientific.

Banyai, L., Hu, Y. Z., Lindberg, M., and Koch, S. W. 1988. Third-order optical nonlinearities in semiconductor microstructures. *Phys. Rev. B* 38: 8142–53.

Banyai, L., Gilliot, P., Hu, Y. Z., and Koch, S. W. 1992. Surface-polarization instabilities of electron-hole pairs in semiconductor quantum dots. *Phys. Rev. B* 45: 14136–42.

Barnes, M. D., Kung, C.-Y., Whitten, W. B., and Ramsey, J. M. 1996. Fluorescence of oriented molecules in a microcavity. *Phys. Rev. Lett.* 76: 3931–34.

Bassani, F., and Parravicini, G. P. 1975. *Electronic States and Optical Transitions in Solids*. Oxford: Pergamon. 1982. Moscow: Mir.

Bawendi, M. G., Carroll, P. J., Wilson, W. L., and Brus, L. E. 1992. Luminescence properties of CdSe quantum crystallites: Resonance between interior and surface localized states. *J. Chem. Phys.* 96: 946–54.

Bawendi, M. G., Kortan, A. R., Steigerwald, M. L., and Brus, L. E. 1989. X-ray structural characterization of larger CdSe semiconductor clusters. *J. Chem. Phys.* 91: 7282–5.

Bawendi, M. G., Steigerwald, M. L., and Brus, L. E. 1990. The quantum mechanics of larger semiconductor clusters ("quantum dots"). *Annu. Rev. Phys. Chem.* 41: 477–96.

Bawendi, M. G., Wilson, W. L., Rothberg, L., Carroll, P. J., Jedju, T. M., Steigerwald, M. L., and Brus, L. E. 1990. Electronic structure and photoexcited-carrier dynamic in nanometer-size CdSe clusters. *Phys. Rev. Lett.* 65: 1623–7.

Beenakker, C. W. J. 1991. Theory of Coulomb-blockade oscillations in the conductance of a quantum dot. *Phys. Rev. B* 44: 1646–50.

Bellegie, L., and Banyai, L. 1991. Population induced nonlinearities in semiconductor nanocrystals. *Phys. Rev. B* 44: 8785–93.

Bellegie, L., and Banyai, L. 1993. Third-order nonlinear susceptibility of large semiconductor microcrystallites. *Phys. Rev. B* 47: 4498–4507.

Benisty, H., Sotomayor-Torres, C. M., and Weisbuch, C. 1991. Intrinsic mechanism for the poor luminescence properties of quantum-box system. *Phys. Rev. B* 44: 10945–8.

Berman, P. R. (ed.). 1994. *Cavity Quantum Electrodynamics*. New York: Academic Press.

Bespalov, V. A., and Kubarev, A. M. 1967. Nonlinear properties of cadmium selenide doped glasses. *Zhurn. Prikl. Spektr.* 7: 349–52.

Bestwick, T. D., Dawson, M. D., Kean, A. H., and Duggan, G. 1995. Uniform and efficient GaAs/AlGaAs quantum dots. *Appl. Phys. Lett.* 66: 1382–4.

Bhargava, R. N. 1996. Doped nanocrystalline materials – physics and applications. *J. Lumin.* 70: 230–8.

Bhargava, R. N., Gallagher, D., Hong, X., and Nurmikko, A. 1994. Optical properties of manganese-doped nanocrystals of ZnS. *Phys. Rev. Lett.* 72: 416–9.

Binggeli, N., and Chelikowsky, J. R. 1994. Langevin molecular dynamics with quantum forces: Application to silicon clusters. *Phys. Rev. B* 50: 11764–70.

Blakemore, J. S. 1985. *Solid State Physics*. Cambridge: Cambridge University Press. 1988. Moscow: Mir.

Blanton, S. A., Hines, M. A., and Guyot-Sionnest, P. 1996. Photoluminescence wandering in single CdSe nanocrystals. *Appl. Phys. Lett.* 69: 3905–7.

Blanton, S. A., Hines, M. A., Schmidt, M. E., and Guyot-Sionnest, P. 1996. Two-photon spectroscopy and microscopy of II-VI semiconductor nanocrystals. *J. Lumin.* 70: 253–68.

Bloembergen, N. 1965. *Nonlinear Optics*. New York: Benjamin Inc.

Blossey, D. F. 1970. Wannier exciton in an electric field. I. Optical absorption by bound and continuum states. *Phys. Rev. B* 2: 3976–90.

Blossey, D. F. 1971. Wannier exciton in an electric field. II. Electroabsorption in direct-band-gap solids. *Phys. Rev. B* 3: 1382–91.

Bobkova, N. M., and Sinevich, A. K. 1984. Study of the color formation in glasses of the selen ruby type. *Sov. Phys. Chem. Glass* 10: 337–44.

Bockelmann, U. 1993. Exciton relaxation and radiative recombination in semiconductor quantum dots. *Phys. Rev. B* 48: 17637–40.

Bockelmann, U., and Bastard, G. 1990. Photon scattering and energy relaxation in two-, one-, and zero-dimensional electron gases. *Phys. Rev. B* 42: 8947–51.

Bockelmann, U., Roussignol, Ph., Filoramo, A., Heller, W., Abstreiter, G., Brunner, K., Bohm, G., and Weimann, G. 1996. Time resolved spectroscopy of single quantum dots: Fermi gas of excitons? *Phys. Rev. Lett.* 76: 3622–5.

Bodnar, I. V., Molochko, A. P., and Solovey, N. P. 1993. Optical properties of silicate glasses activated with ternary semiconductor compounds $CuInS_2$ and $CuInSe_2$. *Inorg. Mat.* 29: 1226–8.

Bogomolov, V. N., and Pavlova, T. M. 1995. Three-dimensional cluster lattices. *Semiconductors* 29: 428–39.

Bogomolov, V. N., Gaponenko, S. V., Kapitonov, A. M., Prokofiev, A. V., Ponyavina, A. N., Silvanovich, N. I., and Samoilovich, S. M. 1996. Photonic band gap in the visible range in a three-dimensional solid state lattice. *Appl. Phys. A* 63: 613–6.

Bogomolov, V. N., Gaponenko, S. V., Germanenko, I. N., Kapitonov, A. M., Petrov, E. P., Gaponenko, N. V., Prokofiev, A. V., Ponyavina, A. N., Silvanovich, N. I., and Samoilovich, S. M. 1997. Photonic band gap phenomenon and optical properties of artificial opals. *Phys. Rev. E* 55: 7619–25.

Bohren, C. F., and Huffman, D. R. 1984. *Absorption and Scattering of Light by Small Particles*. New York: J. Wiley & Sons. 1986. Moscow: Mir.

Bonch-Bruevich, A. M., Razumova, T. K., and Rubanova, G. M. 1967. Experimental study of the optical properties and nonlinear absorption of glass with cadmium selenide. *Sov. Phys. Sol. State* 9: 2265–73.

Bondarenko, V. P., Borisenko, V. E., Dorofeev, A. M., Germanenko, I. N., and Gaponenko, S. V. 1994. Spectral characteristics of visible light emission from porous Si: Quantum confinement or impurity effect? *J. Appl. Phys.* 75: 2727–29.

Born, M., and Wolf, E. 1977. *Principles of Optics*. Oxford: Pergamon. 1979. Moscow: Mir.

Borrelli, N. F., and Smith, D. W. 1994. Quantum confinement of PbS microcrystals in glass. *J. Non-Cryst. Sol.* 180: 25–31.

Borrelli, N. F., Hall, D. W., Holland, H. J., and Smith, D. W. 1987. Quantum confinement effects of semiconducting microcrystallites in glass. *J. Appl. Phys.* 61: 5399–5409.

Bret, G., and Gires, F. 1964. Giant pulse laser and light amplifier using variable transmission coefficient glasses as light switches. *Appl. Phys. Lett.* 4: 175–76.

Brus, L. E. 1983. A simple model for the ionization potential, electron affinity, and aqueous redox potentials of small semiconductor crystallites. *J. Chem. Phys.* 79: 5566–71.

Brus, L. E. 1984. Electron-electron and electron-hole interactions in small semiconductor crystallites: The size dependence of the lowest excited electronic state. *J. Chem. Phys.* 80: 4403–9.

Brus, L. 1986. Electronic wave functions in semiconductor cluster: Experiment and theory. *J. Phys. Chem.* 90: 2555–60.
Brus, L. E. 1991. Quantum crystallites and nonlinear optics. *Appl. Phys. A* 53: 465–74.
Brus, L. E. 1994. Luminescence of silicon materials: chains, sheets, nanocrystals, nanowires, microcrystals, and porous silicon. *J. Phys. Chem.* 98: 3573–81.
Brus, L. 1996. Model for carrier dynamics and photoluminescence quenching in wet and dry porous silicon thin films. *Phys. Rev. B* 53: 4649–56.
Brus, L. E., Szajowski, P. F., Wilson, W. L., Harris, T. D., Shuppler, S., and Citrin, P. H. 1995. Electronic spectroscopy and photophysics of Si nanocrystals. *J. Am. Chem. Soc.* 117: 2915–22.
Bryant, G. W. 1990. Biexciton binding in quantum boxes. *Phys. Rev. B* 41: 1243–46.
Bsiesy, A., and Vial, J. C. 1996. Voltage-tunable photo- and electroluminescence of porous silicon. *J. Lumin.* 70: 310–19.
Butcher, P. N., and Cotter, D. 1990. *The Elements of Nonlinear Optics*. Cambridge: Cambridge University Press.
Butler, L., Redmond, G., and Fitzmaurice, D. 1993. Preparation and spectroscopic characterization of highly confined nanocrystallites of GaAs in decane. *J. Phys. Chem.* 97: 10750–5.
Buzhinskii, I. M., and Bobrova, N. I. 1962. Spectral studies of glasses colored with CdS and CdSe. *Opt. Spectr.* 12: 387–95.
Calcott, P. D. J., Nash, K. J., Canham, L. T., Kano, M. J., and Brumhead, D. 1993. Identification of radiative transition in highly porous silicon. *J. Phys. C* 5: L91–L102.
Camata, R. P., Altwater, H. A., Vahala, K. J., and Flagan, R. C. 1996. Size classification of silicon nanocrystals. *Appl. Phys. Lett.* 68: 3162–4.
Canham, L. T. 1990. Silicon quantum wire fabrication by electrochemical dissolution of wafers. *Appl. Phys. Lett.* 57: 1046–8.
Cardona, M. 1972. *Modulation Spectroscopy*. Berlin: Springer.
Carlsson, N., Georgsson, K., Montelins, L., Samuelson, L., Seifert, W., and Wallenberg, R. 1995. Improved site homogeneity of InP-on-GaInP Stranski-Krastanow islands by growth on a thin GaP interface layer. *J. Cryst. Growth* 156: 23–9.
Chamarro, M. A., Gourdon, C., and Lavallard, P. 1992. Selective excitation of nanocrystals by polarized light. *Sol. Stat. Comm.* 84: 967–70.
Champagnon, B., Andrianasolo, B., Ramos, A., Gandais, M., Allais, M., and Benoit, J-P. 1993. Size of Cd(S,Se) quantum dots in glasses: Correlation between measurements by high-resolution transmission electron microscopy, small-angle x-ray scattering, and low-frequency inelastic Raman scattering. *J. Appl. Phys.* 73: 2775–80.
Chen, W., McLendon, G., Marchetti, A., Rehm, J. M., Freedhoff, M. I., and Myers, C. 1994. Size dependence of radiative rates in the indirect band gap material AgBr. *J. Am. Chem. Soc.* 116: 1585–6.
Chen, X., Uttamchandani, D., Sander, D., and O'Donnell, K. P. 1993. Luminescence decay of porous silicon. *Physica B* 185: 603–7.
Chepic, D. I., Efros, Al. L., Ekimov, A. I., Ivanov, M. G., Kharchenko, V. A., Kudriavtsev, I. A., and Yazeva, T. V. 1990. Auger ionization of semiconductor quantum drops in a glass matrix. *J. Lumin.* 47: 113–27.
Chestnoy, N., Hull, R., and Brus, L. E. 1986. Higher excited electronic states in clusters of ZnSe, CdSe, and ZnS: Spin-orbit, vibronic, and relaxation phenomena. *J. Chem. Phys.* 85: 2237–42.
Cho, K. (ed.) 1979. *Excitons*. Berlin: Springer.

Colvin, V., Cunningham, K. L., and Alivisatos, P. 1994. Electric field modulation of optical absorption in CdSe nanocrystals. *J. Chem. Phys.* 101: 7122–6.
Colvin, V., Schlamp, A., and Alivisatos, P. 1994. Light-emitting diodes made with cadmium selenide nanocrystals. *Nature* 370: 6488–91.
Cotter, D., Burt, M. G., and Girdlestone, H. P. 1990. Electroabsorptive behavior of semiconductor quantum dots in glass. *Semicond. Sci. Technol.* 5: 631–3.
Cotter, D., Girdlestone, H. P., and Moulding, K. 1991. Size-dependent electroabsorptive properties of semiconductor microcrystallites in glass. *Appl. Phys. Lett.* 58: 1455–7.
Cotter, D., Burt, M. G., and Manning, R. J. 1992. Below-band-gap third-order optical nonlinearity of nanometer-size semiconductor crystallites. *Phys. Rev. Lett.* 68: 1200–3.
Craciun, V., Boyd, I. W., Reader, A. H., and Vandenhoudt, D. E. W. 1994. Low temperature synthesis of Ge nanocrystals in SiO_2. *Appl. Phys. Lett.* 65: 3233–5.
Dabbousi, B. O., Bawendi, M. G., Onitsuka, O., and Rubner, M. F. 1995. Electroluminescence from CdSe quantum-dot/polymer composites. *Appl. Phys. Lett.* 66: 1316–18.
Davydov, A. S. 1965. *Quantum Mechanics*. New York: Pergamon.
Davydov, A. S. 1976. *Solid State Theory*. Moscow: Nauka.
Dederen, J. C., Auweraer, M. Van der, and Schryver, F. C. De. 1981. Fluorescence quenching of solubilized pyrene derivatives by metal ions in SDS micelles. *J. Phys. Chem.* 85: 1198–1202.
Delerue, C., Allan, G., and Lannoo, M. 1993. Theoretical aspects of the luminescence of porous silicon. *Phys. Rev. B* 48: 11024–36.
Demtroeder, W. 1995. *Laser Spectroscopy*. Berlin: Springer. 1985. Moscow: Mir.
Dimitrijevic, N. M., and Kamat, P. T. 1988. Oxidation of In_2S_3 and In_2Se_3 colloids as studied by pulse radiolysis. *Radiat. Phys. Chem.* 32: 53–7.
Dissanayake, A. S., Lin, J. Y., and Jiang, H. X. 1995. Quantum-confined Stark effect in CdS_xSe_{1-x} quantum dots. *Phys. Rev. B.* 51: 5457–60.
Dneprovskii, V. S., Efros, Al. L., Ekimov, A. I., Klimov, V. I., Kudriavtsev, I. A., and Novikov, M. G. 1990. Time resolved luminescence of CdSe microcrystals. *Sol. Stat. Com.* 74: 555–7.
Dneprovskii, V. S., Klimov, V. I., Okorokov, D. K., and Vandyshev, Yu. V. 1992. Ultrafast light induced transmission changes and laser emission of semiconductor quantum dots. *Phys. Stat. Sol. (b)* 173: 405–6.
Dneprovskii, V., Eev, A., Gushina, N., Okorokov, D., Panov, V., Karavanskii, V., Maslov, A., Sokolov, V., and Dovidenko, E. 1995. Strong optical nonlinearities in quantum wires and dots of porous silicon. *Phys. Stat. Sol. (b)* 188: 297–306.
Edamatsu, K., Iwai, S., Itoh, T., Yano, S., and Goto, T. 1995. Subpicosecond spectroscopy of the optical nonlinearities of CuCl quantum dots. *Phys. Rev. B* 51: 11205–8.
Efremov, N. A., and Pokutnii, S. I. 1990. Energy spectrum of an exciton in a small spherical semiconductor particle. *Sov. Phys. Sol. State* 32: 955–60.
Efros, A. L. 1992a. Luminescence polarization of CdSe microcrystals. *Phys. Rev. B* 46: 7448–58.
Efros, A. L. 1992b. Optical properties of semiconductor nanocrystals with degenerate valence band. *Superlattices and Microstructures* 11: 167–9.
Efros, Al. L., and Efros A. L. 1982. Interband light absorption in semiconductor sphere. *Semiconductors* 16: 1209–14.
Efros, Al. L., and Rodina, A. V. 1993. Band-edge absorption and luminescence of nonspherical nanometer-size crystals. *Phys. Rev. B* 47: 10005–7.

Efros, Al. L., Rosen, M., Kuno, M., Nirmal, M., Norris, D. J., and Bawendi, M. 1996. Band-edge exciton in quantum dots of semiconductors with a degenerate valence band: Dark and bright exciton states. *Phys. Rev. B* 54: 4843–56.

Einevoll, G. T. 1992. Confinement of excitons in quantum dots. *Phys. Rev. B* 45: 3410–17.

Ekimov, A. I. 1991. Optical properties of semiconductor quantum dots in glass matrix. *Physica Scripta* 39: 217–22.

Ekimov, A. 1996. Growth and optical properties of semiconductor nanocrystals in a glass matrix. *J. Lumin.* 70: 1–20.

Ekimov, A. I., and Onushchenko, A. A. 1982. Quantum-size effect in optical spectra of semiconductor microcrystals. *Semiconductors* 16: 1215–19.

Ekimov, A. I., and Onushchenko, A. A. 1984. Size quantization of electron energy spectrum in semiconductor nanocrystals. *JETP Lett.* 40: 337–40.

Ekimov, A. I., Onushchenko, A. A., and Tsekhomskii, V. A. 1980. Exciton light absorption by CuCl microcrystals in glass matrix. *Sov. Glass Phys. Chem.* 6: 511–12.

Ekimov, A. I., Efros, Al. L., and Onushchenko, A. A. 1985. Quantum size effect in semiconductor microcrystals. *Solid State Comm.* 56: 921–4.

Ekimov, A. I., Onushchenko, A. A., Plukhin, A. G., and Efros, A. A. 1985. Size quantization of excitons and evalution of their energy spectrum parameters in CuCl. *JETP* 88: 1490–501.

Ekimov, A. I., Efros, Al. L., Ivanov, M. G., Onushchenko, A. A., and Schumilov, S. K. 1989. Donor-like exciton in zero-dimension semiconductor structures. *Solid State Comm.* 69: 565–8.

Ekimov, A. I., Efros, Al. L., Shubina, T. V., and Skvortsov, A. P. 1990. Quantum-size Stark effect in semiconductor microcrystals. *J. Lumin.* 46: 97–100.

Ekimov, A. I., Hache, F., Schanne-Klein, M. C., Ricard, D., Flytzanis, C., Kudryavtsev, I. A., Yazeva, T. V., Rodina, A. V., and Efros, Al. L. 1993. Absorption and intensity-dependent photoluminescence measurement on CdSe quantum dots: Assignment of the first electronic transitions. *J. Opt. Soc. Amer. B* 10: 100–10.

Empedocles, S. A., Norris, D. J., and Bawendi, M. G. 1996. Photoluminescence spectroscopy of single CdSe quantum dots. *Phys. Rev. Lett.* 77: 3873–6.

Esch, V., Fluegel, B., Khitrova, G., Gibbs, H. M., Juajin, Xu, Kang, K., Koch, S. W., Liu, L. C., Risbud, S. H., and Peyghambarian, N. 1990. State filling, Coulomb, and trapping effects in the optical nonlinearity of CdTe quantum dots in glass. *Phys. Rev. B* 42: 7450–6.

Eychmüller, A., Hässelbarth, A., Katsikas, L., and Weller, H. 1991. Fluorescence mechanism of highly monodisperse Q-sized CdS colloids. *J. Lumin.* 48 & 49: 745–9.

Fafard, S., Leonard, D., Merz, J. L., and Petroff, P. M. 1994. Selective excitation of the photoluminescence and the energy levels of ultrasmall InGaAs/GaAs quantum dots. *Appl. Phys. Lett.* 65: 1388–90.

Fafard, S., Hinzer, K., Raymond, S., Dion, M., McCaffrey, J., Feng, Y., and Charbonneau, S. 1996. Red-emitting semiconductor quantum dot lasers. *Science* 274: 1350–3.

Faller, P., Kippelen, B., Hoenerlage, B., and Levy, R. 1993. Optical gain in CuCl nanocrystals. *Opt. Mater.* 2: 39–44.

Fauchet, P. M. 1996. Photoluminescence and electroluminescence of porous silicon. *J. Luminescence* 70: 294–310.

Feder, J. 1988. *Fractals*. New York: Plenum Press. 1991. Moscow: Mir.

Feldmann, J., Peter, G., Gobel, E. O., Dawson, P., Moore, K., Foxon, C., and Elliot,

R. J. 1987. Linewidth dependence of radiative exciton lifetimes in quantum wells. *Phys. Rev. Lett.* 59: 2337–40.

Filatov, I. V., and Kuzmitskii, V. A. 1996. Quantum-chemical calculations by CNDO/S technique of the excited states of MgO clusters. *J. Appl. Spectr.* 63: 76–80.

Flugge, S. 1971. *Practical Quantum Mechanics (Part I)*. Berlin: Springer. 1974. Moscow: Mir.

Flurry, R. L. 1983. Quantum chemistry. Englewood Cliffs. N.J.: Prentice-Hall. 1985. Moscow: Mir.

Flytzanis, C., Ricard, D., and Schanne-Klein, M. C. 1996. The role of photodarkening and Auger recombination in the dynamic of the optical response for Cd(S,Se) nanoparticles. *J. Lumin.* 70: 212–21.

Froelich, H. 1937. Die spezifische warme der kleiner metallteilchen. *Physica* 4: 406–10.

Froelich, H. 1958. *Theory of Dielectrics*. Oxford: Clarendon Press.

Froehlich, D., Haselhoff, M., Reinmann, K., and Itoh, T. 1995. Determination of the orientation of CuCl nanocrystals in a NaCl matrix. *Sol. State Comm.* 94: 189–91.

Furukawa, S., and Miyasato, T. 1988. Quantum size effect on the optical band gap of microcrystalline Si: H. *Phys. Rev. B* 38: 5726–9.

Gakamsky, D. M., Goldin, A. A., Petrov, E. P., and Rubinov, A. N. 1992. Fluorescence decay time distribution for polar dye solution with time-dependent fluorescent shift. *Biophys. Chem.* 44: 47–60.

Gallardo, S., Gutierrez, M., Henglein, A., and Janata, E. 1989. Photochemistry and radiation chemistry of colloidal semiconductors. *Ber. Bunsenges. Phys. Chem.* 93: 1080–90.

Gammon, D., Snow, E. S., Shanabrook, B. V., Katzer, D. S., and Park, D. 1996a. Fine structure splitting in the optical spectra of single GaAs quantum dots. *Phys. Rev. Lett.* 76: 3005–8.

Gammon, D., Snow, E. S., Shanabrook, B. V., Katzer, D. S., and Park, D. 1996b. Homogeneous linewidths in the optical spectrum of a single gallium arsenide quantum dot. *Science* 273: 87–90.

Gaponenko, S. V. 1995a. Laser spectroscopy of quantum dots. In *Physics, Chemistry and Application of Nanostructures* Borisenko, V. E., Filonov, A. B., Gaponenko, S. V., and Gurin, V. S. (eds.). Belarusian State University of Informatics and Radioelectronics. pp. 47–9.

Gaponenko, S. V. 1995b. Quasi-zero-dimensional semiconductor structures: Resonant optical nonlinearities and spectral hole-burning. *SPIE Proc.* 2801: 2–10.

Gaponenko, S. V. 1996. Optical processes in semiconductor nanocrystallites (review). *Semiconductors* 30: 577–619.

Gaponenko, S. V., Gribkovskii, V. P., Zimin, L. G., and Nikeenko, N. K. 1982. Bleaching effect in cadmium selenide doped glasses. *Zhurn. Prikl. Spektr.* 37: 863–5.

Gaponenko, S. V., Gribkovskii, V. P., Zimin, L. G., and Nikeenko, N. K. 1984. Influence of recombination via traps on absorption saturation in semiconductors. *J. Appl. Spectr.* 40: 614–8.

Gaponenko, S. V., Zimin, L. G., and Nikeenko, N. K. 1984. Nonlinear optical absorption in semiconductor compound CdS_xSe_{1-x}. *Zhurn. Prikl. Spektr.* 41: 844–6.

Gaponenko, S. V., Germanenko, I. N., Gribkovskii, V. P., Vasiliev, M. I., and Tsekhomskii, V. A. 1992. Nonlinear absorption of semiconducting microcrystallites under quantum confinement: Coexistence of reversible and irreversible effects. *SPIE Proc.* 1807: 65–73.

Gaponenko, S. V., Malinovsky, I. E., Perelman, L. T., and Zimin, L. G. 1992. Light pulse propagation through ZnSe monocrystals under conditions of excitonic nonlinearity and bistability. *J. Lumin.* 52: 225–31.

Gaponenko, S. V., Germanenko, I. N., Gribkovskii, V. P., Zimin, L. G., Lebed, V. Yu., and Malinovskii, I. E. 1993. Nonlinear absorption and photoluminescence of CuCl crystallites under size quantization of excitons. *Physica B* 185: 588–92.

Gaponenko, S. V., Gribkovskii, V. P., Zimin, L. G., Lebed, V. Yu., Malinovskii, I. E., Graham, S., Klingshirn, C., Eyal, M., Brusilovsky, D., and Reisfeld, R. 1993. Nonlinear phenomena of acridine orange in inorganic glasses at nanosecond scale. *Opt. Mater.* 2: 53–8.

Gaponenko, S., Woggon, U., Saleh, M., Langbein, W., Uhrig, A., Muller, M., and Klingshirn, C. 1993. Nonlinear-optical properties of semiconductor quantum dots and their correlation with the precipitation stage. *J. Opt. Soc. Amer. B* 10: 1947–55.

Gaponenko, S. V., Germanenko, I. N., Petrov, E. P., Stupak, A. P., Bondarenko, V. P., and Dorofeev, A. M. 1994. Time-resolved spectroscopy of visibly emitting porous silicon. *Appl. Phys. Lett.* 64: 85–7.

Gaponenko, S. V., Germanenko, I. N., Stupak, A. P., Eyal, M., Brusilovsky, D., Reisfeld, R., Graham, S., and Klingshirn, C. 1994. Fluorescence of acridine orange in inorganic glass matrices. *Appl. Phys. B* 58: 283–8.

Gaponenko, S., Woggon, U., Uhrig, A., Langbein, W., and Klingshirn, C. 1994. Narrow-band spectral hole burning in quantum dots. *J. Lumin.* 60: 302–7.

Gaponenko, S. V., Kononenko, V. K., Petrov, E. P., Germanenko, I. N., Stupak, A. P., and Xie, Y. Z. 1995. Polarization of porous silicon luminescence. *Appl. Phys. Lett.* 67: 3019–21.

Gaponenko, S. V., Germanenko, I. N., Kapitonov, A. M., and Artemyev, M. V. 1996. Permanent spectral hole-burning in semiconductor quantum dots. *J. Appl. Phys.* 79: 7139–42.

Gaponenko, S. V., Petrov, E. P., Woggon, U., Wind, O., Klingshirn, C., Xie, Y. H., Germanenko, I. N., and Stupak, A. P. 1996. Steady-state and time-resolved spectroscopy of porous silicon. *J. Lumin.* 70: 364–76.

Gaponenko, S. V., Germanenko, I. N., Kapitonov, A. M., Woggon, U., Wind, O., and Klingshirn, C. 1997. Three types of spectral hole-burning in CdS nanocrystals. In *Physics, Chemistry, and Application of Nanostructures*. V. E. Borisenko et al. (eds.) pp. 72–7. Singapore: World Scientific.

Gaponenko, S. V., Bogomolov, V. N., Kapitonov, A. M., Prokofiev, A. V., Eychmüller, A., and Rogach, A. L. Electrons and photons in mesoscopic structures: Quantum dots in a photonic crystal (in press).

Geerligs, L. J., Harmans, C. J., and Kouwenhoven, L. P. (eds.) 1993. The physics of few-electron nanostructures. *Physica (Utrecht)* 189B: (1–4). (Special issues)

Gehlen, M. H., and Schryver, F. C. De. 1993. Time-resolved fluorescence quenching in micellar assemblies. *Chem. Rev.* 93: 199–221.

Georgsson, K., Carlsson, N., Samuelson, L., Seifert, W., and Wallenberg, L. R. 1995. Transmission electron microscopy investigation of the morphology in InP Stranski-Krastanow islands grown by metalorganic chemical vapor deposition. *Appl. Phys. Lett.* 67: 2981–2.

Gibbs, H. M. 1985. *Optical Bistability: Controlling Light with Light.* Orlando, Fla.: Academic Press. 1988. Moscow: Mir.

Gibbs, H. M., Olbright, G. R., Peyghambarian, N., Schmidt, H. E., Koch, S. W., and Haug, H. 1985. Kinks: Longtitudinal excitation discontinuities in increasing absorption optical bistability. *Phys. Rev. A* 32: 692–4.

Gilliot, P., Merle, J. C., Levy, R., Robino, M., and Honerlage, B. 1989. Laser induced absorption of CuCl microcrystallites in a glass matrix. *Phys. State Sol. (b)* 153: 403–10.

Gogolin, O., Berosashvili, Yu., Mschvelidze, G., Tsitsishvili, E., Oktjabrski, S., Giessen, H., Uhrig, A., and Klingshirn, C. 1991. CuI microcrystallites embedded in a glass matrix. *Semicond. Sci. Technol.* 6: 401–4.

Golubkov, V. V., Ekimov, A. I., Onushchenko, A. A., and Tsekhomskii, V. A. 1981. Kinetics of CuCl microcrystals growth in a glass matrix. *Sov. Glass Phys. Chem.* 7: 397–401.

Goncharova, O. V., and Sinitsyn, G. V. 1990. Optical characteristics of vacuum deposited quantum-size structures of II-VI compounds. *Izv. Akad. Nauk BSSR* 6: 21–8. (In Russian).

Goncharova, O. V., Sinitsyn, G. V., and Tikhomirov, S. A. 1992. Subpicosecond-scale change in transmission spectrum of vacuum deposited thin film CdS-interferometers. *SPIE Proc.* 1807: 2–13.

Gorokhovskii, A. A., Kaarli, R. K., and Rebane, L. A. 1974. Burning a hole in the curve of purely electronic line in Shpol'skii systems. *JETP Lett.* 20: 216–20.

Grabovskis, V. Ya., Dzenis, Ya., Ekimov, A. I., Kudryavtsev, I. A., Tolstoi, M. N., and Rogulis, U. T. 1989. Photoionization of semiconductor microcrystals in glasses. *Sov. Phys. Sol. Stat.* 31: 149–54.

Gribkovskii, V. P. 1975. *Theory of Light Absorption and Emission in Semiconductors* Minsk: Nauka i Tekhnika.

Gribkovskii, V. P. 1995. Injection lasers. *Progress in Quant. Electron.* 19: 41–88.

Gribkovskii, V. P., Zyulkov, V. A., Kazachenko, A. E., and Tikhomirov, S. A. 1988. Optical nonlinearity of semiconductor microcrystals CdS_xSe_{1-x} under the action of picosecond and nanosecond laser pulses. *Phys. Stat. Sol. (b)* 150: 647–52.

Gribkovskii, V. P., Zimin, L. G., Gaponenko, S. V., Malinovskii, I. E., Kuznetsov, P. I., and Yakushcheva, G. G. 1990. Optical absorption near excitonic resonance of MOCVD-grown ZnSe single crystals. *Phys. Stat. Sol. (b)* 158: 359–66.

Grundmann, M., Christen, J., Ledentsov, N. N., Bohrer, J., Bimberg, D., Ruvimov, S. S., Werner, P., Richter, U., Gosele, U., Heydenreich, J., Ustinov, V. M., Egorov, A. Yu., Zhukov, A. E., Kop'ev, P. S., and Alferov, Zh. I. 1995. Ultranarrow luminescence lines from single quantum dots. *Phys. Rev. Lett.* 74: 4043–6.

Gurevich, S. A., Ekimov, A. I., Kudryavtsev, I. A., Osinskii, A. V., Skopina, V. I., and Chepik, D. I. 1992. Fabrication and study of SiO_2 films activated with CdS nanocrystals. *Semiconductor* 26: 102–6.

Gurin, V. S. 1994. Semiempirical calculation of cadmium chalcogenide clusters. *J. Phys. C* 6: 8691–700.

Gurin, V. S. 1996. Electronic structure of CdS cores in Cd thiolate complexes and clusters. *J. Phys. Chem.* 100: 869–72.

Gurin, V. S., and Grigorenko, N. N. 1995. Photolysis of ultradispersive silver iodide. *J. Phys. Chem. (USSR)* 69: 1863–6.

Gurin, V. S., and Sviridov, V. V. 1995. Optical properties of $CuInS_2$ particles in colloidal solutions. *Colloid. Journ. (USSR)* 57: 313–6.

Haar, D. 1958. *Introduction to the Physics of Many-Body Systems*. New York: Interscience Publishers. 1961. Moscow: Mir.

Hache, F., Ricard, D., and Flytzanis, C. 1989. Quantum-confined Stark effect in very small semiconductor crystallites. *Appl. Phys. Lett.* 55: 1504–7.

Halperin, W. P. 1986. Quantum size effects in metal particles. *Rev. Mod. Phys.* 58: 533–606.

Hanamura, E. 1988a. Rapid radiative decay and enhanced optical nonlinearity of excitons in a quantum well. *Phys. Rev. B* 38: 1228–34.

Hanamura, E. 1988b. Very large optical nonlinearity of semiconductor microcrystallites. *Phys. Rev. B* 37: 1273–9.

Harmans, K. 1992. Next electron, please .... *Physics World* 50–3.

Haroche, S., and Kleppner, D. 1989. Cavity quantum electrodynamics. *Phys. Today* 42: 24–30.

Hatami, F., Ledentsov, N. N., Grundmann, M., Bohrer, J., Heinrichsdorff, F., Beer, M., Bimberg, D., Ruvimov, S. S., Werner, P., Gosele, U., Heydenreich, J., Richter, U., Ivanov, S. V., Meltser, B. Ya, Kop'ev, P. S., and Alferov, Zh. I. 1995. Radiative recombination in type-II GaSb/GaAs quantum dots. *Appl. Phys. Lett.* 67: 656–8.

Haug, H. (ed.). 1988. *Optical Nonlinearities and Instabilities in Semiconductors*. New York: Plenum.

Haug, H., and Koch, S. W. 1990. *Quantum Theory of the Optical and Electronic Properties of Semiconductors*. Singapore: World Scientific.

Haug, H., Ivanov, A. I., and Keldysh, L.V. 1997. *Nonlinear Optical Phenomena in Semiconductors and Semiconductor Microstructures*. Singapore: World Scientific.

Hayashi, R., Yamamoto, M., Tsunemoto, K., Kohno, K., Osaka, Y., and Nasu, H. 1990. Synthesis of Ge nanocrystals in glassy SiO_2 matrix. *Jap. J. Appl. Phys.* 29: 756–9.

Hayashi, S. 1984. Optical study of electromagnetic surface modes in microcrystals. *Jap. J. Appl. Phys.* 23: 665–76.

Hayashi, S., Fujii, M., and Yamamoto, K. 1989. Quantum size effect in Ge microcrystals embedded in SiO_2 thin films. *Jap. J. Appl. Phys.* 28: L1464–7.

Hayashi, S., Kanzawa, Y., Kataoka, M., Nagareda, T., and Yamamoto, K. 1993. Photoluminescence spectra of clusters of group IV elements embedded in SiO_2 matrices. *Z. Phys. D* 26: 144–6.

Hedvig, P. 1975. *Experimental Quantum Chemistry* Budapest: Akademiai Kiado. 1977. Moscow: Mir.

Helwege, K.-H. (ed.) 1982. *Landolt-Börnstein: Numerical Data and Functional Relationships in Science and Technology*. Vol. 17. *Semiconductors*. Berlin: Springer.

Henari, F. Z., Morgenstern, K., Blau, W. J., Karavanskii, V. A., and Dneprovskii, V. S. 1995. Third-order optical nonlinearity and all-optical switching in porous silicon. *Appl. Phys. Lett.* 67: 323–5.

Hendershot, D. G., Gaskill, D. K., Justus, B. L., Fatemi, M., and Berry, A. D. 1993. Organometallic chemical vapor deposition and characterization of indium phosphide nanocrystals in Vycor porous glass. *Appl. Phys. Lett.* 61: 3324–6.

Henglein, A. 1982. Photo-degradation and fluorescence of colloidal cadmium sulfide in aqueous solution. *Ber. Bunsenges. Phys. Chem.* 86: 301–5.

Henglein, A. 1989. Small-particle research: Physicochemical properties of extremely small colloidal metal and semiconductor particles. *Chem. Rev.* 89: 1861–73.

Henneberger, F. 1986. Optical bistability in semiconductors. *Phys. Stat. Sol. (b)* 137: 371–432.

Hill, N. A., and Whaley, K. B. 1993. Electronic structure of semiconductor nanoclusters: A time-dependent theoretical approach. *J. Chem. Phys.* 99: 3707–15.

Hill, N. A., and Whaley, K. B. 1995. Size dependence of excitons in silicon nanocrystals. *Phys. Rev. Lett.* 75: 1130–3.

Hirshman, K. D., Tsybeskov, L., Duttagupta, S. P., and Fauchet, P. M. 1996. Silicon-based visible light-emitting devices integrated into microelectronic circuits. *Nature* 384: 338–40.

Hodes, G., Albu-Yaron, A., Decker, F., and Motisuke, P. 1987. Three-dimensional quantum-size effect in chemically deposited cadmium selenide films. *Phys. Rev. B* 36: 4215–21.

Honea, E. C., Kraus, J. S., and Bower, J. E. 1993. Optical spectra of size-selected matrix-isolated silicon clusters. *Z. Phys. D* 26: 141–3.

Honerlage, B., Levy, R., Grun, J. B., Klingshirn, C., and Bohnert, K. 1985. Excitons and polaritons in semiconductors. *Phys. Rep.* 124: 161–253.

Horan, P., and Blau, W. 1990. Photodarkening effect and the optical nonlinearity in a quantum-confined, semiconductor-doped glass. *J. Opt. Soc. Amer. B* 7: 304–8.

Hsu, S., and Kwok, H. S. 1987. Picosecond carrier recombination dynamics of semiconductor-doped glasses. *Appl. Phys. Lett.* 50: 1782–4.

Hu, Y. Z., Koch, S. W., Lindberg, M.,Peyghambarian, N., Pollock, R., and Abraham, F. F. 1990. Biexcitons in semiconductor quantum dots. *Phys. Rev. Lett.* 64: 1805–07.

Hu, Y. Z., Koch, S. W., and Peyghambarian, N. 1996. Strongly confined semiconductor quantum dots: Pair excitations and optical properties. *J. Lumin.* 70: 185–202.

Hu, Y. Z., Lindberg, M., and Koch, S. W. 1990. Theory of optically excited intrinsic semiconductor quantum dots. *Phys. Rev. B* 42: 1713–23.

Huaxiang, F., Ling, Y., and Xide, X. 1993. Optical properties of silicon nanostructures. *Phys. Rev. B* 48: 10978–82.

Hunsche, S., Dekorsy, T., Klimov, V., and Kurz, H. 1996. Ultrafast dynamics of carrier-induced absorption changes in highly-excited CdSe nanocrystals. *Appl. Phys. B* 62: 3–10.

Hybertsen, M. S. 1994. Absorption and emission of light in nanoscale silicon structure. *Phys. Rev. B* 72: 1515–7.

Ikezawa, M., and Masumoto, Y. 1996. Stochastic treatment of the dynamics of excitons and excitonic molecules in CuCl nanocrystals. *Phys. Rev. B* 53: 13694–9.

Illing, M., Bacher, G., Kümmell, T., Forchel, A., Andersson, T. G., Hommel, D., Jobst, B., and Landwehr, G. 1995. Lateral quantization effects in lithographically defined CdZnSe/ZnSe quantum dots and quantum wires. *Appl. Phys. Lett.* 67: 124–6.

Inoshita, T., and Sakaki, H. 1992. Electron relaxation in a quantum dot: Significance of multiphonon processes. *Phys. Rev. B* 46: 7260–3.

Ishihara, T. 1990. The origin of third-order optical nonlinearity in exciton systems. *Phys. Stat. Sol. (b)* 159: 371–8.

Ishihara, H., and Cho, K. 1990. Cancellation of size-linear terms in the third-order nonlinear susceptibility: Frenkel excitons in a periodic chain. *Phys. Rev. B* 42: 1724–30.

Itoh, T., Iwabuchi, Y., and Kataoka, M. 1988. Study on the size and shape of CuCl microcrystals embedded in alkali-chloride matrices and their correlation with exciton confinement. *Phys. Stat. Sol. (b)* 145: 567–604.

Itoh, T., Iwabuchi, Y., and Kirihara, T. 1988. Size-quantized excitons in microcrystals of cuprous halides embedded in alkali-halide matrices. *Phys. Stat. Sol. (b)* 146: 531–43.

Itoh, T., Furumiya, M., Ikehara, T., and Gourdon, C. 1990. Size-dependent radiative decay time of confined excitons in CuCl microcrystals. *Sol. Stat. Comm.* 73: 271–4.

Itoh, T., Ikehara, T., and Iwabuchi, Y. 1990. Quantum confinement of excitons and their relaxation processes in CuCl microcrystals. *J. Lumin.* 45: 33–6.

Itoh, T., Yano, S., Katagiri, N., Iwabuchi, Y., Gourdon, C., and Ekimov, A. I. 1994. Interface effects on the properties of confined excitons in CuCl microcrystals. *J. Lumin.* 60, 61: 396–9.

Itoh, T., Nishijima, M., Ekimov, A. I., Gourdon, C., Efros, Al. L., and Rosen, M. 1995. Polaron and exciton-phonon complexes in CuCl nanocrystals. *Phys. Rev. Lett.* 74: 1645–8.

Jaskolski, W. 1996. Confined many-electron systems. *Phys. Rep.* 271: 1–66.
Jeffries, C. D., and Keldysh, L. V. (eds.). 1988. *Electron-Hole Droplets in Semiconductors*. Amsterdam: North Holland. 1988. Moscow: Mir.
Jeppesen, S., Miller, M. S., Hessman, D., Kowalski, B., Maximov, I., and Samuelson, L. 1996. Assembling strained InAs islands on patterned GaAs substrates with chemical leam epitaxy. *Appl. Phys. Lett.* 68: 2228–30.
Jesson, D. E., Chen, K. M., and Pennycook, S. J. 1995. Kinetic pathways to strain relaxation in the Si-Ge system. *MRS Bull.* 21/4: 31–7.
Jhe, W., and Jang, K. 1996. Cavity quantum electrodynamics inside a hollow spherical cavity. *Phys. Rev. A* 53: 1126–9.
Jin, C., Yu, J., Qin, W., Zhao, J., Zhou, F., Dou, K., Liu, J., and Huang, S. 1992. Photodarkening effect and optical nonlinearity in CdSSe-doped glasses. *J. Lumin.* 53: 483–6.
Johansson, K. P., McLendon, G., and Marchetti, A. P. 1991. The effect of size restriction on silver bromide. A dramatic enhancement of free exciton luminescence. *Chem. Phys. Lett.* 179: 321–4.
John, G. C., and Singh, V. A. 1995. Porous silicon: Theoretical studies. *Phys. Rep.* 263: 93–152.
John, S. 1987. Strong localization of photons in certain disordered dielectric superlattices. *Phys. Rev. Lett.* 58: 2486–9.
Justus, B. L., Seaver, M. E., Ruller, J. A., and Campillo, A. J. 1990. Excitonic optical nonlinearity in quantum-confined CuCl-doped borosilicate glass. *Appl. Phys. Lett.* 57: 1381–3.
Justus, B. L., Tonucci, R. G., and Berry, A. D. 1992. Nonlinear optical properties of quantum-confined GaAs nanocrystals in Vycor glass. *Appl. Phys. Lett.* 61: 3151–2.
Kagakin, E. I., Petrushina, A. V., Morozov, V. P., Dodonov, V. G., and Pugachev, V. M. 1995. Synthesis of ultradisperse Ag_2S microcrystals. *J. Sci. Appl. Photog.* (in Russian). 40: 63–5.
Kagan, C. R., Murray, C. B., and Bawendi, M. G. 1996. Long-range resonance transfer of electronic excitations in close-packed CdSe quantum-dot solids. *Phys. Rev. B* 54: 8633–43.
Kagan, C. R., Murray, C. B., Nirmal, M., and Bawendi, M. G. 1996. Electronic energy transfer in CdSe quantum dot solids. *Phys. Rev. Lett.* 76: 1517–20.
Kamat, P. V. 1993. Photochemistry on nonreactive and reactive (semiconductor) surfaces. *Chem. Rev.* 93: 267–300.
Kanemitsu, Y. 1995. Light emission from porous silicon and related materials. *Phys. Rep.* 263: 1–92.
Kanemitsu, Y., Hiroshi, U., Masumoto, Y., and Maeda, Y. 1992. On the origin of visible photoluminescence in nanometer-size Ge crystallites. *Appl. Phys. Lett.* 61: 2187–9.
Kanemitsu, Y., Ogawa, T., Shiraishi, K., and Takeda, K. 1993. Visible photoluminescence from oxidized Si nanometer-sized spheres: exciton confinement on a spherical shell. *Phys. Rev. B* 48: 4883–6.
Kang, K. I., McGinnis, B. P., Sandalphon, Hu, Y. Z., Koch, S. W., Peyghambarian, N., Mysyrowicz, A., Liu, L. C., and Risbud, S. H. 1992. Confinement-induced valence-band mixing in CdS quantum dots observed by two-photon spectroscopy. *Phys. Rev. B* 45: 3465–8.
Kang, K. I., Kepner, A. D., Gaponenko, S. V., Koch, S. W., Hu, Y. Z., and Peyghambarian, N. 1993. Confinement-enhanced biexciton binding energy in semiconductor quantum dots. *Phys. Rev. B* 48: 15449–53.

Kang, K. I., Kepner, A. D., Hu, Y. Z., Koch, S. W., Peyghambarian, N., Li, C.-Y., Takada, T., Kao, Y., and Mackenzie, J. D. 1994. Room-temperature spectral hole burning and elimination of photodarkening in sol-gel derived CdS quantum dots. *Appl. Phys. Lett.* 64: 1487–9.
Kanzaki, H., and Tadakuma, Y. 1991. Indirect-exciton confinement and impurity isolation in ultrafine particles of silver bromide. *Sol. Stat. Comm.* 80: 33–6.
Kapitonov, A. M., Gaponenko, N. N., Bogomolov, V. N., Prokofiev, A. V., Samoilovich, S. M., and Gaponenko, S. V. 1997. SiO_2/TiO_2 submicron 3D lattice: A new step towards visible-range photonic crystals. In *Physics, Chemistry, and Application of Nanostructures*. V. E. Borisenko, A. B. Filatov, S. V. Gaponenko, and V. S. Gurin (eds.). pp. 54–7. Singapore: World Scientific.
Karplus, R., and Schwinger, J. 1948. A note on saturation in microwave spectroscopy. *Phys. Rev.* 73: 1020–2.
Karpushko, F. V., and Sinitsyn, G. V. 1978. Bistable optical element for integrated optics based on nonlinear semiconductor interferometer. *J. Appl. Spectr.* 29: 1323–6.
Kastner, M. A. 1993. Artificial atoms. *Physics Today* 46: 24–31.
Katsikas, L., Eychmuller, A., Giersig, M., and Weller, H. 1990. Discrete excitonic transitions in quantum-sized CdS particles. *Chem. Phys. Lett.* 172: 201–4.
Kayanuma, Y. 1986. Wannier exciton in microcrystals. *Solid State Comm.* 59: 405–8.
Kayanuma, Y. 1988. Quantum-size effects of interacting electrons and holes in semiconductor microcrystals with spherical shape. *Phys. Rev. B* 38: 9797–805.
Kayanuma, Y. 1991. Wannier excitons in low-dimensional microstructures: Shape dependence of the quantum size effect. *Phys. Rev. B* 44: 13085–8.
Kharchenko, V. A., and Rosen, M. 1996. Auger relaxation processes in semiconductor nanocrystals and quantum wells. *J. Lumin.* 70: 158–69.
Kharlamov, B. M., Personov, R. I., and Bykovskaya, L. A. 1974. Stable gap in absorption spectra of solid solutions of organic molecules by laser radiation. *Opt. Commun.* 12: 191–4.
Kilin, S. Ya., and Mogilevtsev, D. S. 1992. "Freezing" of decay of a quantum system with a dip in a spectrum of the heat bath-coupling constants. *Laser Physics* 2: 153–63.
Kim, H.-M., Hayashi, S., and Yamamoto, K. 1996. A photoluminescence study of nitrogen doping of gas-evaporated GaP microcrystals. *J. Phys.: Condens. Matter* 8: 2705–14.
Kippelen, B., Levy, R., Faller, P., Gilliot, P., and Bellegie, L. 1991. Picosecond excite and probe nonlinear absorption measurements in CuCl quantum dots. *Appl. Phys. Lett.* 59: 3378–80.
Kittel, Ch. 1986. *Introduction to Solid State Physics*. New York: Willey. 1978. Moscow: Mir.
Klein, M. C., Hache, F., Ricard, D., and Flytzanis, C. 1990. Size dependence of electron-phonon coupling in semiconductor nanospheres: The case of CdSe. *Phys. Rev. B* 42: 11123–32.
Klimov, V., Bolivar, H. P., Kurz, H., Karavanskii, V., Krasovskii, V., and Korkishko, Yu. 1995. Linear and nonlinear transmission of Cu_xS quantum dots. *Appl. Phys. Lett.* 67: 653–5.
Klimov, V., Hunsche, S., and Kurz, H. 1994. Biexciton effects in femtosecond nonlinear transmission of semiconductor quantum dots. *Phys. Rev. B* 50: 8110–3.
Klimov, V. I., Dneprovskii, V. S., and Karavanskii, V. A. 1994. Nonlinear-transmission spectra of porous silicon: Manifestation of size quantization. *Appl. Phys. Lett.* 64: 2691–3.

Klingshirn, C. 1990. Nonlinear optical properties of semiconductors. *Semicond. Sci. Technol.* 5: 457–77.

Klingshirn, C. 1995. *Semiconductor Optics*. Berlin: Springer.

Klingshirn, C., and Gaponenko S. V., 1992. Nonlinear light absorption in semiconductors near the fundamental absorption edge (review). *J. Appl. Spectr.* 56: 512–26.

Knox, R. S. 1963. *Theory of Excitons* New York: Academic. 1966. Moscow: Mir.

Koch, S. W. 1984. *Dynamics of First Order Phase Transitions in Equilibrium and Nonequilibrium Systems*. Berlin: Springer.

Koch, S. W., Hu, Y. Z., Fluegel, B., and Peyghambarian, N. 1992. Coulomb effects and optical properties of semiconductor quantum dots. *J. Cryst. Growth* 117: 592–7.

Kochelap, V. A., Melnikov, L. Yu., and Sokolov, V. N. 1982. Many-valued distribution of nonequilibrium electrons and holes in semiconductors with concentrational nonlinearity of light absorption. *Sov. Phys. Semicond.* 16: 1167–70.

Konnikov, S. G., Sitnikova, A. A., Lipovskii, A. A., Niconorov, N. V., and Kharchenko, M. V. 1995. Silicate glasses doped with microscopic cadmium sulfide-selenide crystals and optical waveguides based on them. *Semiconductors* 29: 767–72.

Kontkiewicz, A. J., Kontkiewicz, A. M., Siejka, J., Sen, S., Nowak, G., Hoff, A. M., Sakthivel, P., Ahmed, K., Mukherjee, P., Witanachchi, S., and Lagowski, J. 1994. Evidence that blue luminescence of oxidized porous silicon originates from SiO_2. *Appl. Phys. Lett.* 65: 1436–38.

Kooij, E., Despo, R., and Kelly, J. 1995. Electroluminescence from porous silicon due to electron injection from solution. *Appl. Phys. Lett.* 66: 2552–4.

Kopelman, P. 1988. Fractal reactions kinetics. *Science* 240: 1620–8.

Kornowski, A., Eichberger, R., Giersig, M., Weller, H., and Eychmüller, A. 1996. Preparation and photophysics of strongly luminescing Cd_3P_2 quantum dots. *J. Phys. Chem.* 100: 12467–71.

Kortan, A. R., Hull, R., Opila, R. L., Bawendi, M. G., Steigerwald, M. L., Carroll, P. J., and Brus, L. E. 1990. Nucleation and growth of CdSe on ZnS quantum crystallites seeds, and vice versa, in inverse micelle media. *J. Am. Chem. Soc.* 112: 1327–32.

Kovalev, D., Ben Chorin, M., Diener, J., Koch, F., Efros, A. L., Rosen, M., Gippius, N. A., and Tikhodeev, S. G. 1995. Porous Si anisotropy from photoluminescence polarization. *Appl. Phys. Lett.* 67: 1585–7.

Kramer, M. A., Tompkin, W. R., and Boyd, R. W. 1986. Nonlinear-optical interactions in fluorescein-doped boric acid glass. *Phys. Rev. A* 34: 2026–31.

Kubo, R. 1962. Electronic properties of metallic fine particles. *J. Phys. Soc. Jap.* 17: 975–80.

Kulinkin, B. S., Petrovskii, V. A., Tsekhomskii, V. A., and Shchanov, M. F. 1988. Hydrostatic pressure effect on exciton absorption spectrum of CuCl microcrystals in photochromic glass. *Sov. Phys. Chem. Glass* 14: 470–1.

Kull, M., Coutaz, J. L., Manneberg, G., and Grivickas, V. 1989. Absorption saturation and photodarkening in semiconductor-doped glasses. *Appl. Phys. Lett.* 54: 1830–2.

Kurochkin, Y. A. 1994. Non-Euclidian quantum mechanics of excitons in semiconductor quantum dots (in Russian). *Proc. Acad. Sci. Belarus* 38: 36–9.

Kuzmitskii, V. A., Gael, V. I. and Filatov, I. V. 1996. Quantum-chemical calculation of electron structure and excited states of ZnS and CdS clusters. *J. Appl. Spectr.* 63: 714–23.

Kuznetsov, S. N., Piculev, V. B., Gardin, Yu. E., Klimov, I. V., and Gurtov, V. A. 1995. Nonradiative processes and luminescence spectra in porous silicon. *Phys. Rev. B* 51: 1601–4.

Landau, L. D., and Lifshitz, E. M. 1982. *Electrodynamics of Continuous Media* Moscow: Nauka.
Landau, L. D., and Lifshitz, E. M. 1988. *Statistical Physics*. Moscow: Nauka.
Landau, L. D., and Lifshitz, E. M. 1989. *Quantum Mechanics*. Moscow: Nauka.
Landsberg, P. T. 1991. *Recombination in Semiconductors*. Cambridge: Cambridge University Press.
Langbein, W., Woggon, U., Gaponenko, S., Uhrig, A., and Klingshirn, C. 1992. Spectrally narrow nonlinear resonances in semiconductor doped glasses. *SPIE Proc.* 1807: 514–21.
Lange, H. 1971. Influence of an electric field on the B, $n = 1$ ground state exciton line of CdS and CdSe single crystals. *Phys. Stat. Sol. (b)* 48: 791–5.
Lannoo, M., Delerue, C., and Allan, G. 1995. Screening in semiconductor nanocrystallites and its consequences for porous silicon. *Phys. Rev. Lett.* 74: 3415–8.
Lavallard, P. and Suris, R. A. 1995. On the origin of polarized luminescence in semiconductor nanocrystals. *Solid State Comm.* 95: 267–9.
Lazarouk, S., Jaguiro, P., Katsouba, S., Masini, G., La Monica, S., Maiello, G., and Ferrari, A. 1996. Stable electroluminescence from reverse biased n-type porous silicon-aluminum Schottky junction device. *Appl. Phys. Lett.* 68: 2108–9.
Lefebvre, P., Richard, T., Mathieu, H., and Allegre, J. 1996. Influence of spin-orbit split-off band on optical properties of spherical semiconductor nanocrystals. The case of CdTe. *Solid State Comm.* 98: 303–6.
Lehmann, V., and Gösele, U. 1991. Porous silicon formation: a quantum wire effect. *Appl. Phys. Lett.* 58: 856–8.
Leonard, D., Krishnamurthy, M., Reaves, C. M., Denbaars, S. P., and Petroff, P. M. 1993. Direct formation of quantum-sized dots from uniform coherent islands of InGaAs on GaAs surfaces. *Appl. Phys. Lett.* 63: 3203–5.
Letokhov, V. S., and Chebotaev, V. P. 1977. *Nonlinear Laser Spectroscopy*. Berlin: Springer. 1990. Moscow: Nauka.
Levy, R., Mager, L., Gilliot, P., and Honerlage, B. 1991. Biexciton luminescence in CuCl microcrystallites. *Phys. Rev. B* 44: 11286–92.
Lianos, P. 1988. Luminescence quenching in organized assemblies as media of noninteger dimensions. *J. Chem. Phys.* 89: 5237–41.
Lifshitz, I. M., and Slyozov, V. V. 1961. Kinetics of diffusive decay of supersaturated solid solutions. *J. Phys. Chem. Sol.* 28: 35–46.
Lifshitz, Z., Yassen, M., Bykov, L., Dag, I, and Chaim, R. 1994. Nanometer-sized particles of PbI_2 embedded in SiO_2 films. *J. Phys. Chem.* 98:1459–63.
Likharev, K. K., and Claeson, T. 1992. Single electronics. *Sci. Am.* 50–5.
Lindberg, M., Koch, S. W., and Haug, H. 1986. Structure, formation, and motion of kinks in increasing-absorption optical bistability. *Phys. Rev. A* 33: 407–15.
Lippens, P. E., and Lanoo, M. 1989. Calculation of the band gap for small CdS and ZnS crystallites. *Phys. Rev. B* 39: 10935–42.
Lipsanen, H., Sopanen, M., and Ahopelto, J. 1995. Luminescence from excited states in strain-induced $In_xGa_{1-x}As$ quantum dots. *Phys. Rev. B* 51: 13868–71.
Lisitza, M. P., Kulish, N. R., Koval, P. N., and Geetz, V. I. 1967. Influence of the laser radiation on the transmission of KS-19 glass. *Sov. Opt. Spectr.* 23: 981–3.
Littau, K. A., Szajowiski, P. J., Miller, A. J., Kortan, A. R., and Brus, L. E. 1993. Luminescent silicon nanocrystal colloids via a high-temperature aerosol reaction. *J. Phys. Chem.* 97: 1224–30.
Liu, L. C., and Risbud, S. H. 1990. Quantum dot size-distribution analysis and precipitation stages in semiconductor doped glasses. *J. Appl. Phys.* 68: 28–32.

Liu, L-C., Kim, M. J., Risbud, S. H., and Carpenter, R. W. 1991. High-resolution electron microscopy and microanalysis of CdS and CdTe quantum dots in glass matricies. *Philosoph. Mag. B* 63: 769–76.

Lockwood, D. J. 1994. Optical properties of porous silicon. *Sol. Stat. Comm.* 92: 101–12.

Loni, A., Simons, A. J., Calcott, P. D. J., and Canham, L. T. 1995. Blue photoluminescence from rapid thermally oxidized porous Si following storage in ambient air. *J. Appl. Phys.* 77: 3557–9.

Lowich, M., Rabe, M., Stegeman, B., Henneberger, F., Grundmann, M., Tuerck, V., and Bimberg, D. 1996. Zero-dimensional excitons in (Zn,Cd)Se quantum structures. *Phys. Rev. B* 54: R11074–7.

Luczka, J., Gadomski, A., and Grzywna, Z. J. 1992. On the diffusion-driven growth: the perturbed sphere problem revisited. *Czech. J. Phys.* 42: 577–82.

Macucci, M., Hess, K., and Iafrate, G. J. 1993. Electronic energy spectrum and the concept of capacitance in quantum dots. *Phys Rev. B* 48: 17357–63.

Madelung, O. 1978. *Introduction to Solid State Theory*. Berlin: Springer. 1980. Moscow: Mir.

Maeda, Y. 1995. Visible photoluminescence from nanocrystallite Ge embedded in a glassy SiO_2 matrix: Evidence in support of the quantum-confinement mechanism. *Phys. Rev. B* 51: 1658–70.

Malhotra, J., Hagan, D. J., and Potter, B. G. 1991. Laser-induced darkening in semiconductor-doped glasses. *J. Opt. Soc. Am. B* 8: 1531–6.

Marchetti, A. P., Johansson, K. P., and McLendon, G. L. 1993. AgBr photophysics from optical studies quantum confinement crystals. *Phys. Rev. B* 47: 4268–75.

Marini, J. C., Stebe, B., and Kartheuser, E. 1993. Influence of the electron-phonon interaction on a donor-like exciton in a semiconductor microsphere. *Solid State Comm.* 87: 435–7.

Martin, E., Delerue, C., Allan, G., and Lannoo, M. 1994. Theory of excitonic exchange splitting and optical Stokes shift in silicon nanocrystallites: Application to porous silicon. *Phys. Rev. B* 50: 18258–67.

Marzin, J.-Y., Gerard, J.-M., Izrael, A., Barrier, D., and Bastard, G. 1994. Photoluminescence of single InAs quantum dots obtained by self-organized growth on GaAs. *Phys. Rev. Lett.* 73: 716–9.

Masumoto, Y. 1996. Persistent hole-burning spectroscopy of semiconductor quantum dots. *J. Lumin.* 70: 386–99.

Masumoto, Y., Wamura, T., and Iwaki, A. 1989. Homogeneous width of exciton absorption spectra in CuCl microcrystals. *Appl. Phys. Lett.* 55: 2535–7.

Masumoto, Y., Kawamura, T., Ohzeki, T., and Urabe, S. 1992. Lifetime of indirect excitons in AgBr quantum dots. *Phys. Rev. B* 46: 1827–30.

Masumoto, Y., Kawamura, T., and Era, K. 1993. Biexciton lasing in CuCl quantum dots. *Appl. Phys. Lett.* 62: 225–7.

Masumoto, Y., Okamoto, S, and Katayanagi, S. 1994. Biexciton binding energy in CuCl quantum dots. *Phys. Rev. B* 50: 18658–63.

Masumoto, Y., Okamoto, S., Yamamoto, T., and Kawazoe, T. 1995. Persistent spectral hole-burning phenomenon of semiconductor quantum dots. *Phys. Stat. Sol. (b)* 188: 209–19.

Mathieu, H., Richard, T., Allegre, J., Lefebvre, P., Arnaud, G., Granier, W., Boudes, L., Marc, J. L., Pradel, A., and Ribes, M. 1995. Quantum confinement effects of CdS nanocrystals in a sodium borosilicate glass prepared by the sol-gel process. *J. Appl. Phys.* 77: 287–93.

Mei, G., Carpenter, S., and Persans, P. D. 1991. Steady-state photomodulation mechanisms in CdS_xSe_{1-x} doped glass. *Sol. Stat. Comm.* 80: 557–61.

Mews, A., Kadavanich, A. V., Banin, U., and Alivisatos, A. P. 1996. Structural and spectroscopic investigations of CdS/HgS/CdS quantum-dot quantum wells. *Phys. Rev. B* 53: R13242–5.

Micic, O. I., Nenadovic, M. T., Peterson, M. W., and Nozik, A. J. 1987. Size quantization in layered semiconductor colloids with tetrahedral bonding: HgI_2. *J. Phys. Chem.* 91: 1295–7.

Micic, O. I., Sprague, J. R., Curtis, C. J., Jones, K. M., Machol, J. L., Nozik, A. J., Giessen, H., Fluegel, B., Mohs, G., and Peyghambarian, N. 1995. Synthesis and characterization of InP, GaP, and $GaInP_2$ quantum dots. *J. Phys. Chem.* 99: 7754–9.

Micic, O. I., and Nozik, A. J. 1996. Synthesis and characterization of binary and ternary III-V quantum dots. *J. Lumin.* 70: 95–107.

Micic, O. I., Sprague, J., Lu, Z., and Nozik, A. J. 1996. Highly efficient band-edge emission from InP quantum dots. *Appl. Phys. Lett.* 68: 3150–2.

Micic, O. I., Cheong, H. M., Fu, H., Zunger, A., Sprague, J. R., Mascarenhas, A., and Nozik, A. 1997. Size-dependent spectroscopy of InP quantum dots. *J. Phys. Chem.* 101:4904–12.

Miguez, H., Fornes, V., Marquez, F., Meseguer, F., and Lopez, C. 1996. Low-temperature synthesis of Ge nanocrystals in zeolite Y. *Appl. Phys. Lett.* 69: 2347–9.

Miller, D. A., and Chemla, D. S. 1986. Mechanism for enhanced optical nonlinearities and bistability by combined dielectric-electronic confinement in semiconductor microcrystallites. *Opt. Lett.* 11: 522–4.

Miller, D. A. B., Chemla, D. S., and Schmitt-Rink, S. 1988. Electroabsorption of highly confined systems: Theory of the quantum-confined Franz-Keldysh effect in semiconductor quantum wires and dots. *Appl. Phys. Lett.* 52: 2154–56.

Minti, H., Eyal, M., Reisfeld, R., and Bercovic, G. 1991. Quantum dots of cadmium sulfide in thin glass films prepared by sol-gel technique. *Chem. Phys. Lett.* 183: 277–82.

Misawa, K., Yao, H., Hayashi, T., and Kobayashi, T. 1991a. Size effects on luminescence dynamics of CdS microcrystallites embedded in polymer films. *Chem. Phys. Lett.* 183: 113–8.

Misawa, K., Yao, H., Hayashi, T., and Kobayashi, T. 1991b. Superradiance quenching by confined acoustic phonons in chemically prepared CdS microcrystallites. *J. Chem. Phys.* 94: 4131–40.

Mitsunaga, M., Shinojima, H., and Kubodera, K. 1988. Laser annealing effect on carrier recombination time in CdSSe-doped glasses. *J. Opt. Soc. Am. B* 5: 1448–52.

Mittleman, D. M., Schoenlein, R. W., Shiang, J. J., Colvin, V. L., Alivisatos, A. P., and Shank, C. V. 1994. Quantum size dependence of femtosecond electronic dephasing and vibrational dynamics in CdSe nanocrystals. *Phys. Rev. B* 49: 14438–47.

Miyoshi, T., Araki, Y., Towata, K., Matsuki, H., and Matsuo, N. 1996. Photoinduced formation of semiconductor nanocrystals in CdS-doped glasses. *Jap. J. Appl. Phys.* 35: 593–4.

Miyoshi, T., and Miki, T. 1992. Time-resolved luminescence and ESR of CdSSe-doped glasses. *Superlat. Microstr.* 12: 243–5.

Moerner, W. E. (ed.). 1988. *Persistent Spectral Hole-Burning: Science and Applications.* Berlin: Springer-Verlag.

Moerner, W. E. 1994a. Examining nanoenvironments in solids on the scale of a single, isolated impurity molecule. *Science* 265: 48–53.
Moerner, W. E. 1994b. Fundamentals of single-molecule spectroscopy in solids. *J. Lumin.* 60–61: 997–1002.
Mohan, V., and Anderson, J. B. 1989. Effect of crystallite shape on exciton energy: quantum Monte Carlo calculation. *Chem. Phys. Lett.* 156: 520–4.
Moison, J. M., Houzay, F., Barthe, F., Leprince, L., Andre, E., and Vatel, O. 1994. Self-organized growth of regular nanometer-scale InAs dots on GaAs. *Appl. Phys. Lett.* 64: 196–8.
Motte, L., Billoudet, F., and Pileni, M. P. 1995. Self-assembled monolayer of nanosized particles differing by their sizes. *J. Phys. Chem.* 99: 16425–9.
Moyo, T., Maruyama, K., and Endo, H. 1992. Photodarkening and photobleaching of CdS microclusters grown in zeolites. *J. Phys. C* 4: 5653–64.
Mueller, M., Lembke, U., Woggon, U., and Rueckmann, I. 1992. Growth of CdSe microcrystallites in a borosilicate glass matrix. *J. Non-Cryst. Sol.* 144: 240–6.
Murray, C. B., Norris, D. J., and Bawendi, M. G. 1993. Synthesis and characterization of nearly monodisperse CdE (E=S, Se, Te) semiconductor nanocrystallites. *J. Am. Chem. Soc.* 115: 8706–15.
Murray, C. B., Kagan, C. R., Bawendi, M. G. 1995. Self-organization of CdSe nanocrystallites into three-dimensional quantum dot superlattices. *Science* 270: 1335–8.
Murray, C. B., and Koch, S. W. 1995. *Semiconductor Lasers*. Berlin: Springer.
Nagamune, Y., Watabe, H., Nishioka, M., and Arakawa, Y. 1995. Observation of a single photoluminescence peak from a single quantum dot. *Appl. Phys. Lett.* 67: 3257–9.
Nair, S. V., Sinha, S., and Rustagi, K. C. 1987. Quantum size effects in spherical semiconductor microcrystals. *Phys. Rev. B* 35: 4098–101.
Nair, S. V., and Takagahara, T. 1996. Weakly correlated exciton pair states in large quantum dots. *Phys. Rev. B* 53: R10516–9.
Nakamura, A., Yamada, H., and Tokizaki, T. 1989. Size-dependent radiative decay of excitons in CuCl semiconducting quantum spheres embedded in glasses. *Phys. Rev. B* 40: 8585–8.
Nakamura, A., Lee, Y. L., Kataoka, T., and Tokizaki, T. 1994. Mesoscopic enhancement of optical nonlinearity in semiconducting quantum dots: CuCl and CuBr microcrystals. *J. Lumin.* 60, 61: 376–9.
Nakano, H., Ishida, Y., and Yanagawa, T. 1991. Energy relaxation and dephasing dynamics of colored filter glasses. *Appl. Phys. Lett.* 59: 3090–2.
Naoe, K., Zimin, L. G., and Masumoto, Y. 1994. Persistent spectral hole burning in semiconductor nanocrystals. *Phys. Rev. B* 50: 18200–10.
Neto, J. A. M., Barbosa, L. C., Casar, C. L., Alves, O. L., and Galembeck, F. 1991. Quantum size effects on $CdTe_xS_{1-x}$ semiconductor-doped glass. *Appl. Phys. Lett.* 59: 2715–7.
Nirmal, M., Murray, C. B., and Bawendi, M. G. 1994. Fluorescence-line narrowing in CdSe quantum dots: Surface localization of the photogenerated exciton. *Phys. Rev. B* 50: 2293–2300.
Nirmal, M., Norris, D. J., Kuno, M., and Bawendi, M. G. 1995. Observation of the "dark exciton" in CdSe quantum dots. *Phys. Rev. Lett.* 75: 3728–31.
Nirmal, M., Dabbousi, B. O., Bawendi, M. G., Macklin, J. J., Trautman, J. K., Harris, T. D., and Brus, L. E. 1996. Fluorescence intermittency in single cadmium selenide nanocrystals. *Nature* 383: 802–4.
Nogami, M., and Abe, Y. 1994. Sol-gel method for synthesizing visible

photoluminescent nanosized Ge-crystal-doped silica glasses. *Appl. Phys. Lett.* 65: 2545–7.
Nogami, M., Nagasaka, K., and Takata, M. 1990. CdS microcrystal-doped silica glass prepared by the sol-gel process. *J. Non-Cryst. Sol.* 122: 101–6.
Nogami, M., Zhu, Y.-Q., and Nagasaka, K. 1991. Preparation and quantum size effect of CuBr microcrystal doped glasses by the sol-gel process. *J. Non-Cryst. Sol.* 134: 71–6.
Nomura, S., and Kobayashi, T. 1990a. Clearly resolved exciton peaks in CdS_xSe_{1-x} microcrystallites by modulation spectroscopy. *Sol. Stat. Comm.* 73: 425–9.
Nomura, S., and Kobayashi, T. 1990b. Variational calculations on electric field dependence of energy of spherical semiconductor microcrystallites. *Sol. Stat. Comm.* 74: 1153–8.
Nomura, S., and Kobayashi, T. 1991. Nonparabolicity of the conduction band in CdSe and CdSSe semiconductor microcrystallites. *Solid State Comm.* 78: 677–80.
Nomura, S., and Kobayashi, T. 1992a. Exciton-LA and -TA phonon couplings in a spherical semiconductor microcrystallite. *Solid State Comm.* 82: 335–40.
Nomura, S., and Kobayashi, T. 1992b. Exciton-LO-phonon couplings in spherical semiconductor microcrystallites. *Phys. Rev. B* 45: 1305–7.
Norris, D. J., and Bawendi, M. G. 1995. Structure in the lowest absorption feature of CdSe quantum dots. *J. Chem. Phys.* 103: 5260–8.
Norris, D. J., and Bawendi, M. G. 1996. Measurement and assignment of the size-dependent optical structure in CdSe quantum dots. *Phys. Rev. B.* 53: 16338–46.
Norris, D. J., Sacra, A., Murray, C. B., and Bawendi, M. G. 1994. Measurement of the size dependent hole spectrum in CdSe quantum dots. *Phys. Rev. Lett.* 72: 2612–5.
Norris, D. J., Efros, Al. L., Rosen, M., and Bawendi, M. G. 1996. Size dependence of exciton fine structure in CdSe quantum dots. *Phys. Rev. B* 53: 16347–54.
Nosaka, Y. 1991. Finite depth spherical well model for excited states of ultrasmall semiconductor particles. An application. *J. Phys. Chem.* 95: 5000–8.
Nosaka, Y., Ohta, N., and Miyama, H. 1990. Photochemical kinetics of ultrasmall semiconductor particles in solution: Effect of size on the quantum yield of electron transfer. *J. Phys. Chem.* 94: 3752–5.
Nötzel, R., Ledentsov, N. N., Däweritz, L., Hohenstein, M., and Ploog, K. 1991. Direct synthesis of corrugated superlattices. *Phys. Rev. Lett.* 67: 3812–5.
Ochoa, O. R., Colajacomo, C., Witkowski, E. J., Simmons, J. H., and Potter, B. J. 1996. Quantum confinement effects on the photoluminescence spectra of CdTe nanocrystallites. *Solid State Comm.* 98: 717–21.
Olbright, G. R., and Peyghambarian, N. 1986. Interferometric measurement of the nonlinear index of refraction of CdS_xSe_{1-x}-doped glasses. *Appl. Phys. Lett.* 48: 1184–6.
Olbright, G. R., Peyghambarian, N., Koch, S. W., and Banyai, L. 1987. Optical nonlinearities of glasses doped with semiconductor microcrystlallites. *Opt. Lett.* 12: 413–5.
Ookubo, N. 1993. Luminescence kinetics of porous silicon. *J. Appl. Phys.* 74: 6375–80.
Pan, J. L. 1992a. Oscillator strengths for optical dipole interband transitions in semiconductor quantum dots. *Phys. Rev. B* 46: 4009–19.
Pan, J. L. 1992b. Reduction of the Auger rate in semiconductor quantum dots. *Phys. Rev. B* 46: 3977–98.
Pan, J., and Ramakrishna, M. V. 1994. Magic numbers of silicon clusters. *Phys. Rev. B* 50: 15431–5.

Pan, J., Xu, X., Ding, S., and Pen, J. 1990. Resonant Raman spectroscopy of CdS crystallites. *J. Lumin.* 45: 45–8.

Pankove, J. I. 1975. *Optical Processes in Semiconductors*. New York: Dover. 1978. Moscow: Mir.

Pascova, R., Gutzow, I., and Tomov, I. 1990. A model investigation of the process of phase formation in photochromic glasses. *J. Mater. Sci.* 25: 914–20.

Pavesi, L. 1996. Influence of dispersive exciton motion on the recombination dynamics in porous silicon. *J. Appl. Phys.* 80: 216–25.

Pavesi, L., and Ceschini, M. 1993. Stretched-exponential decay of the luminescence in porous silicon. *Phys. Rev. B* 48: 17625–8.

Peierls, R. 1979. *Surprises in Theoretical Physics*. Princeton: Princeton University Press. 1988. Moscow: Nauka.

Pellegrini, V., Tredicucci, A., Mazzoleni, C., and Pavesi, L. 1995. Enhanced optical properties in porous silicon microcavities. *Phys. Rev. B* 52: R14328–31.

Perov, P. I., Avdeeva, L. A., Zhdan, A. G., and Elinson, M. I. 1969. Influence of a strong electric field on the exciton absorption of CdS. *Sol. Stat. Phys.* 11: 93–6.

Petrov, E. P., Bogomolov, V. N., Kalosha, I. I., and Gaponenko, S. V. 1998. Spontaneous emission of dye molecules embedded in a photonic crystal. (In press).

Petrov, Yu. I. 1982. *Physics of Small Particles*. Moscow: Nauka.

Peyghambarian, N., Fluegel, B., Hulin, D., Migus, A., Joffre, M., Antonetti, A., Koch, S., and Lindberg, M. 1989. Femtosecond optical nonlinearities of CdSe quantum dots. *IEEE J. Quant. Electron* 25: 2516–22.

Peyghambarian, N., Koch, S. W., and Mysyrowich, A. 1993. *Introduction to Semiconductor Optics*. Englewood Cliffs: Prentice Hall.

Pickering, C. J., Beale, M. I., Robbins, D. J., Pearson, P. J., and Greef, R. 1984. Optical studies of the structure of porous silicon films formed in p-type degenerate and nondegenerate silicon. *J. Phys. C* 17: 6535–52.

Pieranski, P. 1983. Colloidal crystals. *Contemp. Physics* 24: 25–73.

Pilipovich, V. A., and Kovalev, A. A. 1975. *Optical Quantum Generators with Saturable Filters*. (in Russian) Minsk: Nauka i Tekhnika.

Pollock, E. L., and Koch, S. W. 1991. Path-integral study of excitons and biexcitons in semiconductor quantum dots. *J. Chem. Phys.* 94: 6776–81.

Potter, B. G., and Simmons, J. H. 1988. Electronic states of semiconductor clusters: Homogeneous and inhomogeneous broadening of the optical spectrum. *Phys. Rev. B* 37: 10838–45.

Potter, B. G., and Simmons, J. H. 1990. Quantum confinement effects in CdTe-glass composite thin films produced using rf magnetron sputtering. *J. Appl. Phys.* 68: 1218–24.

Priester, C., and Lannoo, M. 1995. Origin of self-assembled quantum dots in highly mismatched heteroepitaxy. *Phys. Rev. Lett.* 75: 93–6.

Purcell, E. M. 1946. Spontaneous emission probabilities at radio frequencies. *Phys. Rev.* 69: 681–7.

Qi, J., and Masumoto, Y. 1996. Spectral hole burning in CdS nanocrystals embedded in polyvinyl alcohol. *Solid State Comm.* 99: 467–72.

Rama Krishna, M. V., and Friesner, R. A. 1991. Quantum confinement effects in semiconductor clusters. *J. Chem. Phys.* 95: 8309–21.

Ramaniah, L. M., and Nair, S. V. 1993. Optical absorption in semiconductor quantum dots: a tight-binding approach. *Phys. Rev. B* 47: 7132–39.

Rashba, E. I., and Sturge, M. D. (Eds.) 1985. *Excitons*. Amsterdam: North Holland. 1985. Moscow: Nauka.

Reed, M., 1993. Quantum dots. *Sci. Am.* 118–23.

Ren, S. Y., and Dow, J. D. 1992. Hydrogenated Si clusters: Band formation with increasing size. *Phys. Rev. B* 45: 6492–5.

Ricard, D., Ghanassi, M., Schanne-Klein, M. C., and Flytzanis, C. 1995. Mechanisms and dynamics of optical nonlinearities in semiconductor nanocrystals. *SPIE. Proc.* 2801: 47–56.

Ridley, B. K. 1982. *Quantum Processes in Semiconductors*. Oxford: Oxford University Press. 1986. Moscow: Mir.

Romanov, S. G., Fakin, A. V., Tretijakov, V. V., Alperovich, V. I., Johnson, N. P., and Sotomayor Torres, C. M. 1996. Optical properties of ordered three-dimensional arrays of structurally confined semiconductors. *J. Crystal Growth* 159: 857–60.

Romestain, R., and Fishman, G. 1994. Exciton wavefunction, correlation energy, exchange energy, and oscillator strength in a cubic quantum dot. *Phys. Rev. B* 49: 1174–81.

Rossetti, R., Nakahara, S., and Brus, L. E. 1983. Quantum size effect in the redox potentials, resonance Raman spectra, and electronic spectra of CdS crystallites in aqueous solution. *J. Chem. Phys.* 79: 1086–88.

Rossetti, R., Ellinson, J. E., Gibson, J. M., and Brus, L. E. 1984. Size effects in the excited electronic states of small colloided CdS crystallites. *J. Chem. Phys.* 80: 4464–9.

Rossman, H., Schulzgen, A., Henneberger, F., and Muller, M. 1990. Quantum confined DC Stark effect in microcrystallites. *Phys. Stat. Sol. (b)* 159: 287–92.

Roussignol, P., Kull, M., Ricard, D., Rougemont, F., Frey, R., and Flytzanis, C. 1987. Time resolved direct observation of Auger recombination in semiconductor doped glasses. *Appl. Phys. Lett.* 51: 1882–4.

Roussignol, P., Ricard, D., Lukasik, J., and Flytzanis, C. 1987. New results on optical phase conjugation in semiconductor-doped glasses. *J. Opt. Soc. Am. B* 4: 5–13.

Roussignol, P., Ricard, D., Flytzanis, C., and Neuroth, N. 1989. Phonon broadening and spectral hole burning in very small semiconductor particles. *Phys. Rev. Lett.* 62: 312–5.

Rubinov, A. N., and Nikolaev, V. I. 1970. Peculiarities of luminescence of the matter in a Fabry-Perot cavity. *Izv. Akad. Nauk USSR* 34: 1308–11.

Ruppin, R. 1989. Optical absorption by excitons in microcrystals. *J. Phys. Chem. Solc.* 50: 877–82.

Sacra, A., Norris, D. J., Murray, C. B., and Bawendi, M. G. 1995. Stark spectroscopy of CdSe nanocrystallites: The significance of transition linewidths. *J. Chem. Phys.* 103: 5236–45.

Saito, S., and Goto, T. 1995. Spatial-confinement effect on phonons and excitons in PbI_2 microcrystallites. *Phys. Rev. B* 52: 5929–34.

Sakamoto, T., Hwang, S. W., Nakamura, Y., and Nakamura, K. 1994. Estimation of electronic confinement in a quantum dot from envelope modulation of Coulomb blockade oscillations. *Appl. Phys. Lett.* 65: 875–7.

Salata, O. V., Dobson, P. J., Hull, P. J., and Hutchison, J. L. 1994. Uniform GaAs quantum dots in a polymer matrix. *Appl. Phys. Lett.* 65: 189–193.

Samuelson, L., Carlsson, N., Castrillo, P., Gustafsson, A., Hessman, D., Lindahl, J., Montelius, L., Petersson, A., Pistol, M., and Seifert, W. 1995. Nano-optical studies of individual nanostructures. *Jap. J. Appl. Phys.* 34: 4392–7.

Saunders, W. A., Vahala, K. J., Atwater, H. A., and Flagan, R. C. 1992.

Resonance-enhanced spontaneous emission from quantum dots. *J. Appl. Phys.* 72: 806–8.

Scamarcio, G., Lugara, M., and Manno, D. 1992. Size-dependent lattice contraction in $CdS_{1-x}Se_x$ nanocrystals embedded in glass observed by Raman scattering. *Phys. Rev. B* 45: 13792–5.

Scamarcio, G., Spagnolo, V., Ventruti, G., Lugara, M., and Righini, G. C. 1996. Size dependence of electron-LO-phonon coupling in semiconductor nanocrystals. *Phys. Rev. B* 53: R10489–92.

Schanne-Klein, M. C., Piveteau, L., Ghanassi, M., and Ricard, D. 1995. The size dependence of the resonant Kerr nonlinearity of CdSSe-doped glasses revisited. *Appl. Phys. Lett.* 67: 579–81.

Schiff, L. I. 1968. *Quantum Mechanics*. New York: McGraw-Hill.

Schmackpfeffer, A., Weber, H. 1967. Absorption saturation in glasses colored by semiconductor microcrystals. *Z. Angew. Physik.* 23: 413–18.

Schmidt, H. M., and Weller, H. 1986. Quantum size effects in semiconductor crystallites: calculation of the energy spectrum for the confined exciton. *Chem. Phys. Lett.* 129: 615–8.

Schmidt, K. H., Patel, R., and Meisel, D. 1988. Growth of silver halides from the molecule to the crystal. A pulse radiolysis study. *J. Am. Chem. Soc.* 110: 4882–4.

Schmidt, M. E., Blanton, S. A., Hines, M. A., and Guyot-Sionnest, P. 1996. Size-dependent two-photon excitation spectroscopy of CdSe nanocrystals. *Phys. Rev. B* 53: 12629–32.

Schmitt-Rink, S., Miller, D. A. B., and Chemla, D. S. 1987. Theory of the linear and nonlinear optical properties of semiconductor microcrystallites. *Phys. Rev. B* 35: 8113–25.

Schoell, E. 1987. *Nonequilibrium Phase Transitions in Semiconductors: Self-Organization Induced by Generation and Recombination Processes*. Berlin: Springer. 1991. Moscow: Mir.

Schooss, D., Mews, A., Eychmüller, A., and Weller, H. 1994. Quantum-dot quantum well CdS/HgS/CdS: Theory and experiment. *Phys. Rev. B* 49: 17072–8.

Segal, G. A. (ed.). 1977. *Semiempirical Methods of Electronic Structure Calculation* New York: Plenum Press; 1980. Moscow: Mir.

Sekikawa, T., Yao, H., Hayashi, T. H., and Kobayashi, T. 1992. Electromodulation spectroscopy of CdS microcrystallites embedded in polymer films. *Solid State Comm.* 83: 969–74.

Sercel, P. C., and Vahala, K. J. 1990. Analytical formalism for determining quantum-wire and quantum-dot band structure in the multiband envelop-function approximation. *Phys. Rev. B* 42: 3690–3710.

Shepilov, M. P. 1992. Calculation of kinetics of metastable liquid-liquid phase separation for the model with simultaneous nucleation of particles. *J. Non-Cryst. Sol.* 146: 1–25.

Shinojima, H., Yumoto, J., and Uesugi, N. 1992. Size dependence of optical nonlinearity of CdSSe microcrystallites doped in glass. *Appl. Phys. Lett.* 60: 298–300.

Siemiarczuk, A., and Ware, W. R. 1989. A novel approach to analysis of pyrene fluorescence decays in sodium dodecylsulfate micelles in the presence of Cu^{2+} ions based on the maximum entropy method. *Chem. Phys. Lett.* 160: 285–9.

Siemiarczuk, A., Wanger, B. D., and Ware, W. R. 1990. Comparison of the maximum entropy and exponential series methods for the recovery of distributions of lifetimes from fluorescence lifetime data. *J. Phys. Chem.* 94: 1661–6.

Silvestri, M. R., and Schroeder, J. 1994. Pressure- and laser-tuned Raman scattering in

II-VI semiconductor nanocrystals: Electron-phonon coupling. *Phys. Rev. B* 50: 15108–12.

Slater, J. C. 1974. *The Self-Consistent Field Theory of Molecules and Solids*. New York: McGraw-Hill; 1978. Moscow: Mir.

Slyozov, V. V., and Sagalovich, V. V. 1987. Diffusive decay of solid solutions. *Sov. Phys. Uspekhi*. 151: 67–106.

Smith, R. A. 1978. *Semiconductors*. Cambridge: Cambridge University Press. 1981. Moscow: Mir.

Spanhel, L., and Anderson, A. 1991. Semiconductor clusters in the sol-gel process: Quantized aggregation, gelation, and crystal growth in concentrated ZnO colloids. *J. Am. Chem. Soc.* 113: 2826–33.

Spanhel, L., Arpac, E., and Schmidt, H. 1992. Semiconductor clusters in the sol-gel process: synthesis and properties of CdS nanocomposites. *J. Non-Cryst. Sol.* 147, 148: 657–62.

Spiegelberg, C., Henneberger, F., and Puls, J. 1991. Spectral hole-burning in quantum confined semiconductor microcrystals. *Superlat. Microstr.* 9: 487–91.

Starukhin, A. N., Lebedev, A. A., Razbirin, B. S., and Kapitonova, L. M. 1992. Anisotropy of radiative transitions in porous silicon. *Sov. J. Tech. Phys. Lett.* 18: 60–2.

Stepanov, B. I., and Gribkovskii, V. P. 1960. Dependence of absorption coefficient on radiation intensity. *Opt. Spectr.* 14: 484–8.

Stepanov, B. I., and Gribkovskii, V. P. 1963. *Introduction to the Theory of Luminescence*. Minsk: Nauka i Tekhnika. 1968. London: ILIFFE Books Ltd.

Stucky, G. D., and Mac Dougall, J. E. 1990. Quantum confinement and host/guest chemistry: probing a new dimension. *Science* 247: 669–78.

Suemoto, T., Tanaka, K., Nakajima, A., and Itakura, T. 1993. Observation of phonon structures in porous Si luminescence. *Phys. Rev. Lett.* 70: 3659–62.

Sviridov, V. V., Branitsky, G. A., Rachmanov, S. K., Stashonok, V. D., and Rogach, A. L. 1987. Silver halide photographic layer as nanostructured system. In *Physics, Chemistry and Application of Nanostructures*. Borisenko, V. E., Gaponenko, S. V., Gurin, V. S., and Filonov, A. B. (eds.). Singapore: World Scientific. pp. 163–6.

Sweeny, M., and Xu, J. 1989. Hole energy levels in zero-dimensional quantum balls. *Solid State Comm.* 72: 301–4.

Szoke, A., Daneu, V., Goldnar, J., and Kurnit, N. A. 1969. Bistable optical device and its applications. *Appl. Phys. Lett.* 15: 376–80.

Takagahara, T. 1987. Excitonic optical nonlinearity and exciton dynamics in semiconductor quantum dots. *Phys. Rev. B* 36: 9293–6.

Takagahara, T. 1989. Biexciton states in semiconductor quantum dots and their nonlinear optical properties. *Phys. Rev. B* 39: 10206–31.

Takagahara, T. 1993a. Effects of dielectric confinement and electron-hole exchange interaction on excitonic states in semiconductor quantum dots. *Phys. Rev. B* 47: 4569–84.

Takagahara, T. 1993b. Electron-phonon interactions and excitonic dephasing in semiconductor nanocrystals. *Phys. Rev. Lett.* 71: 3577–80.

Takagahara, T. 1993c. Enhancement of excitonic optical nonlinearity in a quantum dot array. *Optoelectronics–Devices and Technol.* 8: 545–5.

Takagahara, T. 1993d. Nonlocal theory of the size and tempetature dependence of the radiative decay rate of excitons in semiconductor quantum dots. *Phys. Rev. B* 47: 16639–42.

Takagahara, T. 1996. Electron-phonon interactions in semiconductor nanocrystals. *J. Lumin.* 70: 129–43.

Takagahara, T., and Hanamura, E. 1986. Giant-oscillator-strength effect on excitonic optical nonlinearities due to localization. *Phys. Rev. Lett.* 56: 2533–6.

Takagahara, T., and Takeda, K. 1992. Theory of the quantum confinement effect on excitons in quantum dots of indirect gap materials. *Phys. Rev. B* 46: 15578–83.

Takagahara, T., and Takeda, K. 1996. Exciton exchange splitting and Stokes shift in Si nanocrystals and Si clusters. *Phys. Rev. B* 53: R4205–8.

Takagi, H., Ogawa, H., Yamazaki, Y., Ishizaki, A., and Nakagiri, T. 1990. Quantum size effects on photoluminescence in ultrafine Si particles. *Appl. Phys. Lett.* 56: 2379–80.

Tanaka, K., Nakamura, T., Takamatsu, W., Yamanishi, M., Lee, Y., and Ishihara, I. 1995. Cavity-induced changes of spontaneous emission lifetime in one-dimensional semiconductor microcavities. *Phys. Rev. Lett.* 74: 3380–3.

Tang, Z. K., Nozue, Y., and Goto, T. 1992. Quantum size effects on the excited state of HgI_2, PbI_2 and BiI_3 clusters and molecules in zeolite LTA. *J. Phys. Soc. Japan* 61: 2943–50.

Tennakone, K., Jayatilake, W. D. W., and Ketipearachchi, U. S. 1991. Production of quantum-sized particles of TiO_2 in photohydrolysis of Ti^{3+} ion in aqueous medium. *Chem. Phys. Lett.* 117: 593–6.

Tersoff, J., Teichert, C., and Lagally, M. G. 1996. Self-organization in growth of quantum dot superlattices. *Phys. Rev. Lett.* 76: 1675–8.

Thompson, J. M. 1982. *Instabilities and Catastrophes in Science and Engineering.* Chichester: J. Wiley & Sons. 1985. Moscow: Mir.

Tokizaki, T., Akiyama, H., Takaya, M., and Nakamura, A. 1992. Linear and nonlinear optical properties of CdSe microcrystallites in glasses. *J. Cryst. Growth* 117: 603–7.

Tolbert, S. H., and Alivisatos, A. P. 1995. High-pressure structural transformations in semiconductor nanocrystals. *Annu. Rev. Phys. Chem.* 46: 595–625.

Tolbert, S. H., Herhold, A. B., Johnson, C. S., and Alivisatos, A. P. 1994. Comparison of the quantum confinement effects on the electronic absorption spectra of direct and indirect gap semiconductor nanocrystals. *Phys. Rev. Lett.* 73: 3266–9.

Tomita, M., and Matsuoka, M. 1990. Laser-induced irreversible change of the carrier-recombination process in CdSSe-doped glasses. *J. Opt. Soc. Am. B* 7: 1198–1203.

Tommasi, R., Lepore, M., Ferrara, M., and Catalano, I. M. 1992. Observation of high-index states in $CdS_{1-x}Se_x$ semiconductor microcrystallites by two-photon spectroscopy. *Phys. Rev. B* 46: 12261–5.

Tommasi, R., Lepore, M., and Catalano, I. M. 1993. Two-photon absorption spectrum in CdSSe nanocrystals by nonlinear luminescence technique. *Solid State Comm.* 85: 539–43.

Tran Thoai, D. B., Hu, Y. Z., and Koch, S. W. 1990. Influence of the confinement potential on the eh-pair states in semiconductor microcrystal. *Phys. Rev. B* 42: 11261–67.

Trautman, J. K., Macklin, J. J., Brus, L. E., and Betzig, E. 1994. Near-field spectroscopy of single molecules at room temperature. *Nature* 369: 40–2.

Tsekhomskii, V. A. 1978. Photochromic oxide glasses. *Sov. J. Phys. Chem. Glass* 4: 3–21.

Tsidilkovskii, I. M. 1978. *Band Structure of Semiconductors.* (in Russian). Moscow: Nauka.

Tsu, R., Li, X.-L., and Nicollian, E. H. 1994. Slow conductance oscillations in nanoscale silicon clusters of quantum dots. *Appl. Phys. Lett.* 65: 842–4.

Tsuboi, T. 1980. Absorption spectra of heavily Cu doped KCl, SrCl and KI crystals. *J. Chem. Phys.* 72: 5343–47.

Tu, Y., Levine, H., and Ridgway, D. 1993. Morphology transitions in a mean-field model of diffusion-limited growth. *Phys. Rev. Lett.* 71: 3838–41.

Tyagay, V. A., and Snitko, O. V. 1980. *Electroreflection of Light in Semiconductors.* (in Russian). Kiev. Naukova Dumka .

Ueta, M., Kanzaki, H., Kobayashi, K., Toyozawa, Y., and Hanamura, E. 1986. *Excitonic Processes in Solids.* Berlin: Springer.

Uhrig, A., Banyai, L., Hu, Y. Z., Koch, S. W., Klingshirn, C., and Neuroth, N. 1990. High-excitation photoluminescence studies of CdSSe quantum dots. *Z. Physik B* 81: 385–90.

Uhrig, A., Banyai, L., Gaponenko, S., Worner, A., Neuroth, N., and Klingshirn, C. 1991. Linear and nonlinear optical studies of CdSSe quantum dots. *Z. Physik D* 20: 345–8.

Uhrig, A., Worner, A., Klingshirn, C., Banyai, L., Gaponenko, S., Lacis, I., Neuroth, N., Speit, B., and Remitz, K. 1992. Nonlinear optical properties of semiconductor quantum dots. *J. Cryst. Growth* 117: 598–602.

Valenta, J., Ekimov, A. I., Gilliot, P., Hoenerlage, B., and Levy, R. 1997. Spectral hole-burning and hole-filling in CuBr nanocrystals. *J. Lumin.* 72–74: 386–8.

Valenta, J., Moniatte, J., Gilliot, P., Hoenerlage, B., and Ekimov, A. I. 1997. Hole-filling of persistent spectral holes in the excitonic absorption band of CuBr quantum dots. *Appl. Phys. Lett.* 70: 680–2.

Vandyshev, Y. V., Dneprovskii, V. S., Klimov, V. I., and Okorokov, D. K. 1991. Laser generation in semiconductor quazi-zero-dimensional structure on a transition between size quantization levels. *JETP Lett.* 54: 441–4.

Variano, B. F., Hwang, D. M., Sandroff, C. J., Wiltzius, P., Jing, T. W., and Ong, N. P. 1987. Quantum effects in anisotropic semiconductor clusters: Colloidal suspensions of Bi_2S_3 and Sb_2S_3. *J. Phys. Chem.* 91: 55–8.

Vasiliev, M. I., Grigoriev, N. A., Kulinkin, B. S., and Tsekhomskii, V. A. 1991. Hydrostatic pressure effect on exciton absorption spectra of CuI and CuBr microcrystals in heterogeneous glasses. *Sov. Phys. Chem. Glass* 17: 594–7.

Vasiliev, M. I., Tsekhomskii, V. A., Germanenko, I. N., and Gaponenko, S. V. 1995. Exciton effects in copper halide glasses. *J. Opt. Technol.* 62: 802–11.

Vial, J. C. (ed.). 1995. *Luminescence of Porous Silicon.* Berlin: Springer.

Vial, J. C., Bsiesy, A., Gaspard, F., Herino, R., Ligeon, M., Miller, F., Romestain, R., and Macfarlane, R. M. 1992. Mechanisms of visible-light emission from electro-oxidized porous silicon. *Phys. Rev. B* 45: 14171–6.

Vicsek, T. 1989. *Fractal Growth Phenomena.* Singapore: World Scientific.

Vossmeyer, T., Reck, G., Katsikas, L., Haupt, E. T. K., Schulz, B., and Weller, H.1995. A "double-diamond superlattice" built up of $Cd_{17}S_4(SCH_2OH)_{26}$ clusters. *Science* 267: 1476–79.

Wamura, T., Masumoto, Y., and Kawamura, T. 1991. Size-dependent homogeneous linewidth of Z_3 exciton absorption spectra in CuCl microcrystals. *Appl. Phys. Lett.* 59: 1758–60.

Wang, Y., Suna, A., Mahler, W., and Kasowski, R. 1987. PbS in polymers. From molecules to bulk solids. *J. Chem. Phys.* 87: 7315–22.

Wang, Y., Herron, N., Mahler, W., and Suna, A. 1989. Linear and nonlinear optical properties of semiconductor clusters. *J. Opt. Soc. Am. B* 6: 808–13.

Wang, Y., and Herron, N. 1991. Nanometer-sized semiconductor clusters: material synthesis, quantum-size effect, and photophysical properties. *J. Chem. Phys.* 91: 525–32.

Wang, P. D., Ledentsov, N. N., Sotomayor-Torres, C M., Kop'ev, P. S., and Ustinov, V. M. 1994. Optical characterization of submonolayer and monolayer InAs structures grown in a GaAs matrix. *Appl. Phys. Lett.* 64: 1526–8.

Wang, L., Zhang, J. K., and Bishop, A. R. 1994. Microscopic theory for conductance oscillations of electron tunneling through a quantum dot. *Phys. Rev. Lett.* 73: 585–8.

Weller, H. 1991. Quantum sized semiconductor particles in solution and in modified layers. *Ber. Bunsenges. Phys. Chem.* 95: 1361–5.

Weller, H., Schmidt, H. M., Koch, U., Fojtik, A., Baral, S., Henglein, A., Kunath, W., and Weiss, K. 1986. Photochemistry of colloidal semiconductors. Onset of light absorption as function of size of small CdS particles. *Chem. Phys. Lett.* 124: 557–560.

Wilcoxon, J. P., and Samara, G. A. 1995. Strong quantum-size effects in a layered semiconductor: MoS_2 nanoclusters. *Phys. Rev. B* 51: 7299–7302.

Wilcoxon, J. P., Newcomer, P. P., and Samara, G. A. 1996. Strong quantum confinement effects in semiconductors: FeS_2 nanoclusters. *Solid State Comm.* 98: 581–5.

Williams, R. C., and Smith, K. 1957. A crystallizable insect virus. *Nature* 45: 119–21.

Williams, V. S., Olbright, G. R., Fluegel, B. D., Koch, S. W., and Peyghambarian, N. 1988. Optical nonlinearities and ultrafast carrier dynamics in semiconductor doped glasses. *J. Mod. Opt.* 35: 1979–93.

Woggon, U., and Henneberger, F. 1988. Optical nonlinearities of quantum-confined excitons in CuBr microcrystallites. *J. Physique* 49: C2-255–8.

Woggon, U., and Gaponenko, S. V. 1995. Excitons in quantum dots (review). *Phys. Stat. Sol. (b)* 189: 286–343.

Woggon, U., and Portune, M. 1995. Non-degenerate four-wave mixing in semiconductor quantum dots. *Phys. Rev. B* 51: 4719–22.

Woggon, U., Rueckmann, I., Kornack, J., Mueller, M., Cesnulevicius, J., Kolenda, J., and Petrauskas, M. 1990. Growth, surface passivation, and characterization of CdSe-microcrystallites in glass with respect to their application in nonlinear optics. *SPIE Proc.* 1362: 100–10.

Woggon, U., Rueckmann, I., Kornack, J., Mueller, M., Cesnulevicius, J., Petrauskas, M., and Kolenda, J. 1992. Interface transfer processes and spectral hole burning in CdSe quantum confined clusters. *J. Cryst. Growth* 117: 608–13.

Woggon, U., Bogdanov, S. V., Wind, O., Schlaad, K.-H., Pier, H., Klingshirn, C., Chatziagorastou, P., and Fritz, H. P. 1993. Electro-optic properties of CdS embedded in a polymer. *Phys. Rev. B* 48: 11979–86.

Woggon, U., Gaponenko, S., Langbein, W., Uhrig, A., and Klingshirn, C. 1993. Homogeneous linewidth of electron-hole-pair states in II-VI quantum dots. *Phys. Rev. B* 47: 3684–9.

Woggon, U., Gaponenko, S., Uhrig, A., Langbein, W., and Klingshirn, C. 1994. Homogeneous linewidth and relaxation of excited hole states in II-VI quantum dots. *Adv. Mat. Opt. Electron.* 3: 141–150.

Woggon, U., Saleh, M., Uhrig, A., Portune, M., and Klingshirn, C. 1994. CdS quantum dots in the weak confinement. *J. Cryst. Growth* 138: 988–92.

Woggon, U., Wind, O., Langbein, W., Gogolin, O., and Klingshirn, C. 1994. Luminescence of biexcitons in CuBr quantun dots. *J. Lumin.* 59: 135–41.

Woggon, U. 1996. *Optical Properties of Semiconductor Quantum Dots*. Berlin: Springer.

Woggon, U., Wind, O., Gindele, F., Tsitsishvili, E., and Mueller, M. 1996. Optical transmission in CdSe quantum dots: From discrete levels to broad gain spectra. *J. Lumin* 70: 269–80.

Wooten, F. 1972. *Optical Properties of Solids*. New York: Academic.

Xia, J. B. 1989. Electronic states in zero-dimensional quantum wells. *Phys. Rev. B* 40: 8500–7.

Yablonovitch, E. 1987. Inhibited spontaneous emission in solid-state physics and electronics. *Phys. Rev. Lett.* 58: 2059–62.

Yablonovitch, E. 1994. Photonic crystals. *J. Mod. Opt.* 41: 173–94.

Yamamoto, M., Hayashi, R., Tsunetomo, K., Kohno, K., and Osaka, Y. 1991. Preparation and properties of Si microcrystals embedded in SiO_2 glass films. *Jap. J. Appl. Phys.* 30: 136–42.

Yamanishi, M. 1995. Combined quantum effects for electron and photon systems in semiconductor microcavity light emitters. *Prog. Quant. Electron.* 19: 1–39.

Yano, S., Goto, T., and Itoh, T. 1996. Excitonic optical nonlinearity of CuCl microcrystals in NaCl matrix. *J. Appl. Phys.* 79: 8216–22.

Yano, S., Goto, T., Itoh, T., and Kasuya, A. 1997. Dynamics of excitons and biexcitons in CuCl nanocrystals embedded in NaCl at 2 K. *Phys. Rev B* 55:1667–72.

Yanagawa, T., Sasaki, Y., and Nakano, H. 1989. Quantum-size effects and absorption of microcrystallites in colored filter glasses. *Appl. Phys. Lett.* 54: 1495–98.

Yanagawa, T., Nakano, H., and Sasaki, Y. 1991. Photodarkening and microcrystallite size in colored filter glasses. *Appl. Phys. Lett.* 59: 1690–2.

Yanagawa, T., Nakano, H., Ishida, Y., and Sasaki, Y. 1992. Photodarkening mechanism in colored filter glasses. *Opt. Commun.* 88: 371–5.

Yoffe, A. D. Low-dimensional systems: Quantum size effects and electronic properties of semiconductor microcrystallites (zero-dimensional systems) and quasi-two-dimensional systems. 1993. *Adv. Phys.* 42: 173–266.

Yukselici, H., Persans, P. D., and Hayes, T. M. 1995. Optical studies of the growth of $Cd_{1-x}Zn_xS$ nanocrystals in borosilicate glass. *Phys. Rev. B* 52: 11763–72.

Yumashev. K. V., Mikhailov, V. P., Prokoshin, P. V., Jmako, S. P., and Bodnar, I .V. 1994. CuInSSe-doped glass saturable absorbers for the passive mode-locking of neodymium lasers. *Appl. Phys. Lett.* 65: 2768–70.

Yumoto, J., Fukushima, S., and Kubodera, K. 1987. Observation of optical bistability in CdSSe-doped glasses with 25-psec switching time. *Opt. Lett.* 12: 832–4.

Zhao, X., Schoenfeld, O., Komuro, S., Aoyagi, V., and Sugano, T. 1994. Quantum confinement in nanometer-sized silicon crystallites. *Phys. Rev. B* 50: 18654–7.

Zheludev, N. I., Ruddock, I. S., and Illinworth R., 1987. Generation and amplification of subharmonics in semiconductor doped glass by a modulated argon-ion laser. *J. Mod. Opt.* 34: 1257–64.

Zhu, J. G., Write, C. W., Budai, J. D., Withrow, S. P., and Chen, Y. 1995. Growth of Ge, Si, and SiGe nanocrystals in SiO_2 matrices. *J. Appl. Phys.* 78: 4386–9.

Ziman, J. M. 1982. *Principles of the Theory of Solids.* Cambridge: Cambridge University Press; 1985, Moscow, Mir.

Zimin, L. G., Gaponenko, S. V., and Lebed, V. Yu. 1988. Room-temperature optical nonlinearity in semiconductor-doped glasses. *Phys. Stat. Sol. (b)* 150: 653–6.

Zimin, L. G., Gaponenko, S. V., Lebed, V. Yu., Malinovsky, I. E., Germanenko, I. N., Podorova, E. E., and Tsekhomsky, V. A. 1990. Copper chloride nonlinear optical absorption under quantum confinement. *J. Mod. Opt.* 37: 829–34.

Zimmerman, R. 1988. *Many-Particle Theory of Highly Excited Semiconductors.* Leipzig: Teubner.

Zylberajch, C., Ruaudel-Teixier, A., and Barraud, A. 1989. Properties of inserted mercury sulphide single layers in a Langmuir-Blodgett matrix. *Thin Solid Films* 179: 9–14.

Index